WAYNE Richard Sigleo
Arizona 1969

Extinct Fauna

1) Nature of community occupied
2) was animal gregarious, migrating
3) was animal common, rare, important in biomass
4) grazier, browser omnivore
5)

PLEISTOCENE EXTINCTIONS
The Search for a Cause

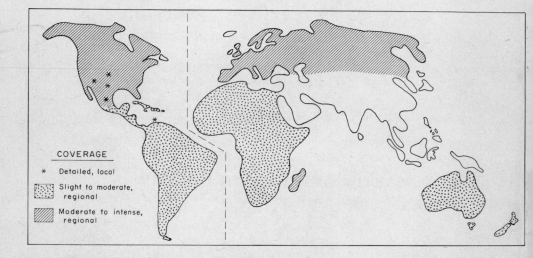

World coverage of Pleistocene extinction as discussed by contributors to this volume.
Unshaded areas are not treated.

PLEISTOCENE EXTINCTIONS

The Search for a Cause

*Volume 6 of the Proceedings of the VII Congress of the
International Association for Quaternary Research*

*Sponsored by the
National Academy of Sciences–National Research Council*

Editors

P. S. MARTIN AND H. E. WRIGHT, JR.

NEW HAVEN AND LONDON, YALE UNIVERSITY PRESS, 1967

Copyright © 1967 by Yale University.
Designed by Beverley Kruger,
set in Times Roman type,
and printed in the United States of America by
Murray Printing Company, Forge Village, Mass.
Distributed in Canada by McGill University Press.
All rights reserved. This book may not be
reproduced, in whole or in part, in any form
(except by reviewers for the public press),
without written permission from the publishers.
Library of Congress catalog card number: 67–24502

PREFACE

Together glaciers and extinct large vertebrates characterize, dramatically, the Quaternary. Alternating expansion and retreat of continental ice sheets affected upper-middle latitudes through much of the Pleistocene. The final reduction and disappearance of a megafauna of proboscideans, gravigrades, artiodactyls, horses, and struthious birds mark its end. By some definitions, the abundant remains of extinct fauna make the beds in which they occur a part of the Pleistocene.

But the commonplace is not necessarily understood. Despite many attempts to clarify it, the cause of extinction has proved elusive to those geologists and biologists most familiar with the evidence. Although the history of Pleistocene ice sheets has captured the interest of many earth scientists, the history of Pleistocene extinct fauna has not. The question of the cause for the expansion and retreat of glaciers has spawned a variety of schools of competing theorists, each seeking to account for all relevant geophysical data in areas ranging from paleoclimatology to astronomy, but the question of ultimate cause or causes of Pleistocene extinction has very often been evaded or even dismissed. There is no closely studied, long-debated Croll-Milankovitch hypothesis or solar–topographic hypothesis to account for the end of the elephants in America, of the diprotodons in Australia, and of the giant tortoises of the Celebes.

Pleistocene geologists are not the only ones who have a good deal at stake in understanding what caused extinction at the end of the glacial period. If climatic change was responsible, then it must have been a change of a magnitude not known previously. Are meteorologists prepared to recognize the possibility of a climatic shock wave of unprecedented dimension within the last 15,000 years? If intensified competition was responsible, if late-glacial climatic changes reduced or altered the range of various species and brought them to competitive crisis, then large mammals may be more susceptible to collapse of niches than is now recognized. Do conservationists consider the possibility that faunal stability in crowded African game parks may be much lower than has been generally assumed? If overhunting by prehistoric man is the underlying cause of extinction, the use of large animals as climatic indicators may be far more speculative than is commonly assumed. Are paleontologists prepared to admit that fossil glyptodonts, tapirs, and giant tortoises are not safe guides to interglacial climates?

Some prudent minds may rebel at the thought that a single factor, either those cited above or any other, might cause such a major event as the extinction of the large vertebrates. Uniformitarians will assert that, after all, extinction did not occur overnight. A combination of causes is so commonly found to determine natural phenomena that focusing on one to the exclusion of others seems naive. Yet the pattern of extinction at the end of the Pleistocene did happen "overnight" in a relative sense—in New Zealand and perhaps also in North America in less than 1,000 years—in roughly 1/300th of the average time estimated for the longevity (duration) of a species of mammals. To focus on a single factor, as several authors have done in the succeeding chapters, can at least be justified in the face of past failures to find a suitable synthetic theory. For a relatively sudden, spectacular, and highly significant geological event, it seems more logical to seek a single simple triggering action, an "essential enzyme," than to assume that many minor factors, each able to reduce animal populations in some small way, suddenly, by chance, conspired to occur at the same time.

Whatever its cause, the fossil record now clearly shows that the Pleistocene faunas of the world, Africa included, were far richer in species than those known in historic times. Ecologists interested in the relations of species or body size to area should take note. The Pleistocene faunas may represent the last natural climax. The pristine range of the American West "where the buffalo roam, where the deer and the antelope play," must contain many empty niches, space once shared by elephants, camels, horses, sloths, extinct bison, and four-horned antelope.

Was late-Pleistocene extinction so effective in upsetting the ecosystem that our National Parks, wilderness areas, and wildlands are an illusion? On a continent where herbivore herds evolved and thrived for tens of millions of years, can there be a natural community without them? Perhaps the high susceptibility of North American "wilderness" to fire, insect invasion, brush expansion, and erosion is the result in large part of depletion of the megafauna.

Finally, the heavy loss of large herbivores in America and Australia may warrant careful experiments with a broad spectrum of exotic large mammals of African and Asian origin. Hippo, African buffalo, or even elephant may hold the key to resource expansion in the Amazon and Orinoco savannas, which have proved to be poor pasture for *Bos*. Camel, kudu, gerenuk, perhaps even giraffe, may hold the solution to vexing problems of brush invasion in arid America. Western man's aversion to desert thorn scrub and tropical woodland may originate in the fact that his major domestic stock—cattle, horses, and sheep—are not browsers and are not well adapted to warm, dry climates. But range ecologists know that brush has its own annual yield of dry matter; paleontologists know the great significance of browsing species in the New World fauna of the late Cenozoic. It remains simply a matter of screening various African browsing or grazing–browsing species in various New World habitats to make animal husbandry and domestication an experimental science. The experiment has much to recommend it on ecological grounds.

Thus in view of the sweeping significance of Pleistocene extinction and in the absence of any such effort previously, it seemed that a symposium on the subject was long overdue. The VIIth Congress of the International Association for Quaternary Research (INQUA), held in Boulder, Colorado, in August of 1965, offered an ideal setting. In preliminary discussions with the Program Chairman, John F. Lance, and with Carl Hubbs, who kindly agreed to serve as chairman of the symposium, it was agreed that no consensus of what caused extinction should, or could, be attempted. For that matter, we could not assume that all symposium participants would regard the question of cause as a proper one. Invitations were extended with an eye toward obtaining some strikingly divergent opinions on the subject, an objective that I soon discovered would have been hard to avoid.

The present book grew out of ten papers (Deevey, Guilday, Hester, Hooijer, Jelinek, Kowalski, Leopold, Mehringer, Slaughter, Martin) presented at the INQUA Congress Symposium on Pleistocene Extinction, Its Cause and Consequences. To these were added papers from the general sessions of the Congress which also treated the problem (Irwin-Williams, Lundelius, Schultz, Vereshchagin). Finally, articles by Edwards, Haynes, Walker, and Battistini and Vérin were accepted afterward as original contributions to the subject.

The outlook of the vertebrate paleontologist, within a zoological setting, is apparent in chapters by Guilday, Hooijer, Kowalski, Lundelius, Schultz, Slaughter, Vereshchagin, and Walker. An anthropological or archaeological viewpoint underlines chapters by Edwards, Hester, Irwin-Willams, Jelinek, Battistini and Vérin. Ecology as it bears on extinction is dealt with in the chapters by Deevey, Martin, and Mehringer; the pattern of plant extinction is offered by paleobotanist Leopold. Haynes covers the radiocarbon chronology of man's arrival in the New World, from the viewpoint of a Pleistocene geologist.

Regarding geographic coverage, articles focused on a single site or local region and starred in the frontispiece are those of Irwin-Williams (Puebla, Mexico), Lundelius (Edwards Plateau and central Texas), Hooijer (Dutch West Indies), Mehringer (the arid southwestern United States), and Schultz (southwest Kansas). North America or North and South America are at least partly covered by Edwards, Guilday, Haynes, Hester, Jelinek, Leopold, Martin, and Slaughter.

Eurasian extinction is treated in papers by Kowalski and Vereshchagin and in part by Leopold. Africa, Australia, and New Zealand receive brief discussion from Guilday and Martin; Madagascar is the subject of chapters by Walker and by Battistini and Vérin. During the symposium, the East Indies was the subject of an outstanding review by Hooijer; unfortunately, Professor Hooijer was unable to prepare another manuscript afterward, and Southeast Asia remains an obvious gap in coverage (see Frontispiece).

I believe the authors of the articles in the present volume have sharpened the issues and, above all, made them more vulnerable to future testing. One may anticipate that before another INQUA Congress, the question of cause of Pleistocene extinction will come to be regarded as no less significant than the question of what controlled glaciation. Is it too much to hope that agreement on a general theory, or perhaps on several, will be an achievement of a future Congress?

Tucson, Arizona   •   Paul S. Martin
May 1966

CONTENTS

PAUL S. MARTIN
and JOHN E. GUILDAY

A BESTIARY
FOR PLEISTOCENE BIOLOGISTS[1]

On the suggestion of E. S. Deevey, Jr., who felt the need for a natural history of animals mentioned in the text, we have assembled an annotated list of all vertebrate genera other than fish. Living genera are barely mentioned, and interested readers should consult Walker's *Mammals of the World* (1964), our primary source. Somewhat longer accounts are presented for the extinct genera or species, animals probably unfamiliar to many readers who may encounter Pleistocene megafauna only through the glass of a museum display case. For each genus marked with a dagger (†) we sought to include available information on distribution, relationship, presumed ecology, time of extinction, the association (if any) with prehistoric man and, in some cases, an illustration by Mr. Harry Clench, Associate Curator of Entomology, Carnegie Museum. The result may qualify as a modern "bestiary"—by no means free from the speculative hazards and droll results one associates with the mediaeval accounts of legendary animals.

Our phylogenetic arrangement mostly follows Simpson's 1945 classification, which lists almost twice the number of extinct Pleistocene mammalian genera treated here. Although much of the literature of vertebrate paleontology is descriptive morphology, often (but not always) aimed at clarifying phylogeny, a good deal of interest in extinct Pleistocene animals comes from other disciplines and is more theoretical.

1. Contribution No. 148, Program in Geochronology, University of Arizona. Various colleagues have assisted in the preparation of this chapter, especially Clayton E. Ray, John F. Lance, and Alan Walker.

Preparation of the manuscript was partly supported by National Science Foundation Grants GB-1959 to Martin and GB-3083 to Guilday.

What was the nature of the community the extinct animal occupied?
Was the animal gregarious, migrating great distances? Was it a grazer,
a browser, an omnivore? Was it common or rare, important in the
biomass of the time, fluctuating in numbers, or fairly closely regulated?
Did it leave an obvious imprint on the vegetation, like the giraffe's
browse-line in certain African game parks? Did it suffer a reduction in
range in the face of a new and superior competitor? How did it react
to the changing climate and vegetation of the Pleistocene? Was it an
easy prey of prehistoric man and an important source of food for him,
or a seldom hunted species largely independent of man's influence?

Zoologists would welcome answers to some of these questions
regarding the living large mammals, not to mention the extinct ones of
thousands of years ago. But if these and similar questions often go
unanswered, they will not go unasked. Through new techniques as well
as more thoughtful application of some old ones, including the strategy
of excavation, the Pleistocene biologist makes his start. Finally, readers
will note that our list is mainly an account of mammals, as it must be.
The major part of Pleistocene generic extinction is confined to this
group, with birds a poor second. Within late-Pleistocene time the
fossil record indicates major displacements but virtually no generic
extinction among other groups of organisms.

CLASS AMPHIBIA

Caudata (order)
    Ambystomidae (family)
        Ambystoma (genus)
Salientia
    Pelobatidae
        Scaphiopus
    Bufonidae
        Bufo
    Ranidae
        Rana

Ambystoma Tschudi, 1838 Spotted salamander, tiger salamander and relatives; North America (N.A.).	CAUDATA Ambystomidae
Scaphiopus Holbrook, 1836 Spadefoot toads.	SALIENTIA Pelobatidae
Bufo Laurenti, 1768 Toads.	SALIENTIA Bufonidae
Rana Linnaeus, 1758 Frogs.	SALIENTIA Ranidae

CLASS REPTILIA

Chelonia (order)
Chelydridae (family)
Chelydra (genus)
Emydidae
Chrysemys
Pseudemys
Testudinidae
Gopherus
Geochelone
Trionychidae
Trionyx
Crocodylia
Crocodilidae
Crocodylus

Chelydra Schweigger, 1812   CHELONIA
Snapping turtles; N.A.   Chelydridae

Chrysemys Gray, 1844   CHELONIA
Painted turtles; N.A.   Emydidae

Pseudemys Gray 1856   CHELONIA
Terrapin; N.A.   Emydidae

Gopherus Rafinesque, 1832   CHELONIA
Four living spp., N.A.; in addition there   Testudinidae
were giant Pleistocene spp.

Geochelone Fitzinger, 1836 (Fig. 1)   CHELONIA
                                                Testudinidae

Pleistocene giant tortoises up to 2 m or even more in length were
found throughout warmer parts of the world, excluding Australia,
where the ecological equivalent of *Geochelone* may have been the
horned tortoise *Melania*. The giants survived into modern times only
on remote oceanic islands such as the Galapagos and the Seychelles,
which apparently were never inhabited by prehistoric man. With the
possible exception of Lewisville and Friesenhahn Cave, there are no
American deposits suggesting the coexistence of *Geochelone* and
prehistoric man. The giant tortoise has become a paleoclimatic
thermometer in the hands of some American paleontologists. Its
remains are regarded as indicating a warm, frost-free climate without
a winter freeze-up, yet the inference that this group was exterminated
by the onset of continentality at the end of the Pleistocene is biologic-
ally implausible. Giant tortoises disappeared from tropical parts of the
West Indies (W.I.). In the case of Mono Island, the fresh and un-
mineralized nature of the bones indicates relatively recent (postglacial)
extinction. Surely even in glacial times freezing temperatures did not
extend through the Caribbean.

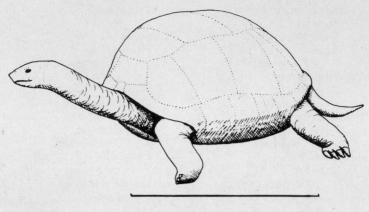

1 m

Fig. 1. *Geochelone*

Trionyx Geoffroy St. Hilaire, 1809	CHELONIA
Soft-shelled turtles.	Trionychidae
Crocodylus Laurenti, 1768	CROCODILIA
Worldwide; the crocodiles	Crocodilidae
escaped Pleistocene extinction.	

CLASS AVES

Aepyornithiformes (order)
  Aepyornithidae (family)
    †*Aepyornis* (genus)
Dinornithiformes
  Emeidae
    †*Emeus*
    †*Megalapteryx*
    †*Pachyornis*
    †*Euryapteryx*
    †*Zelornis*
    †*Anomalopteryx*
  Dinornithidae
    †*Dinornis*
Procellariiformes
  Hydrobatidae
    Oceanodroma
Anseriformes
  Anatidae
    Branta
    †*Camptorhynchus*

Falconiformes
  Falconidae
    Polyborus
  Teratornithidae
    †*Teratornis*
Galliformes
  Tetraonidae
    Tetrao
    Tympanuchus
Galliformes
  Meliagridae
    Meleagris
Gruiformes
  Gruidae
    Grus
  Rallidae
    Fulica
  Phorusrhacidae
    †*Titanis*

Charadriiformes
  Scolopacidae
    Numenius
  Alcidae
    †*Pinguinus*
Columbiformes
  Columbidae
    †*Ectopistes*
Psittaciformes
  Psittacidae
    †*Conuropsis*
Strigiformes
  Strigidae
    †*Ornimegalonyx*

Tytonidae
  Tyto
Piciformes
  Picidae
    Campephilus
    Colaptes
Passeriformes
  Drepanidae
    Drepanis
  Troglodytidae
    Thyromanes
  Parulidae
    Dendroica
  Fringillidae
    Pipilo

---

†*Aepyornis* Geoffroy, 1850   AEPYORNITHIFORMES
                                                  Aepyornythidae

The extinct giant, flightless birds of Madagascar include 4 spp. of *Aepyornis* and 3 of *Mullerornis*. The largest, *A. maximus*, was somewhat smaller than an ostrich, heavier set, and presumably much less cursorial, as one might expect on an island free of large carnivorous mammals. *Aepyornis* eggs exceeded 1-ft length and 2-gallon capacity (up to 31.5 × 24.5 cm, 11,035.8 cc); broken pieces litter coastal dunes of southwestern Madagascar where Battistini has named the Aepyornian Formation for them. Battistini reports a radiocarbon date of about A.D. 1200 on shell fragments from Diego Suarez. Remains of the giant birds are found in many archaeological sites. The species apparently persisted into early historic time.

†*Emeus* Reichenbach, 1852   DINORNITHIFORMES
                                                  Emeidae

A genus of small New Zealand (N.Z.) moas, 1.37–1.68 m tall. Brodkorb recognizes 2 spp. Of the 150 moas excavated from Pyramid Valley about 40% were *Emeus*. One contained an eggshell measuring 179 × 134 mm. A C^{14} date on another was 3,920 B.P.

†*Megalapteryx* Haast, 1886   DINORNITHIFORMES
                                                  Emeidae

Of the extinct moas, this is one of the smallest, a significant fact, for the bird apparently survived long after extinction of *Dinornis* and the larger forms. If man were the major cause of extinction, one would expect the smaller species to survive longer, because they are more difficult to find and kill. Most of the N.Z. extinct birds disappeared in the time of the Moa Hunters, the East Polynesian predecessors of the Maori. However, *Megalapteryx* apparently was known and exterminated by the Maori themselves. *M. benhami* is known from two localities only; *M. didinus* was much more common and also occurred on North Island. Its remains from relatively dry caves in the Otago District of South Island include heads and necks with ligaments, skin, eyes, and a huge leg with skin and feathers attached (Greenway, 1958).

†*Pachyornis* Lydekker, 1891   DINORNITHIFORMES
                                                                Emeidae

This genus of 5–7 spp. differed from *Anomalopteryx* and *Megalopteryx* in having a relatively longer tibia compared with the femur and metatarsus. These moas and *Pachyornis* "must have looked quite incredible . . . like a forty-gallon drum supported on knee-length gum boots" (Duff, 1952, p. 22). Their bones are found on both North and South islands of N.Z., often associated with the Moa Hunters.

†*Euryapteryx* Haast, 1874   DINORNITHIFORMES
                                                                Emeidae

Squat, massive moas of 4 spp. all found on North Island, two found on South Island, N.Z. Their bones were found in great quantities in human ovens; at the Shag River site moa bones (mainly *Euryapteryx*) were removed by boat and train, filling several railway trucks, before being dispatched to Dunedin bone-mills. "At Wairau it is not possible to dig anywhere over an area of fifteen to twenty acres without striking bones of moas which have been killed and eaten." Other massive kill sites are enumerated in Duff (1956, p. 67–68) who could not believe that without bows, arrows, or long-range weapons the Moa Hunters would have been capable of exterminating such an abundant bird. An egg of *E. curtus* from North Island measured 120×91 mm.

†*Zelornis* Oliver, 1949   DINORNITHIFORMES
                                                                Emeidae

Two spp. of N.Z. moas formerly included with *Eurapteryx*, one known only from North Island, the other from South Island.

†*Anomalopteryx* Reichenbach, 1852   DINORNITHIFORMES
                                                                Emeidae

A. didiformis from the Quaternary of New Zealand is found at many localities in both North and South islands. Those in the latter were slightly smaller. A second moa, *A. antiquus*, is known from the Upper Miocene or Lower Pliocene of N.Z.

†*Dinornis* Owen, 1843   DINORNITHIFORMES
                                                                Dinornithidae

The alkaline peat of Pyramid Valley, N.Z., is estimated to contain 800 fossil moas/acre. In more acid peats in N.Z., only beaks and gizzard stones remain. Depending on its mode of body carriage, *D. maximus*, the largest extinct member of this genus, would have been 3.0–3.5 m tall. *Dinornis* eggs, 25×18 cm and exceeding 4,000 cc in capacity, are only half the volume of the egg of *Aepyornis* from Madagascar. Brodkorb (1963) recognizes 8 spp. of *Dinornis*, 5 from North Island and 3 from South Island of N.Z.

Oceanodroma Reichenbach, 1852   PROCELLARIFORMES
Petrels.   Hydrobatidae

Branta Scopoli, 1769   ANSERIFORMES
Canada geese and allies.   Anatidae

†*Camptorhynchus* Bonaparte, 1838   ANSERIFORMES
                                             Anatidae

Labrador duck. Males especially rare in early 19th century before the last specimen was taken alive on Long Island in 1875.

Polyborus Meriam, 1826   FALCONIFORMES
Caracaras.   Falconidae

†*Teratornis merriami* Miller, 1909   FALCONIFORMES
                                     Teratornithidae

This great vulture stood over 75 cm high and had a wing span of 3.6 m. It ranks among the largest known birds of flight, possibly weighing 50 pounds. The feet were surprisingly small and weak when compared with the great size of the body, wings, and head (Stock, 1961). It was undoubtedly a carrion feeder, and remains are very common in the Rancho la Brea tar pits of California.

Tetrao Linnaeus, 1758   GALLIFORMES
Old World grouse.   Tetraonidae

Tympanuchus Linnaeus, 1758   GALLIFORMES
Prairie chickens, N.A.   Tetraonidae

Meleagris Linnaeus, 1758   GALLIFORMES
Turkey, N.A.   Meleagridae

Grus Pallas, 1766   GRUIFORMES
Cranes; 2 spp. near extinction.   Gruidae

Fulica Linnaeus, 1758   GRUIFORMES
Coots.   Rallidae

†*Titanis* Brodkorb, 1963   GRUIFORMES
                                     Phorusrhacidae

A flightless bird larger than the African ostrich recently described from Pleistocene deposits on the Santa Fe River in Florida. Similar giant birds are known from S.A.

Numenius Brisson, 1760   CHARADRIIFORMES
Various spp. of curlew;   Scolopacidae
the Eskimo curlew, *N. borealis*,
is practically extinct.

†*Pinguinus* Bonnaterre, 1790   CHARADRIIFORMES
                                     Alcidae

The Great Auk is a famous case of extinction by modern man. The last specimen was taken off the coast of Iceland in 1844; doubtful sighting records come from the Grand Banks in 1852. Great Auk bones appear in a variety of postglacial midden deposits from Scandinavia, W. Europe (to Italy), and along the Atlantic coast of

America south to Florida. Evidently western man delivered the coup
de grace to a flightless species that had been much reduced in its
natural range by prehistoric hunters; sometimes it is included within
Alca Linnaeus, 1758.

†*Ectopistes* Swainson, 1827   COLUMBIFORMES
Passenger pigeon. The species   Columbidae
declined from billions
in the early 19th century
to extinction by 1914.

†*Conuropsis* Salvadori, 1891   PSITTACIFORMES
                                                     Psittacidae

Carolina paroquet; the last one died in captivity in the Cincinnati
Zoological Gardens in 1914. Their habits made them especially
vulnerable to hunters. There is no reason to believe extinction
followed from habitat changes.

†*Ornimegalonyx* Arredondo, 1961   STRIGIFORMES
A giant owl from Cuba, the   Strigidae
largest known.

Tyto Bilberg, 1828   STRIGIFORMES
Barn owls, worldwide. Extinct   Tytonidae
species found on various islands,
especially in the W.I.

Campephilus Gray, 1840   PICIFORMES
Ivory-billed woodpeckers; N.A.   Picidae

Colaptes Vigors, 1826   PICIFORMES
Flickers.   Picidae

Drepanis Temminck, 1820   PASSERIFORMES
One of several genera of   Drepanidae
Hawaiian honey-creepers.

Thryomanes Sclater, 1862   PASSERIFORMES
Bewick's wren and its allies.   Troglodytidae

Dendroica Gray, 1842   PASSERIFORMES
Wood warblers; N.A., S.A.   Parulidae

Pipilo Bieillot, 1816   PASSERIFORMES
Towhees.   Fringillidae

CLASS MAMMALIA

Marsupialia (order)   Macropodidae
    Didelphidae (family)   †*Palorchestes*
      Didelphis (genus)   *Macropus*
    Phalangeridae   Diprodontidae
      Phascolarctos   †*Nototherium*
                                    † *Diprotodon*

Insectivora
  Solenodontidae
    Solenodon
  Soricidae
    Notiosorex
    Sorex
    Blarina
    Cryptotis
    Neomys
    Suncus
    Crocidura
  Talpidae
    Desmana
    Talpa
    Scalopus
  Nesophontidae
    †*Nesophontes*
Chiroptera
  Phyllostomidae
    Ardops
    Ariteus
    Stenoderma
    Phyllonycteris
    Phyllops
    Monophyllus
    Reithronycteris
  Natalidae
    Natalus
  Vespertilionidae
    Lasiurus
    Euderma
Primates
  Lemuridae
    Hapalemur
    Lemur
    Lepilemur
    Cheirogaleus
    Microcebus
    Phaner
    †*Megaladapis*
    †*Archaeolemur*
    †*Hadropithecus*
  Indridae
    Avahi
    †*Palaeopropithecus*
    †*Mesopropithecus*
    †*Neopropithecus*
    Propithecus
    Indri
    †*Archaeoindris*

  Daubentoniidae
    Daubentonia
  Cercopithecidae
    Macaca
    †*Simopithecus*
    Papio
    Cercopithecus
    Erythrocebus
  Pongidae
    Pongo
Edentata
  Megalonychidae
    †*Nothropus*
    †*Nothrotherium*
    †*Acratocnus*
    †*Megalocnus*
    †*Megalonyx*
  Megatheriidae
    †*Eremotherium*
    †*Megatherium*
  Mylodontidae
    †*Glossotherium*
    †*Paramylodon*
    †*Mylodon*
    †*Lestodon*
    †*Scelidotherium*
  Dasypodidae
    Dasypus
    †*Pampatherium*
  Glyptodontidae
    †*Neothoracophorus*
    †*Hoplophorus*
    †*Brachyostracon*
    †*Panochthus*
    †*Glyptodon*
    †*Boreostracon*
Lagomorpha
  Ochotonidae
    Ochotona
  Leporidae
    Lepus
    Sylvilagus
    Brachylagus
Rodentia
  Sciuridae
    Sciurus
    Marmota
    Cynomys
    Citellus
    Callospermophilus

Geomyidae
Geomys
Thomomys
Heterogeomys
Heteromyidae
Perognathus
Dipodomys
†*Prodipodomys*
Castoridae
Castor
†*Castoroides*
†*Trogontherium*
Cricetidae
†*Megalomys*
†*Bensonomys*
Sigmodon
Reithrodontomys
Peromyscus
Ondatra
Onychomys
Oryzomys
Neotoma
Zygodontomys
Arvicola
Cricetulus
Dicrostonyx
Lemmus
Microtus
Pitymys
Synaptomys
†*Mimomys*
†*Pliolemmus*
†*Pliophenacomys*
†*Pliopotamys*
Muridae
Rattus
Mus
Zapodidae
Zapus
Hystricidae
Hystrix
Hydrochoeridae
Hydrochoerus
†*Neochoerus*
Heptaxodontidae
†*Elasmodontomys*
Dinomyidae
†*Eumegamys*
Dasyproctidae
Dasyprocta

Capromyidae
Capromys
Geocapromys
Plagiodontia
†*Hexolobodon*
†*Isolobodon*
Echimyidae
†*Brotomys*
Cetacea
Delphinidae
Delphinus
Tursiops
Carnivora
Canidae
†*Theriodictis*
Canis
Vulpes
Nyctereutes
Cuon
Ursidae
†*Arctotherium*
†*Arctodus*
Tremarctos
Selenarctos
Ursus
Euarctos
Procyonidae
Procyon
†*Parailurus*
Mustelidae
Mustela
Gulo
†*Trigonictis*
†*Canimartes*
Martes
Meles
Taxidea
Mephitis
Lutra
Hyaenidae
Crocuta
Hyaena
Felidae
Felis
Leo
Lynx
Panthera
Uncia
†*Smilodon*
†*Dinobastis*

Phocidae
   Phoca
Liptoterna
  Macraucheniidae
    †*Macrauchenia*
Notoungulata
  Toxodontidae
    †*Toxodon*
Proboscidea
  Gomphotheriidae
    †*Stegomastodon*
    †*Cuvieronius*
    †*Notiomastodon*
  Mammutidae
    †*Mammut*
  Elephantidae
    †*Loxodonta*
    †*Palaeoloxodon*
    †*Archidiskodon*
    Elephas
    †*Mammuthus*
Sirenia
  Dugongidae
    †*Hydrodamalis*
Perissodactyla
  Equidae
    †*Stylohipparion*
    †*Nannippus*
    †*Hippidion*
    †*Onohippidium*
    †*Plesippus*
    Equus
  Tapiridae
    Tapirus
  Rhinocerotidae
    Dicerorhinus
    Rhinoceros
    †*Coelodonta*
Artiodactyla
  Leptochoeridae
    †*Leptochoerus*
  Suidae
    †*Stylochoerus*
    †*Nesochoerus*
    †*Tapinochoerus*
    Sus
  Tayassuidae
    †*Platygonus*
    †*Mylohyus*
    Tayassu

Hippopotamidae
  Hippopotamus
Camelidae
  †*Camelops*
  †*Titanotylopus*
  †*Palaeolama*
  †*Tanupolama*
  Camelus
Cervidae
  Moschus
  †*Megaceros*
  Dama
  Cervus
  Odocoileus
  †*Charitoceros*
  †*Cervalces*
  Alces
  Rangifer
  †*Sangamona*
  Capreolus
Giraffidae
  †*Libytherium*
Antilocapridae
  †*Breameryx*
  †*Capromeryx*
  †*Stockoceros*
  †*Tetrameryx*
  Antilocapra
Bovidae
  Taurotragus
  Bos
  Poephagus
  Bison
  †*Spirocerus*
  Gazella
  Saiga
  Oreamnos
  †*Myotragus*
  Rupicapra
  †*Bootherium*
  †*Symbos*
  †*Euceratherium*
  †*Preptoceras*
  Ovibos
  Capra
  Ovis

Didelphis Linnaeus, 1758   MARSUPIALIA
N.A., S.A.; opossums.   Didelphidae

Phascolarctos Blainville, 1816   MARSUPIALIA
Koalas, 1 sp., wt. 4–15 kg, living   Phalangeridae
singly or in small groups;
Australia.

†*Palorchestes* Owen, 1873   MARSUPIALIA
                                                    Macropodidae

Perhaps the largest of the extinct kangaroos of Australia, *Palorchestes*
had a skull 384 mm long and an estimated ht. of 3.5 m.

Macropus Shaw, 1790   MARSUPIALIA
                                                    Macropodidae

The great gray kangaroo of E. Australia, *M. giganteus*, is over 2 m
tall; more common inland in open country is the red kangaroo, *M.
rufus*, 1.7 m tall and weighing 23–70 kg. Kangaroos may attain
48 kilometers/hour for short distances.

1 m

Fig. 2. *Diprotodon*

†*Nototherium* Owen, 1845   MARSUPIALIA
One of the 6 genera in an extinct   Diprotodontidae
Pleistocene family, Australia.

†*Diprotodon* Owen, 1838 (Fig. 2)
The largest of the extinct Australian
Pleistocene marsupials, *Diprotodon*
was rhinoceros-sized—3.2 m long.

MARSUPIALIA
Diprotodontidae

Solenodon Brandt, 1833

INSECTIVORA
Solenodontidae

Two W.I. species, one on Hispaniola, one on Cuba; mainly nocturnal,
the size of a house cat.

Notiosorex Coues, 1877
The desert shrew, arid
N.A.; 2 spp.

INSECTIVORA
Soricidae

Sorex Linnaeus, 1758
Long-tailed shrews, 40 spp.
N. Hemisphere.

INSECTIVORA
Soricidae

Blarina Gray, 1838
Greater N.A. short-tailed shrews.

INSECTIVORA
Soricidae

Cryptotis Powell, 1848
Shrews, N.A. to northern S.A.;
many species.

INSECTIVORA
Soricidae

Neomys Kaup, 1829
Old World water shrews, 2 spp.

INSECTIVORA
Soricidae

Suncus Ehrenberg, 1832

INSECTIVORA
Soricidae

Musk shrews, ca. 20 spp., Old World; *S. etruscus* of the Mediterranean region in Africa is the world's smallest mammal.

Crocidura Wagler, 1832
White-tooth shrews of Eurasia,
many spp.

INSECTIVORA
Soricidae

Desmana Güldenstädt, 1777

INSECTIVORA
Talpidae

Russian desmans, 1 sp., aquatic. The desman is a large, muskrat-size
aquatic mole found only in the Volga Basin of U.S.S.R. A closely
related genus, *Galemys*, occurs in the Spanish Pyrenees.

Talpa Linnaeus, 1758
Old World moles, 4 spp., Eurasia.

INSECTIVORA
Talpidae

Scalopus Geoffroy, 1803
N.A.; eastern American mole.

INSECTIVORA
Talpidae

†*Nesophontes* Anthony, 1916

INSECTIVORA
Nesophontidae

Extinct W.I. shrews, 6 spp., found with bones of *Rattus* and *Mus* in
cave deposits. No study skins have reached museums.

Ardops Miller, 1906   CHIROPTERA
Tree bats; 4 spp. confined to the   Phyllostomidae
Lesser Antilles.

Ariteus Gray, 1838   CHIROPTERA
Jamaican fig-eating bats, rare in   Phyllostomidae
collections.

Stenoderma E. Geoffroy, 1818   CHIROPTERA
Red fruit bat of the W.I.   Phyllostomidae

Phyllonycteris Gundlach, 1860   CHIROPTERA
Four spp. of small fruit bat; W.I.   Phyllostomidae

Phyllops Peters, 1865   CHIROPTERA
Falcate-winged bats; W.I.   Phyllostomidae

Monophyllus Leach, 1821   CHIROPTERA
Six spp., W.I., small, cave-inhabiting   Phyllostomidae
fruit bats.

Reithronycteris Miller, 1898   CHIROPTERA
A rare Jamaican bat, probably   Phyllostomidae
a species of the W.I.
group *Phyllonycteris*.

Natalus Gray, 1838   CHIROPTERA
Funnel-eared bats, tropical America.   Natalidae

Lasiurus Gray, 1831   CHIROPTERA
Hairy-tailed bats, ca. 12 spp.;   Vespertilionidae
N.A., S.A.

Euderma H. Allen, 1892   CHIROPTERA
Spotted bats, southwestern U.S.A.   Vespertilionidae

Hapalemur I. Geoffroy, 1851   PRIMATES
Gentle lemurs, two species in   Lemuridae
Madagascar forests, solitary in habits.

Lemur Linnaeus, 1758   PRIMATES
                                                   Lemuridae

The extinct Madagascar spp. of this genus, *L. insignis*, *L. jullyi*, and
L. majori, were all slightly larger in cranial features than the half-
dozen living species.

Lepilemur I. Geoffroy, 1851   PRIMATES
Sportive lemurs, the single species is   Lemuridae
secretive and elusive, nocturnal,
arboreal, and bears a single young.

Cheirogaleus E. Geoffroy, 1812   PRIMATES
Dwarf lemurs; 3 nocturnal spp.   Lemuridae
occupying wooded areas of
Madagascar.

Microcebus E. Geoffroy, 1828   PRIMATES
The 2 spp. of this tiny Madagascar   Lemuridae
primate are the smallest members of
the order, wt. 45–85 g.

Phaner Gray, 1870   PRIMATES
The single sp. of squirrel lemur   Lemuridae
occurs in N.W. Madagascar,
chiefly nocturnal.

†*Megaladapis* Major, 1893   PRIMATES
                                                 Lemuridae

The extinct "giant" lemurs of Madagascar are sometimes compared
with the giant ground sloths of S.A., but the only species of comparable size were the dwarf W.I. sloths. The largest herbivores
of Madagascar were the giant birds and the extinct hippopotamus.
The largest of the "gigantic" subfossil lemuroids, *M. edwardsi*, had a
maximum skull length of 300–310 mm. The genus is characterized by
disproportionate elongation of the facial region of the skull, a small
neurocranium with a brain only half the dimensions of *Archaeolemur*,
and small orbits indicating a diurnal habit. The extinction of *Megaladapis* has apparently occurred since the arrival of prehistoric man in
Madagascar; no radiocarbon dates directly associated with the giant
lemurs have been reported to date.

†*Archaeolemur* Filhol, 1895   PRIMATES
                                                 Lemuridae

This extinct group of 3 spp. of lemurs had a skull about 50 % longer
than that of the larger living indrids. Its bones appear in late postglacial deposits of Madagascar.

†*Hadropithecus* Lorenz, 1899   PRIMATES
                                                 Lemuridae

One of the rarest of the extinct lemurs, *Hadropithecus* is known only
from caves of Andrahomana and adjacent deposits in the extreme
southern part of Madagascar. It is similar to *Archaeolemur* but with
pectoral limbs longer than pelvic, with different cheek teeth, and with
a profile more simian than that of any other lemur.

Avahi Jourdan, 1834   PRIMATES
                                                 Indriidae

The woolly lemur has a head and body length of 300–450 mm, and
a tail almost as long. The fur is thick and woolly rather than silky as in
the indris and sifaka. Woolly lemurs are generally solitary, nocturnal,
and arboreal. On the ground they will stand upright.

†*Palaeopropithecus* Grandidier, 1899   PRIMATES
                                                 Indriidae

Two species of this extinct lemur are known from central and western
Madagascar, both associated with prehistoric man. The group differs
from its relatives in the general form of the skull, which is depressed

and elongated, an unusual shape for a primate. The orbits are small, indicating a diurnal habit. Its teeth suggest a leafy diet.

†*Mesopropithecus* Standing, 1908   PRIMATES
                                                            Indriidae

This extinct lemur is known only from the type locality, the famous Madagascar marsh of Ampasambazimba. Its skull was about 100 mm long and its orbits proportionately smaller than existing indrids.

†*Neopropithecus* Lambert, 1936   PRIMATES
                                                            Indriidae

Two spp. of extinct lemurs, *N. globiceps* and *N. platyfrons*, are known from well-preserved skulls and mandibular rami in western and central Madagascar. In their smaller size, more graceful build, wider occipital region, and in the absence of a sagittal crest, they differ from *Mesopropithecus* and are more closely related to the living sifaka (*Propithecus*).

Propithecus Bennett, 1832   PRIMATES
                                                        Indriidae

Sifakas of Madagascar, 2 spp., large, arboreal, diurnal lemurs travelling in family groups of 6–10. They easily leap 10–12 m.

Indri E. Geoffroy, 1796   PRIMATES
                                                     Indriidae

These large lemurs live in family groups, are diurnal and almost strictly arboreal, rarely coming to the ground and rarely seen, although their weird dog-like howls are often heard. Like howler monkeys, indris feeds mainly on leaves, fruits, nuts, bark, and flowers; head and body length, 610–712 mm, tail 50–64 mm.

†*Archaeoindris* Standing, 1908   PRIMATES
                                                            Indriidae

Approaching *Megaladapis* in skull size, this extinct Madagascar lemur is known only from 1 sp. collected in the late postglacial deposit of Ampasambazimba. Its mandible was massive and powerfully built. The orbits were large and close together leading Lamberton to think that it was nocturnal; however, the ratio of orbit to skull length is not great enough to equal that of the living nocturnal lemurs and Walker (this volume) believes it was diurnal like the other giant lemurs.

Daubentonia E. Geoffroy, 1795   PRIMATES
                                                           Daubentoniidae

Subfossil remains of living *D. madagascariensis* and extinct *D. robusta* are known from dry S.W. Madagascar where aye-ayes do not occur today. The living species persists in more heavily forested areas. It is nocturnal, silent and solitary and is famous to the zoologist for its long, attenuate middle finger and large rodent-like incisors with which it extracts wood-boring larvae. In the length of its radius and ulna *D. robusta* was about 50% larger than the living aye-aye.

Macaca Lacépède, 1799   PRIMATES
Cercopithecidae

Macaques, ca. 12 spp., throughout S. Asia, Indonesia, and the Mediterranean region. Large males weigh up to 13 kg.

†*Simopithecus* Andrews, 1916   PRIMATES
Cercopithecidae

The extinct giant baboons of E. Africa were apparently related to the Gelada of Abyssinia (*Theropithecus*). In the largest species, *S. jonathani*, the female mandible was as large or slightly more massive than that of an average adult male gorilla. A smaller species, *S. oswaldi*, was as large as or slightly larger than modern baboons (*Papio*). Other extinct species and genera in the baboon subfamily are known from S. Africa, especially in cave breccias associated with the *Australopithecus* fauna.

Papio Erxleben, 1777   PRIMATES
Baboons   Cercopithecidae

Cercopithecus Linnaeus, 1758   PRIMATES
Guenons, ca. 12 spp. in   Cercopithecidae
Africa, wt. ca. 7 kg.

Erythrocebus Trouessart, 1897   PRIMATES
Cercopithecidae

The red hussar monkey, *E. patas*, occurs in central Africa. It is terrestrial, with the males attaining 26 kg body wt.

Pongo Lacépède, 1799   PRIMATES
Pongidae

Orangutan, the single living sp. found in Sumatra and Borneo; females weigh 40 kg, males 75–100 kg.

†*Nothropus* Burmeister, 1882   EDENTATA
Pleistocene S.A. ground sloths   Megalonychidae
related to *Nothrotherium* of N.A.

†*Nothrotherium* Lydekker, 1889 (Fig. 3)   EDENTATA
Megalonychidae

Apart from the W.I. species, the shasta ground sloth (*Nothrotherium shastense*) of arid America was the smallest of the N.A. genera, barely 1 m high at the hindquarters. The remarkable dung deposits attributed to *Nothrotherium* in Nevada, Arizona, and New Mexico provide an invaluable guide to its food habits (Martin et al., 1963). Apparently it browsed on various species of desert shrubs including Joshua tree (*Yucca brevifolia*), Mormon tea (*Ephedra*), and creosote bush (*Larrea tridentata*). Unless contaminated by twigs or urine of trade rats (*Neotoma*), which also occupied sloth caves, the dung of these animals should be an ideal material for radiocarbon dating. The youngest dung date thus far, C-222 from Gypsum Cave, Nev., roughly 8,000 years, has not been duplicated by subsequent dates on other deposits, which are several thousand years older. Semipopular writing on the subject of ground sloth extinction leaves the impression that the animals survived until only a few thousand or even a few hundred years ago. Such a possibility must be rejected not only on the basis of radiocarbon dating of the dung heaps but also in the absence of sloth remains from archaeological refuse, flood-plain alluvium and other widespread postglacial deposits in the Southwest.

1 m

Fig. 3. *Nothrotherium*

†*Acratocnus* Anthony, 1918   EDENTATA
                                                     Megalonychidae

Of the four W.I. sloths recognized in this genus, two occur in Puerto Rico and one each in Hispaniola and Cuba, all in cave deposits. They have not been dated by radiocarbon, but a late postglacial survival seems probable. Their bones are associated with those of pre-historic men (Miller, 1929). The dwarf ground sloths of the W. I. may have been ecologically equivalent to the extinct giant lemurs of Madagascar.

†*Megalocnus* Leidy, 1868   EDENTATA
                                                     Megalonychidae

About the size of a black bear, this was the largest of the W. I. ground sloths; there were 2 spp. Most of the known specimens came from Baños de Ciego Montero, thermal springs near Cienfuegos, Cuba. Approximate time of extinction is unknown, as none of the W.I. extinct fauna, including sloths, has been directly dated by radiocarbon. Related genera of smaller size are *Mesocnus* Matthew, 1931, and *Microcnus* Matthew, 1931.

†*Megalonyx* Harlan, 1825 (Fig. 4)   EDENTATA
                                                   Megalonychidae

Jefferson's ground sloth is considered a woodland or forest species. Known in N.A. since the Pliocene, *Megalonyx* apparently moved against the tide of invading Holarctic mammals. With the forming of the Panamanian isthmus it spread northward to the subarctic. Fossil remains of this ground sloth have been found near Great Slave Lake and at Fairbanks, Alaska (one proximal phalanx). The genus is represented by several species in the late Pleistocene and possibly the early postglacial if the Evansville, Ind., date (W-418, 9400 ± 250) is a valid terminal record. The dated wood from the *Megalonyx* beds was collected in 1870, before critical questions of site stratigraphy were fully appreciated.

1 m

Fig. 4. *Megalonyx*

†*Eremotherium* Spillman, 1948 (Fig. 5)   EDENTATA
                                                   Megatheriidae

The giant browsing ground sloth reached a height about equivalent to that of the tallest giraffe, over 5 m; 1 sp. occurred in southeastern U.S., another ranged through tropical America, including the tar seeps of Peru. At least 8 giant megatheriids were excavated at El

Hatillo in Panama by Gazin (1956). For 150 years this genus was confused with *Megatherium*, which has 4 rather than 3 well-developed toes on the forefoot.

†*Megatherium* Cuvier, 1796   EDENTATA
                                                               Megatheriidae

An elephant-sized ground sloth from the Pleistocene of S.A.; see *Eremotherium*.

1 m

Fig. 5. *Eremotherium*

†*Glossotherium* Owen, 1840   EDENTATA
                                                               Mylodontidae

A small S.A. genus of late-Pleistocene ground sloth. At Lost Hope Inlet Cave in Patagonia, skin, dung, and bones of this sloth are well preserved. They appear to have been contemporaneous with Early Man.

†*Paramylodon* Brown, 1903 (Fig. 6)   EDENTATA
                                                               Mylodontidae

This genus is the most common sloth in the Rancho la Brea tar seeps of California; Marcus (1960) estimates 76 individuals from all the excavations. It was characterized by dermal ossicles and lobate grinding teeth, indicating a grazing habit. About 130 cm high over its back, it was roughly 20 cm higher than *Nothrotherium* but of a much heavier build, with an ox-like chest. It is in uncertain association with wood dated at 14,500 years at Rancho la Brea and 10,500 B.P. at Big Bone Lick, Ky. There are no Early Man kill-sites.

†*Mylodon* Owen, 1840   EDENTATA
                                                          Mylodontidae

S.A. giant sloths that survived into the late Pleistocene. The name *Mylodon* is sometimes applied to N.A. ground sloths of the genus *Paramylodon*.

1 m

Fig. 6. *Paramylodon*

†*Lestodon* Gervais, 1855   EDENTATA
                                                          Mylodontidae

A tapir-sized ground sloth closely related to *Mylodon*, with a broad muzzle and front teeth modified into short tusks.

†*Scelidotherium* Owen, 1840   EDENTATA
                                                          Mylodontidae

A ground sloth with the bulk of a rhinoceros but with an elongate tubular skull and weak dentition. It is found in the tar seeps of Peru, in Ecuador, and elsewhere in S.A.; 1 sp. has been described by Duges (1892) from Guanajuato, Mexico.

Dasypus Linnaeus, 1758   EDENTATA
                                                          Dasypodidae

About 5 spp. of living armadillos occur in N. and S.A. The common species weighs 4–8 kg. Extinct *D. bellus* was about three times larger in body size; its scutes are now recognized as a common element in the late-Pleistocene faunas of Texas. It reached the U.S. in the early Rancholabrean (Slaughter 1966, p. 90) and was extinct at the end of the late-glacial or slightly later if the radiocarbon date of 9,550 (SM-532) on the Ben Franklin Local Fauna is in direct association.

†*Pampatherium* Gervais and Ameghino,   EDENTATA
1880   Dasypodidae

A group of giant armadillos, some 1 m high and 2 m long, probably
weighing 90 kg or more, with a shorter, deeper skull than modern
armadillos. The single N.A. species has been listed under *Holmesina*
Simpson, 1930, and *Chalmytherium* Lund, 1838. It was more
numerous apparently than the glyptodonts with essentially the same
late-Pleistocene range in the U.S.—Florida and S. and E. Texas.
There is one record from Oklahoma. In S.A. pampatheres are known
from the late Pliocene to the late Pleistocene.

†*Neothoracophorus* Ameghino, 1889   EDENTATA
                                                         Glyptodontidae

A late-Pleistocene glyptodont found on the pampas of Argentina and
Uruguay, probably herbivorous, differentiated from *Hoplophorus* by
dentition, details of armor, and shorter tail.

†*Brachyostracon* Brown, 1912   EDENTATA
                                                         Glyptodontidae

This genus belongs to a large subfamily well developed in the late
Tertiary of S.A. and distinct from *Boreostracon* and *Glyptodon*.
In N.A. it occurs only in Mexico, from Nuevo León to Veracruz.
Hibbard (1955) reports a number of scutes in Upper Becerra beds of
Tequixquiac, Mexico, one associated with fragments of a giant tor-
toise (*Geochelone*). Maldonado and Aveleyra report an association
with artifacts.

†*Panochthus* Burmeister, 1867   EDENTATA
                                                         Glyptodontidae

Armadillo-like in build, the S.A. glyptodont *Panochthus* had a body
and upper limbs covered with a tortoise-like bony shield, a bony head
casque, and a long tail ringed with bony plates and armed with long,
bony spikes. There is no evidence of contemporaneity with man.

†*Glyptodon* Owen, 1838 (Fig. 7)   EDENTATA
                                                         Glyptodontidae

Glyptodon was armored in a solid carapace composed of a mosaic of
bony polygons covered by a horny scale-like epithelium, as in the
armadillo. The head was covered by a similar bony casque, the long
tail completely encased in rings of bony plates, and the short
tortoise-like limbs were heavily clawed. This genus is known from
the Blancan of N.A., and the late Pleistocene of S.A.

†*Hoplophorus* Lund, 1838   EDENTATA
                                                         Glyptodontidae

Giant armored edentates, closely resembling *Glyptodon*, but lower
and longer and with a long armored tail, armed with two pairs of
bony plaques; S.A.

†*Boreostracon* Simpson, 1929   EDENTATA
                                        Glyptodontidae

Boreostracon floridanus is known from 6 localities in Florida only; 3 spp. occur in S.A. Slaughter and McClure (1965) report one from Sims Bayou near Houston, Texas, in association with a 23,000-year-old radiocarbon date. If valid, this is a very interesting record of an allegedly tropical animal persisting in the U.S. near the time of the Wisconsin glacial maximum. As in the case of the giant tortoise (*Geochelone*), certain paleontologists regard the glyptodonts as strictly tropical in adaptation, and they assign an interglacial or interstadial age to any deposits containing their distinctive scutes. The onset of continental climates with intense winter cold is said to account for their extinction at the end of the Pleistocene. Unless one is willing to postulate freezing temperatures across the equator, such an explanation clearly begs the question of their extinction in tropical America. At present there are no indisputable associations of this genus with remains of Early Man.

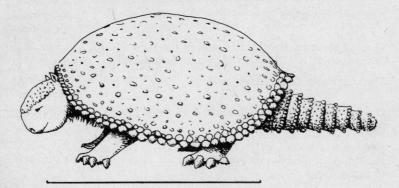

1 m

Fig. 7. *Glyptodon*

Ochotona Link, 1795   LAGOMORPHA
                                        Ochotonidae

Pikas, ca. 12 spp., rat-sized, short-legged relatives of the rabbit, Holarctic. In eastern N.A., formerly occurring in the Appalachian Mts. where they are found in Pleistocene cave deposits.

Lepus Linnaeus, 1758   LAGOMORPHA
The hares comprise ca. 26 spp.   Leporidae
found in Africa, Eurasia, and
N.A.; body wt. 1.3–7.0 kg.

Sylvilagus Gray, 1868   LAGOMORPHA
Cottontail rabbits; N.A., S.A.,   Leporidae
13 spp., wt. to 2.3 kg. The genus
Brachylagus is here included.

Brachylagus. See Sylvilagus.

Sciurus Linnaeus, 1758   RODENTIA
Eurasian and American tree-squirrels,   Sciuridae
55 spp., first found in the Pliocene or
even earlier.

Marmota Blumenbach, 1779   RODENTIA
Marmots, ca. 16 spp., 3.0–7.5 kg   Sciuridae
adult wt.

Cynomys Rafinesque, 1817   RODENTIA
                                                              Sciuridae

Colonial ground squirrels of the American West, formerly occurring
in colonies numbering thousands of individuals, now greatly
reduced in numbers; 5 spp.; wt. 0.9–1.4 kg.

Citellus Oken, 1816   RODENTIA
                                                              Sciuridae

Ground squirrels, sousliks; 14 spp. in the New World, 7 in Eurasia.
Some authors refer them to the genus *Spermophilus*.

Callospermophilus Merriam, 1897   RODENTIA
                                                              Sciuridae

Golden-mantled ground squirrels, 2 or 3 spp.; some authors include
them in *Citellus*.

Geomys Rafinesque, 1817   RODENTIA
Eastern American pocket gophers   Geomyidae
or tuzas, 4 spp.

Thomomys Wied-Neuwied, 1839   RODENTIA
Western pocket gophers, 6 spp., N.A.   Geomyidae

Heterogeomys Merriam, 1895   RODENTIA
Tropical American pocket gophers.   Geomyidae

Perognathus Wied-Neuwied, 1839   RODENTIA
Pocket mice, 25 spp.; drier   Heteromyidae
parts of N.A.

Dipodomys Gray, 1841   RODENTIA
Kangaroo rats, 22 spp., nocturnal   Heteromyidae
rodents of arid N.A.

†*Prodipodomys* Hibbard, 1939   RODENTIA
An extinct Blancan genus closely   Heteromyidae
related to the kangaroo rat
(*Dipodomys*).

Castor Linnaeus, 1758
Beavers. Probably only 1 sp.,
averaging 13 kg, occasionally to
32 kg in wt.

RODENTIA
Castoridae

†*Castoroides* Foster, 1838 (Fig. 8)

RODENTIA
Castoridae

The giant beaver is the only extinct late-Pleistocene genus of its order known from northern N.A. Its near relative, *Trogontherium*, is the only extinct rodent known from the late Pleistocene of Europe. From its fossil distribution, which extends north of the maximum extent of the Wisconsin ice sheet, it has long been known that *Castoroides* endured into the late-glacial, an age interpretation recently verified by radiocarbon dating (Y-526).

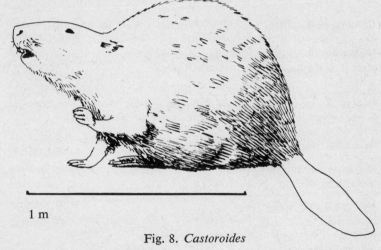

1 m

Fig. 8. *Castoroides*

†*Trogontherium* Fischer, 1809

RODENTIA
Castoridae

The giant beaver of the Upper Pliocene and Pleistocene of Europe and Asia. At least 2 spp. have been described. Presence of *Trogontherium* in the caves of Choukoutien, near Peking, suggests contemporaneity with *Homo erectus*.

†*Megalomys* Trouessart, 1881

RODENTIA
Cricetidae

A muskrat-sized rodent closely related to the rice rat, *Oryzomys*. Only a few museum specimens are known. One sp. was described from Martinique, one from Barbuda (bones only), and one from Santa Lucia, all in the Lesser Antilles.

†*Bensonomys* Gazin, 1942   RODENTIA
An extinct mouse of the   Cricetidae
early-Pleistocene, N.A.

Sigmodon Say and Ord, 1825   RODENTIA
Cotton rats, N.A. and northern S.A.   Cricetidae

Reithrodontomys Giglioli, 1873   RODENTIA
American harvest mice, at least   Cricetidae
17 spp., N.A. and S.A.

Peromyscus Gloger, 1841   RODENTIA
                                                      Cricetidae

White-footed mice, N.A. south to Colombia; ca. 55 spp., the most
common and widespread of all N.A. mammals, of wide ecological
tolerance.

Ondatra Link, 1795   RODENTIA
                                                      Cricetidae

Muskrats, N.A., 2 spp. This semiaquatic rodent has spread rapidly
after its introduction into western Eurasia.

Onychomys Baird, 1857   RODENTIA
Grasshopper mouse, 2 spp., western N.A.   Cricetidae

Oryzomys Baird, 1857   RODENTIA
Rice rats; 100 spp., N.A., S.A.   Cricetidae

Neotoma Say and Ord, 1825   RODENTIA
Wood rats, 22 spp., N.A.   Cricetidae

Zygodontomys Allen, 1897   RODENTIA
Cane mice, ca. 17 spp.,   Cricetidae
medium-sized, ground dwelling;
tropical America.

Arvicola Lacépède, 1799   RODENTIA
                                                      Cricetidae

The water vole, *A. terrestris*, with a head and body length usually
over 150 mm, is the largest native microtine rodent in the Old World.

Cricetulus Milne-Edwards, 1867   RODENTIA
Rat-like hamsters of S.E. Europe to   Cricetidae
Siberia, 7 spp. in open country.

Dicrostonyx Gloger, 1841   RODENTIA
                                                      Cricetidae

Collared lemmings, 2 spp. restricted to arctic tundra and turning white
in winter. *Dicrostonyx* occurs far south of its present range in the
Pleistocene of both New and Old World.

Lemmus Link, 1795   RODENTIA
Cricetidae

True lemmings, found in arctic tundra except N.E. Canada; this animal is responsible for the spectacular "lemming migrations."

Microtus Schrank, 1798   RODENTIA
Meadow voles, about 44 spp.,   Cricetidae
holarctic.

Pitymys McMurtrie, 1831   RODENTIA
N. Hemisphere; pine mouse.   Cricetidae

Synaptomys Baird, 1857   RODENTIA
Lemming mice, 2 spp. in N.A.   Cricetidae

†*Mimomys* Major, 1902   RODENTIA
A primitive vole of the early   Cricetidae
Pleistocene of Europe and N.A.

†*Pliolemmus* Hibbard, 1937   RODENTIA
A small primitive vole from   Cricetidae
the early Pleistocene, N.A.

†*Pliophenacomys* Hibbard, 1937   RODENTIA
A small primitive vole from   Cricetidae
the early Pleistocene of N.A.

†*Pliopotamys* Hibbard, 1938   RODENTIA
A rat-sized vole of the Blancan   Cricetidae
(early Pleistocene) of N.A.

Rattus Fischer, 1803   RODENTIA
Muridae

Typical rats; the largest genus of mammals with about 570 named spp.; originally confined to the Old World but now cosmopolitan.

Mus Linnaeus, 1758   RODENTIA
House mice; native of the Old World;   Muridae
worldwide as introduced.

Zapus Coues, 1875   RODENTIA
Zapodidae

Jumping mice, 2 spp., known from N.A.; a close relative, *Eozapus*, occurs in E. Asia.

Hystrix Linnaeus, 1758   RODENTIA
Old World porcupines of a dozen   Hystricidae
spp., wt. 17–27 kg.

Hydrochoerus Brisson, 1762   RODENTIA
Hydrochoeridae

At 50 kg the pig-sized capybaras are the largest living species of rodents. They range from S.A. north to Panama, are semiaquatic, and live in family groups or bands.

†*Neochoerus* Hay, 1926 (Fig. 9)   RODENTIA
                                                          Hydrochoeridae

These extinct capybaras are known from the late Pleistocene of
Florida and S.A. They have not been dated by radiocarbon or found
associated with Early Man.

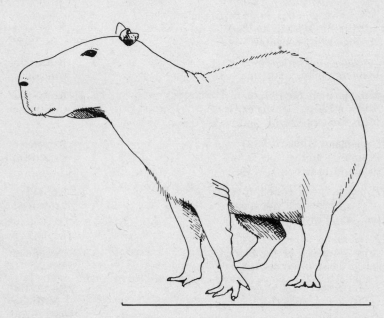

1 m

Fig. 9. *Neochoerus*

†*Elasmodontomys* Anthony, 1916   RODENTIA
                                                          Heptaxodontidae

Rabbit-sized W.I. rodents closely related to the modern hutias known
only from cave remains in Puerto Rico and Hispaniola.

†*Eumegamys* Kraglievich, 1926   RODENTIA
                                                          Dinomyidae

The largest known rodent, the size of a small rhino, this S.A.
herbivore of the Plio-Pleistocene has no known association with
prehistoric man.

Dasyprocta Illiger, 1811   RODENTIA
Agoutis, of tropical America,   Dasyproctidae
wt. 1.3–4.0 kg. Diurnal except
where molested.

Capromys Desmarest, 1822   RODENTIA
                                                          Capromyidae

Cuban hutias of 3 spp. including one on the Isle of Pines. Adults
weigh 4–7 kg, live in pairs, are diurnal, and are regarded as a delicacy
by the natives.

Geocapromys Chapman, 1901 RODENTIA
Capromyidae

The Jamaican hutia, *G. brownii*, occurs both in Jamaica, where it is the only surviving mammal other than bats, and on Little Swan Island. It is hunted and eaten by man and survives only in the most inaccessible areas.

Plagiodontia Cuvier, 1836 RODENTIA
Hispaniolan hutias, more often found Capromyidae
subfossil in kitchen middens
than as living rodents.

†*Hexolobodon* Miller, 1929 RODENTIA
Capromyidae

The Haitian relative of Cuban hutias, known only from isolated teeth and bone fragments from caves; apparently became extinct in early historic times.

†*Isolobodon* Allen, 1916 RODENTIA
Capromyidae

An extinct W.I. rodent related to the living hutias; 2 spp. are known from fragmentary remains in caves, one from Puerto Rico, the other from the Dominican Republic.

†*Brotomys* Miller, 1916 RODENTIA
Echimyidae

The "mohuy" of Oviedo, 2 subfossil spp. common on Hispaniola in caves and prehistoric Indian middens. *Brotomys* Miller, 1916, was described from caves and Indian middens of Cuba.

Delphinus Linnaeus, 1758 CETACEA
Delphinidae

Common dolphin. Small beaked whales up to 2.5 m in length and 75 kg in wt. Worldwide in temperate and warm seas.

Tursiops Gervais, 1855 CETACEA
Bottle-nosed dolphins, 2 spp. Delphinidae

†*Theriodictis* Mercerat, 1891 CARNIVORA
A canid of the S.A. Pleistocene. Canidae

Canis Linnaeus, 1758 CARNIVORA
Canidae

The dogs, wolves, coyotes, and dire wolves. Perhaps the most interesting of the N.A. extinct species was *Canis dirus*, formerly treated as a separate genus, *Aenocyon* (Merriam, 1918). This was the most abundant mammal found at Rancho la Brea in California where 1,646 individual dire wolves were recorded (Marcus, 1960). They were also common in the Upper Pleistocene tar seeps of Peru. There is no direct association of dire wolves with prehistoric man, but late-glacial deposits containing cultural material such as those in Ventana Cave, Jaguar Cave, and Gypsum Cave also contain remains of dire wolves. While it is very likely that they survived to the end of the late-glacial in N.A. and S.A., there is no critical stratigraphic or radiometric determination of their time of extinction. Although as large as a timber wolf, the dire wolf was differently proportioned, with relatively shorter limb segments and a massive, hyena-like dentition. A smaller extinct member of the genus, *C. orcutti*, is represented by 239 individuals at Rancho la Brea.

Vulpes Brisson, 1762   CARNIVORA
Red foxes, to 9 kg. *V. vulpes* is   Canidae
Holarctic in range; the kit fox,
V. macrotus, is endemic to N.A.

Nyctereutes Temminck, 1838–39   CARNIVORA
Raccoon dogs, 1 sp., Eurasia,   Canidae
up to 7.5 kg.

Cuon Hodgson, 1837   CARNIVORA
Indian dhole or red wolf; the single   Canidae
sp. *C. alpinus* occurs in S.E. Asia
to Siberia; wt. 14–21 kg.

†*Arctotherium* Bravard, 1857   CARNIVORA
See *Arctodus*.   Ursidae

†*Arctodus* Leidy, 1854 (Fig. 10)   CARNIVORA
                                      Ursidae

Taxonomic treatment of the giant short-faced bears has suffered from uncertainty surrounding the type, an unworn second molar illustrated

1 m

Fig. 10. *Arctodus*

by Leidy. It is probable that *Arctodus, Arctotherium,* and *Tremarctotherium* are congeneric. The short-faced bear from Rancho la Brea was almost 30 cm taller than the extinct California grizzly (*Ursus horribilis*) and probably about twice its weight. Stock considers the tremarctotheres as more carnivorous than the true bears and comparable in size to the Kodiak bear. The group was widespread over N. and S.A. and is closely related to the spectacled bear (*Tremarctos*) of the S.A. Andes. Although known from various deposits that postdate the last glacial maximum of 20,000 years ago, such as Friesenhahn Cave, Potter Creek Cave, and Burnet Cave, there are no critical, stratigraphically controlled C¹⁴ dates and no indisputable associations of the giant short-faced bear with artifacts.

Tremarctos Gervais, 1855   CARNIVORA
Ursidae

Spectacled bears, a single herbivorous sp., *T. ornatus*, of the Andean forests, occurring up to 3,000 m; wt. 140 kg. Extinct *T. floridanus* was approximately twice as large; it has been found in late-Pleistocene cave deposits of Mexico and eastern U.S. north to Tennessee.

Selenarctos Heude, 1901   CARNIVORA
Ursidae

Asiatic black bears, 1 sp., wt. to 120 kg; considered a subgenus of *Ursus* by some authors.

Ursus Linnaeus, 1758   CARNIVORA
Brown bears and grizzly bears, the   Ursidae
former weighing up to 780 kg.

Euarctos Gray, 1864   CARNIVORA
Ursidae

American black bears, wt. 120–150 kg. The single sp. is often considered a subgenus of *Ursus*. Its N.A. fossil record dates from the Illinoian glaciation.

Procyon Storr, 1780   CARNIVORA
Raccoons; N.A., S.A.   Procyonidae

†*Parailurus* Schlosser, 1899   CARNIVORA
Extinct pandas, Europe.   Procyonidae

Mustela Linnaeus, 1758   CARNIVORA
Weasels, ermines, ferrets, mink,   Mustelidae
ca. 15 spp.; Eurasia, N.A., S.A.

Gulo Pallas, 1780   CARNIVORA
Mustelidae

A single sp., the wolverine of the taiga and forest-tundra, prefers carrion but will prey on living animals in winter; wt. 14–27.5 kg.

†*Trigonictis* Hibbard, 1941   CARNIVORA
                                                         Mustelidae

A weasel-like mammal from the Blancan deposits of N.A., closely related to the present-day Central and S.A. grison (*Grison vittatus*).

†*Canimartes* Cope, 1892   CARNIVORA
                                                         Mustelidae

An extinct genus of marten-like carnivore of the early Pleistocene in western N.A.

Martes Pinel, 1792   CARNIVORA
                                                         Mustelidae

Martens and fishers, ca. 8 spp., semiarboreal carnivores usually confined to northern coniferous forests.

Meles Brisson, 1762   CARNIVORA
The single sp. is the Old World   Mustelidae
badger, weighing up to 20 kg.

Taxidea Waterhouse, 1839   CARNIVORA
American badgers, 3.5–10 kg.   Mustelidae

Mephitis G. Cuvier, 1800   CARNIVORA
Striped and hooded skunks, N.A.,   Mustelidae
2 spp.

Lutra Brisson, 1762   CARNIVORA
River otters, ca. 12 spp., wt. 4.5–14 kg.,   Mustelidae
found on all major land masses
except Australia.

Crocuta Kaup, 1828   CARNIVORA
                                                         Hyaenidae

The one living sp., *C. crocuta*, is the largest member of the family, weighing 59–82 kg. The closely related extinct cave hyena of Europe was common in the Pleistocene; its coprolite is often found.

Hyaena Brisson, 1762   CARNIVORA
Striped hyenas; 2 spp. in Af. and   Hyaenidae
cent. Asia, wt. 27–54 kg.

Felis Linnaeus, 1758   CARNIVORA
                                                         Felidae

Cats, mountain lions, servals, etc., ca. 25 spp. in all. The genus *Panthera* is sometimes included with *Felis*. Of the large felines Simpson recognizes only 3 groups from the Pleistocene of N.A.: (1) the jaguar, *Panthera onca*, which may average slightly larger in the Pleistocene than the largest of the present living races; (2) pumas or mountain lions, of which the Pleistocene forms reach a little larger than the living species; and (3) *Panthera atrox*, a large cat that is not a lion and might be called a giant jaguar; it is close to but fairly different from the living jaguar.

Leo. See *Panthera.*

Lynx Kerr, 1792   CARNIVORA
                                                               Felidae

Lynx, bobcat, caracal; N. Hemisphere into Africa and S.E. Asia,
included by Simpson in the genus *Felis.*

Panthera Oken, 1816   CARNIVORA
                                                              Felidae

In addition to living lions, tigers, leopards, and jaguars, this genus
or subgenus of *Felis* includes the extinct jaguar of N.A., *P. atrox,*
and the cave lion, *P. spelaea,* of Europe and Asia. The extinct jaguar
was the most formidable predatory felid, larger than both the living
cats of Eurasia and the extinct saber-tooth. It ranged from Alaska
to Mexico. Other late-glacial age occurrences include Wilson Butte
Cave, Idaho, and Ventana Cave, Ariz. At least 76 individuals of
P. atrox were recovered in the excavations from Rancho la Brea.

Uncia Gray, 1867   CARNIVORA
The snow leopard of central Asia,   Felidae
23–41 kg; some authorities regard it
as a subgenus of *Felis.*

†*Smilodon* Lund, 1842 (Fig. 11)   CARNIVORA
                                                             Felidae

Best known from the tar pits of Rancho la Brea, Calif., where it is
second in abundance to the dire wolf, the saber-tooth cat *Smilodon*
was first described from the caves of Brazil. It had approximately the
size of an African lion but had heavier forequarters, lighter hindquarters, and a bob-tail, implying it was not so fleetfooted. This plus
its long saber-like upper canines make it probable that it preyed upon
the larger, more slowly moving animals such as ground sloths and

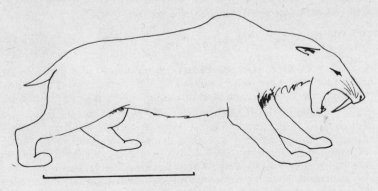

1 m

Fig. 11. *Smilodon*

young proboscideans. The Rancho la Brea bones indicate many pathological lesions especially in the limb bones and lumbar region, consistent with an active and violent life. Numbers of species have been described; it is not clear how many biological species they represent. The time of extinction in the late Pleistocene is still to be determined critically. *Smilodon* bones were packed around the tree at Pit 3 at Rancho la Brea, radiocarbon dated at about 14,500 years, but the possibility of a secondary association in the case of tar-pit stratigraphy cannot be discounted. No direct associations with Early Man are known. If one accepts the ecological proposition that a reduction in number of prey species would directly result in extinction of certain predators, carnivores, and commensals, the extinction of *Smilodon* is self-explanatory.

†*Dinobastis* Cope, 1893

CARNIVORA
Felidae

Dinobastis is a specialized saber-tooth cat, structurally closer to *Smilodon* than any other genus. It stood conspicuously high in the forequarters and had serrated canines but also many skeletal features similar to those of true cats. A nearly complete skeleton and also much other material are known from Friesenhahn Cave, Texas, where its chief diet was young proboscideans.

Phoca Linnaeus, 1758
Harbor seals, 1 sp., *P. vitulina;*
found along coasts of the Holarctic.

CARNIVORA
Phocidae

†*Macrauchenia* Owen, 1840 (Fig. 12)

LIPTOTERNA
Macraucheniidae

This late-Pleistocene camel-like animal with a long neck has been known since the time of Darwin. The reduction of liptoterns from 10 genera in both the Oligocene and Miocene to 6 in the Pliocene and 2 in the Pleistocene is thought to be the result of competition with artiodactyls entering S.A. in the late Tertiary. It is likely that at least this genus survived into the time of Early Man.

†*Toxodon* Owen, 1837 (Fig. 13)

NOTOUNGULATA
Toxodontidae

A rhinoceros-sized giant herbivore of an order that evolved in isolation in, and was always confined to, S.A., *Toxodon* was built low and long, was hornless, and equal to a hippopotamus in girth. Some authorities have surmised it was semiaquatic.

†*Stegomastodon* Pohlig, 1912

PROBOSCIDEA
Gomphotheriidae

According to Hibbard and Dalquest (1966) there are only two localities in which *Stegomastodon* and *Mammuthus* occur in the same fauna: "*Mammuthus* rapidly displaced *Stegomastodon*, which was chiefly a grazer and semibrowser. *Mammut* and *Cuvieronius* were better adapted to browsing than was *Stegomastodon* and it appears that the chief habitat of *Stegomastodon* was taken over by *Mammuthus*:" In N.A., *Stegomastodon* is regarded as a Blancan guide fossil and its extinction seems to be simply a matter of phyletic replacement by a superior Holarctic invader. But in S.A., which was not penetrated by *Mammuthus* or *Mammut*, *Stegomastodon* survives into the late Pleistocene, apparently in association with Early Man at Muaco, Venezuela.

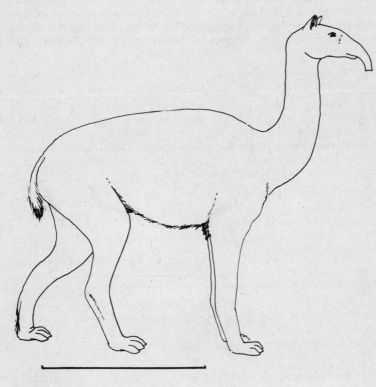

1 m

Fig. 12. *Macrauchenia*

†*Cuvieronius* Osborn, 1923   PROBOSCIDEA
                                              Gomphotheriidae

While commonly called mastodons, *Stegomastodon* and *Cuvieronius* are not in the same family as *Mammut*. Hibbard and Dalquest (1966) recently reported *Cuvieronius* from the Seymour Formation (Irvingtonian) of Texas. It is also reported from Middle Pleistocene gravels near Las Cruces, N.M. Considering only its distribution in temperate

N.A., one might conclude that *Cuvieronius* disappeared because of natural causes long before the arrival of Upper Paleolithic hunters. But in tropical America its remains are associated with human artifacts.

†*Notiomastodon* Cabrera, 1929   PROBOSCIDEA
                                                                 Gomphotheriidae

This rather rare S.A. late-Pleistocene elephant is distinguished from *Stegomastodon* by an enamel band running along its tusks. Basic stratigraphic information, including approximate time of extinction, is unknown, but there is no reason to believe that the Pleistocene gomphotheres of S.A. disappeared before the arrival of Early Man.

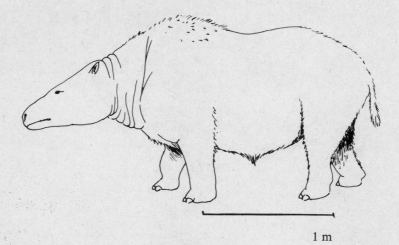

1 m

Fig. 13. *Toxodon*

†*Mammut* Blumenbach, 1799 (Fig. 14)   PROBOSCIDEA
                                                                 Mammutidae

These large, heavily built proboscideans occurred throughout the N. Hemisphere in the late Tertiary. They survived the Pleistocene only in N.A. where the American mastodon is known from at least 1,000 localities in deposits mainly postdating the last glacial maximum. Radiocarbon dates on mastodon bones are as young as 6,000–8,000 B.P., but it is not certain that the dated samples were uncontaminated by humic acids or other intrusive carbon. From associated pollen it is certain the American mastodon occupied late-glacial boreal forest or woodland, characterized by spruce and other coniferous trees. However, mastodons are known as far south as central Florida in Pleistocene deposits far beyond the southern limit of boreal forest even during the time of its maximum Wisconsin age displacement. Mastodons were certainly not confined to coniferous-forest habitats. From both its bunodont molars and its Pleistocene center of distribution in forested regions it is apparent that mastodons were more strictly browsing animals than were mammoths. In Texas the ratio of mastodon to mammoth bones is an indication of the position of the forest-prairie boundary at the time. In the Southwest, late-Pleistocene mastodons are known only from certain montane forest "islands" such as the Sandia Mts. of New Mexico, where they reached an elevation of approximately 2,550 m.

1 m

Fig. 14. *Mammut*

Many names have been proposed, but there is no certainty that they represent more than one good biological sp. This animal stood over 2 m high, the Warren mastodon attaining over 3 m. If its disappearance was caused or hastened by the arrival of prehistoric man, the kill sites must have been mainly or entirely in upland areas where bones were not preserved.

Loxodonta F. Cuvier, 1827 PROBOSCIDEA
Elephantidae

The single living sp., *L. africana*, is the largest living terrestrial mammal, wt. 5,000–7,500 kg. Females may bear 4–5 young in a lifetime.

†*Palaeoloxodon* Matsumoto, 1924   PROBOSCIDEA
According to Simpson, a synonym   Elephantidae
of *Loxodonta*, the genus of the
African elephant.

†*Archidiskodon* Pohlig, 1888   PROBOSCIDEA
A synonym of *Mammuthus* Burnett,   Elephantidae
1830.

Elephas Linnaeus, 1758   PROBOSCIDEA
          Elephantidae

The erratic use of generic names in the subfamily Elephantinae
including *Loxodonta, Paleoloxodonta, Mammutus, Archidiskodon,*
and *Elephas,* plus half a dozen less common synonyms, is likely to
discourage a nonpaleontologist. Perhaps the main morphological
feature of the group is a complicated tooth with enamel plates set
close together. Unlike mastodons, elephant teeth are adapted to
grazing. Dwarf elephants are known from certain nearshore islands
in the Mediterranean, the Channel Islands of California, and the
Celebes. In Sicily their remains were found in great numbers with a
degenerate stag, *Bos, Bison, Hyaena,* and partly worked flint
implements. The elephants of Sardinia and Malta were remarkably
small—1 m or less. The population of the Celebes, an *Archidiskodon*
of Chinese origin, was associated with fossils of the giant tortoise.
In each case the island elephants seemed to have arrived with the
help of sea-level recession during the glacial periods. In no case is it
certain that the insular forms disappeared before prehistoric man
arrived. The New World elephants did not range south of Mexico,
apparently being blocked by the tropical forests of Central America.
There are numerous archaeological records of elephants (*Mammutus*) in the late-glacial of N.A., summarized by Haynes (1964).
Apart from a few doubtful radiocarbon dates on bone or tusk, there
is no good evidence that elephants survived into the early postglacial
of N.A., Europe, or Eurasia. From both the abundance of elephants
in undisturbed parts of Africa and the number of fossil bones of this
group found in late-Pleistocene beds of N.A. and Eurasia, it seems
likely that they were important animals in the fauna of their time—
a species that the Early Hunters would have encountered repeatedly.

†*Mammuthus* Burnett, 1829 (Fig. 15)   PROBOSCIDEA
          Elephantidae

In a broad sense the mammoths of the N. Hemisphere may be
considered part of the genus *Elephas.* The four widely distributed,
morphologically distinct, N.A. late-Pleistocene species, in order of
increasing size, were the woolly mammoth (*M. primigenius,* 2.8 m),
Jefferson's mammoth (*M. jeffersoni,* 3.4 m), the Columbian mammoth
(*M. columbi,* 3.6 m), and the imperial mammoth (*M. imperator,*
4.0 m), the latter twice the shoulder height of the dwarf mammoths
(*M. exilis*) known from Santa Rosa Island, and considerably larger
than the largest African elephant (3.52 m). *M. imperator* was nevertheless dwarfed by the Mosbach elephant, *Elephas* (*Mammuthus*)

trogontherii of Eurasia, estimated to have been 4.5 m tall at the shoulder and considered the predecessor of the woolly mammoth. While not so common as mastodons east of the Mississippi, mammoth remains are extremely numerous in the Pleistocene of western N.A. Slaughter (1960, p. 486) reports 15 semiarticulated skeletons of Columbian elephants in a single pit along the Trinity River, Dallas, where he believes that not more than one in a thousand mammoth bones are recovered. In Europe, Cuvier had to prove anatomically that the fossil elephants found in Britain were not

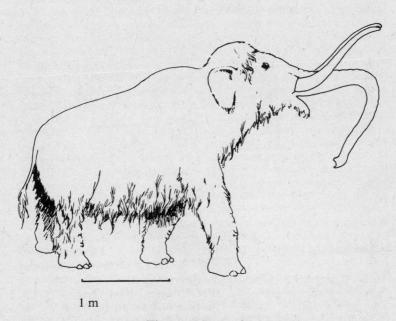

1 m

Fig. 15. *Mammuthus*

brought there by the Romans. Thousands of teeth of woolly mammoths were dredged up off Norfolk, England, in the early 19th century. It is estimated that 117,000 mammoths have been discovered in Russia during the last 250 years; many were removed for their ivory. Some frozen remains of woolly mammoth from Siberia are at or beyond the range of radiocarbon dating. The last woolly mammoth lingered into the late-glacial and coexisted with Upper Paleolithic hunters. In America, mammoths invaded in the Irvingtonian, approximately dated by potassium argon at 1.5 million years. *M. columbi* and *M. imperator* have been found associated with Clovis fluted points radiocarbon-dated at 11,200 B.P.

†*Hydrodamalis* Retzius, 1794 (Fig. 16)   SIRENIA
                                                          Dugongidae

The Rhytina or Stellar's sea cow is of particular interest as the only genus of large mammal known to have become extinct in the last 500 years. It was a whale-sized (8 m long) manatee-like marine herbivore. Living sea cows are much smaller (2 to 3 m) and are tropical. Formerly known from the shallows of two small islands of the Komandorskie group in the Bering Straits of the North Pacific, it was obliterated between 1742 and 1768 by Vitus Bering's crew and other seafarers. Fossils of *Hydrodamalis* have been reported from the Tertiary and late-Pleistocene of the Pacific Coast of N.A., the latter dated by radiocarbon at 19,000 B.P. (SI-115).

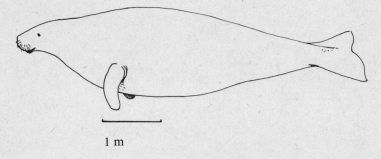

1 m

Fig. 16. *Hydrodamalis*

†*Stylohipparion* van Hoepen, 1932   PERISSODACTYLA
                                                          Equidae

In E. Africa and S. Africa, *Stylohipparion* is clearly associated with Acheulean-age cultures—as at Olduvai Gorge (Bed IV) and Olorgesailie. In N. Africa, it is regarded as an indicator of the Villafranchian. Possibly the extinction pattern matches that of the New World where certain genera disappeared in the early Pleistocene of N.A., yet survived into the late Pleistocene in S.A. On the other hand, the lack of any major geographic barrier, such as the Central American isthmus, makes a transgressive wave of extinction within the African continent seem unreasonable.

†*Nannippus* Matthew, 1926   PERISSODACTYLA
                                                          Equidae

A small 3-toed horse of the Pliocene, last recorded in the Blancan of N.A. It is not impossible that this genus survived to a later time in Mexico.

†*Hippidium* Owen, 1869 (Fig. 17)   PERISSODACTYLA
                                                          Equidae

Three to four spp. of this horse are known from the Lower Pleistocene of Argentina. They differ from *Equus* in their shorter, heavier limbs, very large, heavy-looking heads, and extremely slender, splint-like nasal bones.

†*Onohippidium* Moreno, 1891   PERISSODACTYLA
                                                       Equidae

A horse found associated with prehistoric man in the Eberhart Cave, Ultima Esperanza, of the S.A. Pleistocene.

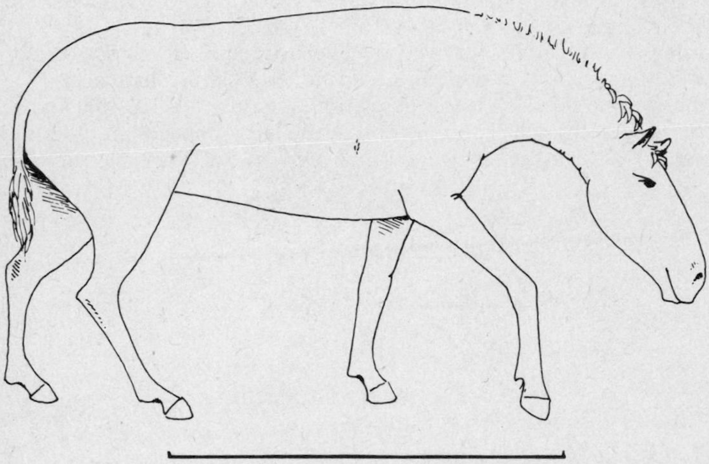

1 m

Fig. 17. *Hippidium*

†*Plesippus* Matthew, 1924   PERISSODACTYLA
Zebrine horses of the Blancan of   Equidae
N.A., sometimes treated as a
subgenus of *Equus*.

Equus Linnaeus, 1758   PERISSODACTYLA
                                                       Equidae

Numerous fossil horses have been described from the late Pleistocene of N.A., ca. 12 from Texas alone. No more than 2 or occasionally 3 distinct fossil populations can be recognized in any single horizon or quarry. Horse bones are among the most common Pleistocene fossils in N.A.; 163 identifiable bone and tooth fragments were found in the conglomerate layer of Ventana Cave, with 23 more associated with prehistoric man in the volcanic debris. A large horse was the most common of the fossils in Burnet Cave, N.M. At least 130 individual *Equus occidentalis* are reported by Marcus from Rancho

la Brea. Despite stratigraphic association with Early Man, there are
no obvious kill sites similar to those found in Europe. Of the approximately 8 living spp., horses were domesticated mainly from the
tarpan of eastern Europe and Asia. Although *Equus* had been extinct
throughout the Americas for at least 8,000 years at the time of
European colonization, introduced horses escaped captivity and
multiplied phenomenally. An estimated· 18,000–34,000 feral horses
and 5,500–13,000 feral burros still occupy Anglo-America, including
the driest of the desert regions (McKnight, 1964).

Tapirus Brisson, 1762 (Fig. 18)   PERISSODACTYLA
                                                     Tapiridae

Four spp. of tapirs survived the Pleistocene, 3 in tropical America
and 1 in the Old World. In addition, N.A. alone had at least 3–4
Pleistocene spp. The larger living tapirs may weigh 225–300 kg. As in
the case of most Pleistocene faunas, the largest species in the genus is
extinct—the subgenus *Megatapirus augustus* of Yunnan, Indochina,

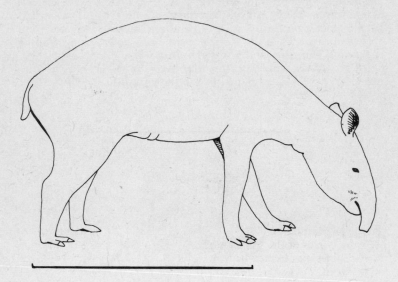

1 m

Fig. 18. *Tapirus*

and perhaps Java. Extinct tapirs survived in N.A. through the late-
glacial until 9,400 years ago if the Evansville, Ind., date (W-418) is
accepted. On the assumption that tapirs can live only in humid areas,
their occurrence in Ventana Cave in the Sonoran desert region in
southern Arizona seems a perplexing problem. But the Ventana Cave
record may be no more remarkable than the occurrence in equally

arid regions of African large mammals as elephant, rhinoceros, and hippopotamus.

Dicerorhinus Gloger, 1841   PERISSODACTYLA
                                             Rhinocerotidae

Smallest of the living rhinos, 1.1–1.5 m tall at the shoulders. The disappearance of Eurasian members of this genus with the survival of a tropical species (in Sumatra) matches the worldwide pattern of temperate-zone extinction and tropical survival seen in many large mammals.

Rhinoceros Linnaeus, 1758   PERISSODACTYLA
One-horned rhinos, 1 sp. in   Rhinocerotidae
northern India, another in Java;
wt. 2,000–4,000 kg.

†*Coelodonta* Bronn, 1831 (Fig. 19)   PERISSODACTYLA
                                             Rhinocerotidae

The woolly rhinoceros of the Eurasian Upper Pleistocene steppe-tundra was hunted by Paleolithic man. Its physical appearance is well known both from European cave paintings and carvings and from preserved carcasses in the Siberian permafrost and the salt deposits of Starunia, formerly in S.E. Poland. Remarkably, the woolly rhino failed to follow woolly mammoths over the Bering Bridge into America where the family perished in the Pliocene.

1 m

Fig. 19. *Coelodonta*

†*Stylochoerus* van Hoepen and   ARTIODACTYLA
van Hoepen, 1932   Suidae

The tusks of this giant relative of the bush pig were mistaken for those of mastodon in the earlier work at Olduvai Gorge. The genus is called *Afrochoerus* by Leakey. In both E. and S. Africa it occurs associated with Middle Pleistocene faunas and Acheulean sites including Vaal River, Cornelia, Florisbad, and Vlakkraal. Cooke (1963) notes it is almost impossible to guess the habits of the giant African pigs beyond the fact that the brachyodont forms were probably thicket dwellers, and the more hypsodont types were grass eaters.

†*Meschoerus* Shaw and Cooke, 1941   ARTIODACTYLA
An extinct Middle Pleistocene   Suidae
pig of Africa.

†*Tapinochoerus* van Hoepen and   ARTIODACTYLA
van Hoepen, 1932   Suidae

A large relative of the bush pig common in beds at Olduvai Gorge and from other Middle Pleistocene deposits south of the Sahara.

Sus Linnaeus, 1758   ARTIODACTYLA
Wild boars, pigs; at least 5 spp.,   Suidae
Old World.

†*Platygonus* Le Conte, 1848 (Fig. 20)   ARTIODACTYLA
                                                Tayassuidae

Widespread through N. and S.A., the common Pleistocene "flat-headed peccary" ranged in size from that of a large European wild boar down to the modern white-lipped peccary. Longer limbed, longer tailed, and longer snouted than living peccaries, they are

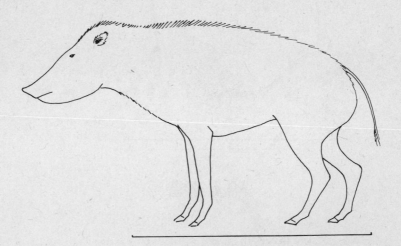

1 m

Fig. 20. *Platygonus*

believed to have been primarily a prairie–plains form in contrast to
Mylohyus, which seemingly preferred more wooded country. Records
of *Platygonus* east of the Mississippi River may in some cases be
associated with late- or full-glacial treeless habitats along the ice
margin. They were probably as gregarious as modern peccaries, and
herds are often found in the same deposit, under situations that imply
mass mortality. A 4,000-year C^{14} date (M-1516) from the Midwest
must be regarded as contaminated. There are no good associations
of *Platygonus* with Early Man.

†*Mylohyus* Cope, 1889 (Fig. 21)   ARTIODACTYLA
                                                    Tayassuidae

Compared with *Platygonus* and the living peccaries *Mylohus*, the
"long-nosed peccary" of the N.A. Pleistocene, was longer legged, had
a much longer, slimmer snout, and was probably a woodland

1 m

Fig. 21. *Mylohyus*

browser and omnivore. It ranged throughout eastern U.S. south of
42°N. Lat. from the Atlantic Coast west to the 100th Meridian. In
Texas the faunal change from glacial to interglacial ages was characterized by a change from *Mylohyus* to *Platygonus*. In the northeast

it is possible *Platygonus* occupied the open, grassy, ice-margin environment of full- or late-glacial times in Pennsylvania, succeeded by *Mylohyus* with the return of forest. From associated pollen and small vertebrates in New Paris Sinkhole No. 4, it is apparent that *Mylohyus* occupied a cold woodland during deglaciation (Guilday et al., 1964). The mistaken notion that fossil peccaries indicate warm climate is based on the tropical and warm-temperate distribution of the white-lipped and collared peccaries. In both Pennsylvania and Texas, *Mylohyus* remains are associated with radiocarbon dates of late-glacial or possibly early postglacial age. No direct archaeological associations are known, although both *Mylohyus* and cultural material were removed from Hartman Cave, Pa., during the 1800s.

Tayassu Fischer, 1814   ARTIODACTYLA
                                                          Tayassuidae

Peccaries; N.A., S.A. There are two species in N.A., one of them (collared peccary) sometimes included in a separate genus *Dicotyles*.

Hippopotamus Linnaeus, 1758   ARTIODACTYLA
                                                         Hippopotamidae

The only living species, *H. amphibus*, is found in Africa. It occurs in interglacial faunas from England to India. The dwarf extinct Madagascar hippo, *H. lemerlei*, an abundant fossil, has been radiocarbon-dated at around 1,000 years B.P.

†*Camelops* Leidy, 1854 (Fig. 22)   ARTIODACTYLA
                                                           Camelidae

Its limbs were 20% longer, its joints even knobbier, and its head was longer and narrower; otherwise, *Camelops* was the morphological equivalent of the dromedary of S.W. Asia. The tall neural spines and lack of lumbar spines suggest a single mid-dorsal hump. Webb (1965) finds 18 valid species names referred to *Camelops*. But typically one finds only 1 sp. in a well-studied single fossil horizon, such as Rancho la Brea, where all 36 individuals recovered were referred to *C. hesternus*. Three to 5 biological spp., not all contemporaneous, would seem to be a generous estimate for the entire Pleistocene fauna. *Eschatius* is probably a synonym. Association with prehistoric man is claimed in many cases, such as at Sandia Cave, Burnet Cave, and Clovis, N. M.; Paisley Cave, Ore.; the Lindenmeier site, Colo.; Tule Springs, Nev.; and Double Adobe, Ariz. But no kill sites are known. As in the case of the shasta ground sloth, mummified remains of camels have been found in caves (Fillmore, Utah), with dried muscle still attached to the skull. Despite conjecture about a late survival, both stratigraphic and C^{14} evidence point to late-glacial extinction. The Cenozoic evolution of the family in N.A., plus the fact that introduced dromedaries (*Camelus*) found native southwestern shrubs such as mesquite and creosote bush entirely to their liking, provoke the question of why *Camelops* disappeared. The case is an outstanding example of late-Pleistocene extinction without replacement.

†*Titanotylopus* Barbour and Schultz, 1934 (Fig. 23)   ARTIODACTYLA
Camelidae

This great browser, called *Gigantocamelus* in earlier literature, stood over 3.5 m tall. Its skull was over 855 mm long. Webb regards *Titanotylopus* as closer to *Camelus* than the Pleistocene camels of N.A. North of Mexico it is found only in deposits of Blancan age; however, Maldonado and Aveleyra (1949) report it from the Upper Becerra of Tequixquiac. If correct, this record would indicate a late-Pleistocene survival south of the Rio Grande. Was *Titanotylopus* the New World ecological equivalent of the giraffe?

1 m

Fig. 22. *Camelops*

†*Palaeolama* Gervais, 1867   ARTIODACTYLA
Camelidae

A group of perhaps 6 spp. of camel larger than *Lama* but closely related, found throughout the late Pleistocene of S.A. and presumably associated with ancient man at Muaco, Venezuela.

1 m

Fig. 23. *Titanotylopus*

†*Tanupolama* Stock, 1928 (Fig. 24)   ARTIODACTYLA
                                                      Camelidae

Webb lists 10 recognized spp. of llama-like camels but he was unable
to indicate how many may be synonyms. Nor is it proven that
Tanupolama is generically distinct from *Paleolama* of S.A. (Webb,
1965, p. 35). Older literature records of *Camelus americanus* presumably refer to *Tanupolama*. Its remains are associated with a
17,000-year radiocarbon date at Grossbeck Creek and with an
8,500-year date at Gypsum Cave (C-222). The latter, however, was on

.1 m

Fig. 24. *Tanupolama*

sloth dung and appears too young in terms of subsequent information on the age of *Nothrotherium*. According to Stock (1931), the Gypsum Cave *Tanupolama* came from a layer of burned dung above and not more than 6 ft from dart points. The camel remains included charred and broken limb bones.

Camelus Linnaeus, 1758   ARTIODACTYLA
                                                       Camelidae

The one-humped camel, *C. dromedarius*, is known only as a domestic form. It was introduced into N.A. unsuccessfully and is a feral species in Australia. The two-humped or Bactrian camel, *C. bactrianus*, weighs 450–690 kg. A wild population is said to persist in Chinese Turkestan and Mongolia.

Moschus Linnaeus, 1758   ARTIODACTYLA
                                                  Cervidae

Musk deer, 1 sp. only, in central Asia through Siberia; upper canine developed as a tusk in males. The species is generally solitary; wt. 9–11 kg.

†*Megaloceros* Brookes, 1828   ARTIODACTYLA
(*Megaceros* Owen,1844)   Cervidae

The giant Irish elk and its relatives were widespread throughout the late Pleistocene of Europe and Asia and penetrated N. Africa. Apparently none survived the late-glacial. The usual explanation for extinction of the giant Irish elk has been the selective disadvantage of supporting its enormous antlers—an obligation supposedly imposed by allometric growth rates. This hypothesis would be more plausible if the genus had not disappeared simultaneously with many others.

Dama Frisch, 1775   ARTIODACTYLA
                                                  Cervidae

Some authors treat the fallow deer as a subgenus of *Cervus*. Adults reach 1 m in shoulder height and weigh 40–80 kg. They are gregarious when domesticated but occur in small groups in the wild.

Cervus Linnaeus, 1758   ARTIODACTYLA
                                                  Cervidae

Perhaps because of smaller size and more inconspicuous habits, proportionately more cervids than bovids survived the Pleistocene. In the Americas, 9 of 14 genera survived, but only 5 of 14 bovids. In Eurasia, 11 of 14 cervid genera survived along with only 20 of 41 bovids (genera mainly from Simpson, 1945). Among the largest of the surviving deer is *C. canadensis. C. elaphus*, the red deer of Europe, and 13 additional spp. range through Europe, Asia, the East Indies, and the Philippines. The red deer weighs 100–250 kg; the American elk or wapiti may attain 350 kg. It is both a grazer and browser.

Odocoileus Rafinesque, 1832   ARTIODACTYLA
                                                  Cervidae

A New World deer of both N.A. and S.A., from 15 to 200 kg in body wt., dependent on the species. White-tailed deer, *O. virginianus*, and mule deer, *O. hemonius*, occur in temperate N.A., the former into tropical lowlands.

†*Choritoceros* Hoffstetter, 1963   ARTIODACTYLA
A small Pleistocene deer from   Cervidae
Bolivia, characterized by peculiar
antler form.

†*Cervalces* Scott, 1885 (Fig. 25) ARTIODACTYLA
Cervidae

E. A. Hibbard (1958) concluded that 4 described spp. of giant deer assigned to *Cervalces* represented antler variations no greater than that known in the closely related modern moose (*Alces alces*). The recently described *Libralces* of Europe was a close relative. *Cervalces* ranged through east-central U.S. and Alaska. While neither C[14] dates nor association with artifacts have been reported, the record from marl at Berrian Springs, Mich., and other post-Cary deposits leaves little doubt that *Cervalces* survived into the late-glacial.

1 m

Fig. 25. *Cervalces*

Alces Gray, 1821   ARTIODACTYLA
                                                      Cervidae

Called moose in N.A. and elk in Europe; the largest member of its
family, weighing up to 825 kg, is a solitary browser of boreal forests
and marshy regions through the northern regions. *Alces* appears in
Danish late-glacial deposits in the Allerød and it replaces reindeer
at the beginning of the postglacial as the dominant Cervid.

Rangifer H. Smith, 1827   ARTIODACTYLA
                                                      Cervidae

The reindeer and caribou were placed by Banfield (1961) in a single
cosmopolitan species. *Rangifer* is first found associated with
Chellean–Acheulean sites of middle-Pleistocene age. The genus is the
largest open-country, gregarious cervid to survive the Pleistocene.
The only larger genera are found in heavier cover and are somewhat
less gregarious (*Cervus*) or solitary (*Alces*). Adult caribou weigh
up to 318 kg.

†*Sangamona* Hay, 1920   ARTIODACTYLA
                                                      Cervidae

An extinct deer about the size of the modern caribou, *Sangamona*
was widely distributed in N.A. Remains have been found in both
western and eastern U.S.A., but there is no evidence of contemporaneity with man. The type specimen is a single tooth. While additional
fragmentary remains appear to uphold the validity of this late-
Pleistocene deer, a definitive description has yet to appear.

Capreolus Gray, 1821   ARTIODACTYLA
                                                      Cervidae

The single species, the roe deer (*C. capreolus*), is widespread through
Eurasia except N.W. India and the extreme Arctic. Adults weigh
15–59 kg, up to 93 cm in shoulder height. They travel in family
groups, avoid thick forests, and may sometimes gather in herds of 40.
Wolves are their main enemy.

†*Libytherium* Pomel, 1892 (Fig.26)   ARTIODACTYLA
                                                      Giraffidae

The short-legged, antlered giraffes of Africa, formerly regarded as
Sivatherium, are now believed to be an independent development
from a Pliocene invader (Cooke, 1963, p. 107). *Libytherium* is
associated with many Acheulean sites in S. and E. Africa. Records
from N. Africa have been considered no younger than Villafranchian.
At Olduvai Gorge, the presence of *Libytherium* and other extinct
mammals of large size has led Leakey (1965, p. 76) to interpret the
beds in which they occur as deposited under a humid climate with
extinction caused by drought. The answer begs the question of why
the large species did not endure in more favorable humid habitats
elsewhere.

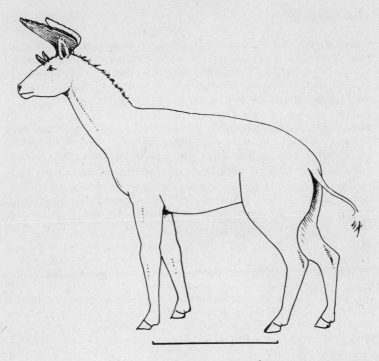

1 m

Fig. 26. *Libytherium*

†*Breameryx* Furlong, 1946   ARTIODACTYLA
                                               Antilocapridae

The dwarf pronghorn antelopes sometimes assigned to this genus barely exceed 2 ft in height, equivalent in size and possibly in ecology to the dik-diks and duikers of central Africa. But they are related to the modern pronghorn (*Antilocapra*), from which they differed not only in smaller size but also in having a pair of prongs, the front one much shorter and triangular in cross section. On both morphological and geological grounds Hibbard and Taylor (1960) propose that *Breameryx* be synonymized with *Capromeryx*. The group is notable as one of very few cases, perhaps the only one in N.A., of an extinct Pleistocene artiodactyl that is of *smaller* size than its closest living relative. Except possibly at Midland, Texas, there is no well-known association of its bones with a kill site. Radiocarbon dates, not necessarily in association with the dwarf antelope bones, are available from Rancho la Brea, where at least 34 individuals were recovered in excavations of the tar pits, and from Blackwater Draw in the Brown Sand Wedge.

†*Capromeryx* Matthew, 1902   ARTIODACTYLA
                                                              Antilocapridae

Generic differences among members of the Antilocapridae are based
largely on the form of horn cores. Individual teeth and isolated
skeletal parts are hard to identify. The type description of *Capromeryx*
from Hay Springs, Kans., did not include horn cores, thus its relationship to *Breameryx* remains uncertain. Scarcely larger than a jackrabbit, the long-crowned teeth of *Capromeryx* suggest grazing
habits. The group occurred in western N.A. from the Rocky Mts.
south to Mexico. No reliable associations with Early Man and no
indisputable radiocarbon dates are known; Hester (this volume)
lists 11,000 B.P. for *Capromeryx* at the Scharbauer Site.

†*Stockoceros* Frick, 1937 (Fig. 27)   ARTIODACTYLA
                                                              Antilocapridae

Bones of extinct four-horned "pronghorns" somewhat larger than
Antilocapra are found from a number of southwestern caves presumably of late-glacial age, such as Burnet Cave and Ventana Cave.
None of the remains has been critically dated by radiocarbon. The
chronology and ecology of this genus is not likely to be resolved until
the status of *Tetrameryx* is better known.

1 m

Fig. 27. *Stockoceros*

†*Tetrameryx* Lull, 1921   ARTIODACTYLA
                                                           Antilocapridae

An extinct four-horned "pronghorn", heavier bodied than modern
Antilocapra. In Shelter Cave, N.M., *T. conklingi* is represented by
many broken bones; associated dung is supposedly also of this genus.

Antilocapra Ord, 1818   ARTIODACTYLA
                                                           Antilocapridae

Unlike the other bovids, members of this family annually shed their
horn sheaths. The swiftest of the New World mammals, antelope are
able to run as fast as 65 km/hour. They may gather in herds of up to
100 in winter. Adult males weigh 36–60 kg.

Taurotragus Wagner, 1855   ARTIODACTYLA
African eland, 2 spp., wt. to 900 kg.   Bovidae

Bos Linnaeus, 1758   ARTIODACTYLA
                                                           Bovidae

There are 7 spp. of cattle, including the yak, which some authors
include in a separate genus, *Poephagus;* adult wt. is 450–1,000 + kg.
The long-horned wild auroch, *B. primigenius*, was probably an
ancestor of domestic cattle (*Bos taurus*).

Poephagus Gray, 1843   ARTIODACTYLA
                                                           Bovidae

See genus *Bos*.

Bison H. Smith, 1827 (Fig. 28)   ARTIODACTYLA
                                                           Bovidae

In the early postglacial of N.A., Paleo-Indians hunted *B. antiquus*
and the slightly smaller *B. occidentalis*. At the Olsen-Chubbuck Site
in Colorado a herd of 200 was driven into an arroyo and killed by a
band of perhaps 150 hunters using points of the Scottsbluff type.
The hunters butchered only part of the animals (Wheat, 1967).
The living N.A. spp., *B. bison*, and the European wisent, *B. bonasus*,
weigh 450–1,350 kg. The latter is a woodland form, feeding on
leaves, twigs, and bark, rather than grazing. The early 19th century
herd of N.A. bison was estimated by Seaton to be approximately
50 million. Extinct bison were apparently equally abundant; 1,766
jaws and 4,838 metapodials of *B. crassicornis* are reported from
Alaskan muck deposits near Fairbanks. The most common herbivore
from Rancho la Brea was *B. antiquus*, represented by 159 individuals
(Marcus 1960). Following the Wood Committee, the genus *Bison* has
become a guide fossil to the Rancholabrean. *B. latifrons* and another
species with extremely long horns, *B. alleni*, have been assumed to be
extinct by the end of the Sangamon. There is evidence to indicate that
the long-horned bison survived into the early Wisconsin, but there
is not yet clear proof that they endured until the time of extinction of
most of the genera of N.A. large mammals. The common interglacial long-horned spp. of Europe and Asia was *B. priscus*.

†*Spirocerus* Boule and Teilhard de
Chardin, 1928
An extinct antelope from the
Upper Pliocene–Pleistocene of Asia.

ARTIODACTYLA
Bovidae

Gazella Blainville, 1816

ARTIODACTYLA
Bovidae

Perhaps a dozen valid spp. of gazelles, found in drier parts of the
Old World; adult wt. 14–75 kg.

1 m

Fig. 28. *Bison*

Saiga Gray, 1843

ARTIODACTYLA
Bovidae

This Asiatic antelope is characterized by a remarkable inflated
proboscis-like nose. Adults weigh 36–69 kg. In the Pleistocene a
population penetrated Alaska but failed to survive the late-glacial.
Through careful management, the Russian herd of saigas has been
restored to over a million individuals.

Oreamnos Rafinesque, 1817

ARTIODACTYLA
Bovidae

The single living species of mountain goat occupies high country of
western N.A. Adults weigh 75–140 kg; herds may shelter together

in caves in severe weather. An extinct fossil species known from cave deposits of lower elevations in western N.A., *O. harringtoni*, was approximately two thirds the size of *O. americanus*. The type came from Smith Creek Cave, Nev., associated with *Equus*, *Camelops*, and *Ovis* at a depth of 3–4 ft. Radiocarbon dates from Rampart Cave are not in direct association with bones of *O. harringtoni*.

†*Myotragus* Bate, 1909   ARTIODACTYLA
                                         Bovidae

An extinct Pleistocene goat-antelope from some of the islands of the Mediterranean, related to the chamois and mountain goat.

Rupicapra Blainville, 1816   ARTIODACTYLA
The chamois of Europe and Asia   Bovidae
Minor; wt. 24–50 kg.

†*Bootherium* Leidy, 1852 (Fig. 29)   ARTIODACTYLA
                                         Bovidae

This extinct woodland musk-ox lacks the inflated frontal development of *Symbos*. The latter may simply be the male of the same species. *Bootherium* ranged from Alaska to the east coast of N.A. south to Texas. An unpublished pollen count by W. Benninghoff of endo-cranial matrix from Saltville, Va., is dominated by spruce and pine, suggesting a boreal woodland habitat. While known from post-Cary

1 m

Fig. 29. *Bootherium*

deposits as far north as the "Mason-Quimby Line" in Michigan and almost certainly a contemporary of Clovis hunters in the New World, no direct association of bones and artifacts has been reported.

†*Symbos* Osgood, 1905 (Fig. 30)   ARTIODACTYLA
                                                                  Bovidae

The woodland musk-ox was taller and more slender than living bison. It is known from various late-glacial deposits in eastern N.A. west to Alaska. From pollen associated with the fossil remains, it apparently occupied boreal forest or woodland of spruce, pine, and fir. The radiocarbon date on bone of the Scotts specimen, N-1402, 11,100 ± 400, is supported by a late-glacial pollen record. There is no obvious association of *Symbos* with Early Man.

1 m

Fig. 30. *Symbos*

†*Euceratherium* Furlong and   ARTIODACTYLA
Sinclair, 1904   Bovidae

The shrub oxen are related to the goats (Caprinae) perhaps as closely as to the musk-oxen (Ovibovinae). They were larger than mountain sheep and probably occupied lower slopes of hilly country. Teeth identified as *Euceratherium* were found with charcoal, a stone flake, and remains of mammoth in Potter Creek Cave, Calif. Others occur in post-Blancan faunas mainly in western N.A. from Mexico to Oregon, east to Cumberland Cave in Maryland, where shrub oxen remains were initially misidentified as the African eland (*Tauro-tragus*). Another synonym is *Aftonius* of Hay. Stock and Furlong suggested that *Euceratherium* and *Preptocerus* are sexually dimorphic forms of the same genus. According to Simpson (1963), "The occurrence in a single deposit of two separate species of such closely similar herbivores of almost exactly the same size would be extremely unusual. At present the three genera, six species, and two varieties named in this group are without taxonomic or biological significance."

†*Preptoceras* Furlong, 1905 (Fig. 31)   ARTIODACTYLA
                                           Bovidae

The shrub oxen were larger than mountain sheep; this genus may simply be the female of *Euceratherium*. The rich late-Pleistocene fauna of Burnet Cave is doubtfully dated by C-873 at 7,900 ± 300

1 m

Fig. 31. *Preptoceras*

B.P. Although fossil bones were found interbedded with charcoal, the radiocarbon date may be questioned. An artifact regarded as a Clovis point was also collected in the cave. Hester (this volume) would accept the Clovis point as a better age estimate for the associated fauna than the radiocarbon date—that is, the Clovis point should be ca. 11,000 years in age. *Preptoceros* and *Bison antiquus* were found directly under a large rock in a hearth along with the grooved spear point in question.

Ovibos Blainville, 1816   ARTIODACTYLA
                                                                 Bovidae

Next to bison, the musk-ox is the largest bovid to survive the Pleistocene in N.A. It reaches 430–510 mm in shoulder height; adults weigh 320–410 kg, are gregarious, and occur in herds of up to 100. The survival of the species in the American Arctic and its extinction in the late-glacial of Eurasia invite an explanation. Living musk-oxen are admirably suited for tundra life and do not reproduce well even in captivity as far south as Vermont. The center of evolution of tundra and cold woodland vertebrates, the Eurasian subarctic, should certainly have provided suitable habitat for musk-oxen throughout the postglacial as well as earlier. The pattern of survival suggests a refuge in unglaciated areas in N.A. as Greenland and the eastern Canadian Arctic. These areas contain no evidence of Paleolithic tools and it seems possible that *Ovibos* survived in and later expanded its range beyond those regions not penetrated by Upper Paleolithic hunters. In historic time a major conservation effort has saved the species.

Capra Linnaeus, 1758   ARTIODACTYLA
Goats, ibexes, etc., 5 spp. from   Bovidae
Spain to Siberia; males weigh
75–120 kg.

Ovis Linnaeus, 1758   ARTIODACTYLA
                                                                 Bovidae

The mountain sheep have proportionately larger horns than any of the other bovids. Approximately 6 spp. occupy Asia and N.A.; adults weigh 75–200 kg. They live in dry upland and mountainous areas, especially in rough country.

References

Banfield, A. W. F., 1961, A revision of the reindeer and caribou, genus *Rangifer: Nat. Mus. Canada Bull. 177*, 137 p.

Brodkorb, Pierce, 1963, Catalogue of fossil birds: *Florida State Mus. Biol. Sci. Bull.*, v. 7, p. 179–293

Cooke, H. B. S., 1963, Pleistocene mammal faunas of Africa, p. 65–117, *in* F. C. Howell and F. Bourliere, *Editors, African ecology and human evolution:* Chicago, Aldine

Duff, Roger, 1952, *Pyramid Valley:* Christchurch, Pegasus Press, 48 p.

———— 1956, *The Moa-Hunter period of Maori culture:* Wellington, N.Z., Owen, 400 p.

—— 1963, *The problem of moa extinction:* Thomas Cawthron Memorial Lecture no. 38, Cawthron Inst., Nelson, N.Z., Stiles, 27 p.

Duges, A., 1892, Nota sobre un fosil de Arperos, Estado de Guanajuato: *El Minero Mexicano*, v. *19*, p. 233–35

Gazin, C. L., 1956, Exploration for the remains of giant ground sloths in Panama: Smithsonian Inst. Ann. Rept., publ. 4272, p. 341–54

Greenway, J. C., Jr., 1958, Extinct and vanishing birds of the world: New York, *American Comm. Int. Wild Life Protection, Spec. Bull.*, v. *13*, 518 p.

Guilday, J. E., Martin, P. S., and McCrady, A. D., 1964, New Paris No. 4, A Pleistocene cave deposit in Bedford Co., Penna.: *Nat. Speleol. Soc. Bull.*, v. *26*, p. 121–94

Haynes, C. V., Jr., 1964, Fluted projectile points, their age and dispersion: *Science*, v. *145*, p. 1408–13

Hibbard, C. W., 1955, Pleistocene vertebrates from the Upper Becerra: *Univ. Michigan Mus. Paleont. Contrib.*, v. *12*, pp. 47–96

Hibbard, C. W., and Dalquest, W. W., 1966, Fossils from the Seymour Formation: *Univ. Michigan Mus. Paleont. Contrib.*, v. *21*, p. 1–66

Hibbard, C. W., and Taylor, D. W., 1960, Two late Pleistocene faunas from Southwestern Kansas: *Univ. Michigan Mus. Paleont. Contrib.*, v. *16*, p. 1–223

Hibbard, E. A., 1958, Occurrence of the extinct moose, *Cervalces*, in the Pleistocene of Michigan: *Michigan Acad. Sci., Arts Lett.*, v. *43*, p. 33–37

Leakey, L. S. B., 1965, *Olduvai Gorge 1951–1961:* Cambridge Univ. Press, 118 p.

McKnight, Tom, 1964, Feral livestock in Anglo-America: *Univ. California Publ. Geog.*, v. *16*, 78 p.

Maldonado-Koerdell, M., and Aveleyra Arroyo de Anda, L., 1949, Nota preliminar sobre dos artefactos del Pleistoceno Superior hallados en al region de Tequixquiac, Mexico: *El Mex. Antiquo*, v. 7, p. 154–161

Marcus, L. F., 1960, A census of the abundant large Pleistocene mammals from Rancho la Brea: *Los Angeles Co. Mus. Contrib. Sci.*, no. *38*, p. 1–11

Martin, P. S., Sabels, B. E., and Shutler, Dick, Jr., 1961, Rampart Cave coprolite and ecology of the shasta ground sloth: *Amer. Jour. Sci.*, v. *259*, p. 102–27

Miller, G. S., 1929, A second collection of mammals from Caves near St. Michel, Haiti: *Smithsonian Inst. Misc. Coll.*, v. *81*, no. 9, 30 p.

Simpson, G. G., 1945, The principles of classification and a classification of mammals: *Amer. Mus. Nat. Hist. Bull.*, v. *85*, 350 p.

—— 1963, A new record of *Euceratherium* or *Preptoceras* (extinct Bovidae) in New Mexico: *Jour. Mamm.*, v. *44*, p. 583–84

Slaughter, B. H., 1960, A new species of *Smilodon* from a late Pleistocene alluvial terrace deposit of the Trinity River: *Jour. Paleont.*, v. *34*, p. 486–92

—— 1966, The Moore Pit local fauna; Pleistocene of Texas: *Jour. Paleont.*, v. *40*, p. 78–91

Slaughter, Bob H., and McClure, W. L., 1965, The Sims Bayou local fauna; Pleistocene of Houston, Texas: *Texas Jour. Sci.*, v. *17*, p. 404–17

Stock, Chester, 1931, Problems of antiquity presented in Gypsum Cave, Nevada: *Sci. Monthly*, v. *32*, p. 22–32

——— 1961, Rancho la Brea: *Los Angeles Co. Museum Sci. Ser.*, no. 20, 81 p.

Walker, E. P., 1964, *Mammals of the World*, Vols. I and II: Baltimore, Johns Hopkins Press, 1,500 p.

Webb, S. D., 1965, The osteology of *Camelops: Los Angeles Co. Mus. Sci. Bull.*, no. 1, 54 p.

Wheat, J. B., 1967, A Paleo-Indian bison kill: *Sci. American*, v. *216*, p. 44–52

EDWARD S. DEEVEY, Jr.

Yale University
New Haven, Connecticut

INTRODUCTION

This book, like so much else of international significance, began in a sidewalk cafe in Warsaw. It was a sunny afternoon in August 1961. At that very moment (so the waiter told us) two sisters of President Kennedy were touring Chopin's birthplace. The VI INQUA Congress had convened, and we were relaxing between sessions. In this heady postwar atmosphere, a good deal more reminiscent of Hemingway than of Conrad, Paul Martin joined us over lemonade. The good talk continued; Papa would have loved it. It was all about bulls, and hunting, and war in the trenches, and the green hills of Africa. The hand of Neolithic and industrial man lies heavy on parts of Poland, and the ugly trenches still scar the landscape in the Masurian lake district. Yet, on the way back from Mikolajki, some of us—who had never seen a bison outside a zoo—had caught sight of a herd of wisent, roaming free in the forest. Nearby, we saw deposits (either muddy or dusty, like the Pleistocene of other countries) that had yielded mammoth, woolly rhino, aurochs, and Chinese elm.

Disturbance, extinction, and the survival of elephants in Africa were much on Paul Martin's mind. Long before he had worked in Arizona, he had thought of the three problems as interlocked. After firsthand study in the field, the professional skill of the Clovis mammoth hunters had impressed him, though anyone could see—and Martin's pollen stratigraphy was supposed to demonstrate—that climatic change, not hunting, had driven the big Pleistocene game from Arizona. But the timing of the climatic change and of extinction seemed oddly out of balance, and in Arizona, at any rate, the climate seemed not to have changed so much. He was also thinking about the mastodons of the northeastern states. They flourished for millennia in wooded country, and elected extinction after the climate changed for the better, while their range, and that of the men who hunted them, was expanding. And no one, at least after Duff's work, needed climatic change to explain

the disappearance of moas from New Zealand. Could direct action,
i.e. killing, have been responsible for the loss of the ground sloths and
the marsupial lion, as we know it was for the dodo? Or, if the action was
indirect, by "disturbance of the habitat?" And what, exactly, does that
mean? What was disturbed, when, where, and how much?

It is important to notice that Martin was not asking a series of questions
of fact. As any historian knows, reliable tools for finding the facts are
much more useful than the facts themselves. "Who killed Cock Robin"—
why this or that animal became extinct—is a problem for local anti-
quaries and other romantics to puzzle over. No one is immune to romantic
issues but, in a world full of fascinating problems, scientists try to keep
romanticism under discipline. The conjectures most worth pursuing are
those that seem *productive*, in the sense that an apparently unprovable
theorem in mathematics is productive. Paul Martin's suggestion, that
all the major late-Pleistocene extinctions resulted from human influence,
was suddenly seen, that August day in Warsaw, to be productive in
that way. For one thing, what he was really asking turned on a methodological question. It was one we all had a stake in, because, up to then,
we thought we knew the answer. It was this: *How, by the methods
favored by historical ecologists, could one distinguish the effects of
disturbance from those of climatic change?* If that question could not be
answered, there was no profit in debating the causes of extinction.
But if it could, and if Paul could document his case, a whole new view of
"disturbance" and "habitat" would result, with major implications for
management as well as for theory.

I will need a few more pages to explain the question, and why 1961 was
the right year, and an INQUA congress the right place, to raise it.
Anyway, we (Estella Leopold, James Griffin, John Lance, my wife,
and I) thought so. We encouraged Paul to organize a conference (at the
next INQUA congress in Boulder, for of course it was too late to do it
at Warsaw), and see where his conjecture might lead. This book is one
of the first results. It's only a beginning, for any reader will discover
that the questions raised are mostly still open, and that there are a lot of
new ones. But that is what is meant by *productive*.

THE PREVALENCE OF POLLEN

Historical ecologists favor a great many methods, and the standard ones
that also belong to physical geology—those having to do with glaciation,
for instance—are *not* at issue here. What concerns us, as biologists,
is the use of animals and plants, living and fossil, to help define a climate.
Even today the more delicate points in climatology, such as the distinction

between steppe and desert, are better decided by vegetation than by instruments. As clues to more ancient climates, a hippopotamus near London, or a caribou near Philadelphia, can be remarkably informative. The great thing about Pleistocene fossils, of course, is that most of them are still extant, somewhere. So inferences about climate, drawn from the present range, can usually be checked against physiology. Even though mammals are warm-blooded, a surprising amount of Pleistocene climatology is deduced from mammalian fossils—natural ones, that is, those found in middens being disqualified. If, now, it is alleged that the reason reindeer no longer thrive near Biarritz is not that the climate has changed, but that the Magdalenians killed them, some of the well-prepared ground begins to shift under one's feet.

The matter gets stickier where plants are involved. As everyone knows that plants are more reliably stationary than mammals, a challenge to their testimony is that much more threatening. And the idea of human disturbance, as an alternative to climatic change, is an insidious challenge. What might have been disturbed, for a much longer time than anyone realized, is *vegetation*—as abstract a noun as any, but one composed of plants. Disturbance, as a complementary abstraction, does not shout that all the evidence from plants is wrong, or based on circular arguments, but only whispers that much of it could be misleading.

To see why, we need a review of events since 1916, when custody of Pleistocene vegetation passed to experts on pollen statistics. Like other atmospheric pollutants, pollen can settle anywhere that moving air can carry it; it lands in caves and middens, on Antarctic ice, and on the floor of the deep sea. Most especially, fossil pollen turns up in countable proportions (which immediately bridge the mental gap between plants and vegetation) in lake mud or peat that was built up at roughly four inches a century. Von Post's simple idea, that a series of changes in pollen proportions in accumulating peat was a four-dimensional look at vegetation, must rank with the double helix as one of the most productive suggestions of modern times.

It proved astonishingly easy to test. Some well-planned studies of modern pollen fallout showed that different vegetations in fact shed different assemblages of pollen. All the expected biases were found— against herbs and in favor of wind-pollinated trees, among others—and it was a pity that *species* of birch, pine, oak, and so on, which might have lived in different regions, could not be told apart by their pollen. But no one haggled much over details of any one vegetation when it became clear that every spot in northern Europe had experienced at least six different kinds, with all gradations between them. In Sweden, successive layers of peat are dominated by pollen of birch, pine, hazel,

oak, and spruce, each in its turn reaching percentages over fifty. Nothing
but climatic change, of course, could account for a sequence so repro-
ducible, over so wide an area. Starting with the moraine underneath,
and counting a smudge of mud just over it with pollen of arctic herbs
but little or none of trees, the sequence (to be exact, the first five parts
of it) said "postglacial warming of climate" in terms that could not
be doubted.

Not that anyone had doubted it before, what with leaves of arctic
willow underlying pine stumps, and reindeer giving way to red deer in
the middens, and all that marine mollusks had told about cold and
warm stages of the Baltic. But these bigger, rarer fossils were too far
apart, stratigraphically, to convey much information, unless (as sometimes happened when a curator neglected to wash them) they could be
fitted into their exact place in a pollen sequence. "Exact," in this context,
meant "to within about a centimeter," or something on the order of a
decade. Locally, one could do even better than that. As between sections
in different regions, connected by nothing but pollen-laden air masses,
events could not be ordered much more closely than to the nearest
century. With that kind of precision available, pollen stratigraphy
became much too important to be left to botanists. Geophysicists, for
example, were interested in more than climates; as the glacially unloaded
crust rebounded under Scandinavia, lakes were isolated from the salty
Baltic and North Sea, at times ingeniously specified by the pollen in
their first freshwater layers. Archaeologists learned to prize the pollen,
in middens and cave-earth, more than the artifacts. And such primitives
as the Bronze Age Lake Dwellers, who had had the foresight to settle
on the bogs and to die *in* them, became far more popular among the
cognoscenti than their rich relations on Crete.

In Scandinavia and Britain, where postglacial events moved with
some of the demoralizing swiftness of an arctic springtime, the composition of a particular vegetation, as inferred from proportions of
pollen, was not the central problem, even for botanists. It was easier
and much more objective to look beyond pollen to climate, the signal
that was transmitted through vegetation, regarded as a noisy channel.
If climatologists did not understand, at first, what was meant by an
oak-pollen climate, they soon learned (as anyone could who had first
grasped the concept of a hippopotamus climate). In situations where
precise dating was less essential, though—and these included older rocks,
from Paleozoic coal measures to interglacial peats—pollen statistics
were used in a more purely botanical way to reconstruct former environments. The idea that a pollen assemblage is a true though fossil vegetation, having definable meaning once the obvious biases are allowed for,

is backed by captains of the petroleum industry, among others. Though mixed up initially with the galloping forests of Scandinavia, and the self-evident northward trend of their movement, it rests on other inferences that were rather less obvious. If we tease these out of their stratigraphic matrix, and remember that what radiocarbon confirmed so gloriously was the stratigraphy, not the botanical inferences, we find them to be of two sorts. One has to do with human disturbance of the vegetation, beginning (in Scandinavia) in the Mesolithic, and evident almost everywhere in Europe by 3000 B.C. The other has to do with advancing glaciers and worsening climate, especially around 9000 B.C. I bracket them together, for the moment, because both have to do with *deteriorating* vegetation, i.e. with disorder progressing with time, a concept that is as nonobvious in biology as it is obvious in physics.

Around the margin of the Scandinavian ice sheet, where it paused in its long retreat, and readvanced for several centuries, the Upper Dryas flora (containing leaves of the arctic–alpine *Dryas octopetala*) had long been known, or suspected, to be the nonglacial equivalent of the Fennoscandian readvance. As the second of two cold times, following a somewhat warmer time, the Allerød, the upper Dryas zone was then found to have higher percentages of herb, or nonarboreal pollen (NAP), normally including some of Dryas or its arctic associates. On the evidence of pollen alone, and regardless of which came first in time (the prediction was first made in Germany by Firbas), a tundra vegetation could be distinguished from something as like it as a steppe–tundra. Like a standing wave, the Allerød sequence—high NAP, lower NAP, high NAP— was then propagated by pollen stratigraphers throughout northern Europe, north of the Alps, and even to France, Spain, and Italy. Needless to say, it was absent on top of Fennoscandian moraine *and its correlatives;* the stratigraphic dividends were immense. As one of the correlatives was the Valders drift in Wisconsin, and as radiocarbon confirmed the datings with uncanny regularity, NAP came to be invested with semimagical properties. Among the problems exposed around the margin of the Wisconsin ice sheet in New England, beginning about 1957, prominent place must now be given to my own belief that NAP (most of which was grass and sedge pollen) probably meant tundra.

More precisely, it was tree pollen mixed with NAP, equated (as in Europe) with trees advancing on tundra, that caused misunderstanding. A pollen percentage, as Margaret Davis insisted, cannot tell the difference between *adding* tree-pollen grains to NAP, from any distance, and *substituting* tree pollen for tundra pollen. Yet the idea of an Allerød sequence, absent heavier fossils like Dryas, rested on that ambiguity. One need not dispute the treelessness of Europe, between two ice sheets, or deny that some NAP comes from tundra vegetation, to grant that percentages might be more misleading in North America. There, with a single ice sheet in the north, all the trees were growing somewhere south of the ice.

When fallout of tree pollen and NAP was measured separately, against time (using rates of settling computed from radiocarbon dates of mud), Mrs. Davis found, as she expected, that as postglacial reforestation proceeded, tree grains were first added, and then substituted. No trace of an Allerød sequence was found (and I lost a five-dollar bet). It was easier, now, to see why the trees that had seemed to advance on the tundra, and had then, debatably, retreated, were not mainly birches but spruce, pine, and oak. If palms and magnolias had lighter pollen grains, these, too, might have turned up on Connecticut's tundra.

More was lost by this display of virtuosity than steppe–tundra in New England; fifty years' credulous faith in pollen statistics was shaken. Steppe–tundra is hard to tell from tundra under the best of circumstances, but some other more fundamental distinctions, as between forest and prairie, had also rested on confrontation of tree pollen and NAP. Not entirely, of course, for fossils as heavy as horses can be left behind when savanna gives way to forest. But even in Minnesota, where the post-glacial prairie-forest border had been mapped in four dimensions, a fair amount of fuzziness could be blamed on pollen. In the rougher Southwest, where forest, steppe, and desert are intermixed, Paul Martin had given much thought to separating them, also in four dimensions. In a country not noted for its peat bogs, he had measured modern pollen fallout in every cattle tank and playa he could find, and the percentages *were* different under different vegetations. Here, though, as some of the ambiguity was deleted from the pollen, the clarity of postglacial climatic changes seemed to disappear with it. Pleistocene climates had moved up and down the mountainsides by as much as four thousand feet, but the shifting mosaic of rainfall belts was rather like today's, and not at all like temperature zones arranged around an ice sheet. The farther one moved from the margin of Wisconsin ice, in fact, the harder it became to distinguish what was glacial from what was interglacial or postglacial. The pollen from highland Mexico is notably indecisive on this point. As we began to work in lowland Guatemala—this project has not yet reached anything glacial—we found pine and oak pollen confronting other mixtures, from rainforest and from tropical thorn scrub, that lacked the resilience of NAP and became almost invisible. What *was* visible, in the NAP, was maize and weed pollen, resulting from Maya agriculture.

Meanwhile, in a context where all savannas are certainly man-made,

NAP was reintroduced into Europe along with Neolithic technology. The botanical details, many of them derived from an experimental plot in Jutland, are too rich to be reviewed here; but it is worth noticing that slash-and-burn, or swidden agriculture, is better understood as it was practiced in Denmark in 2750 B.C. than it is in tropical Africa or America, where millions of lives depend on it. As one result, ecologists everywhere are considerably more respectful of fire, as an agent of disorder, than I was in 1939 when I wrote of "aboriginal pyromania" with youthful scorn.

After each episode of burning—and fire, as we know now, touched every part of northern Europe within four centuries of 3200 B.C.—the forest was allowed to regenerate. According to the degree of disturbance and the quality of the local soils, reforestation was partial or nearly complete. Even before the first clearance, an early Gallic habit of lopping elm branches (to bed the cattle) suppressed the flowering of elm. Though later documented by all the techniques of archaeology, all this is plainly shown by proportions of tree pollen as mixed with NAP. What is dated, by radiocarbon, at 3000 ± 200 B.C., is the first drop in percentage of elm pollen, which preceded the first rise of NAP by some fifty years. Admitting, once again, that *if* one knows how to read them, pollen percentages can tell one vegetation from another, we have to recognize something else. A famous climatic event, the shift from moist, elm-rich Atlantic climate to drier, elm-poor Sub-Boreal climate, has somehow evaporated. It was lopped away, in fact, by Neolithic disturbance.

Vegetation can deteriorate for many reasons, including volcanic eruptions, and I do not mean to set human influence against climatic change as if two rival explanations competed for historical truth. Neither do I intend to pit European against American virtuosity in reading pollen percentages. What I have been trying to show is why, in asking questions about late-Pleistocene extinction, Paul Martin reopened some chapters in the history of vegetation and asked that someone, preferably their authors, re-read them. To the extent that post-Pleistocene climatic change had been inferred from extinction of big game, the inference might be entirely circular. And if one appealed to vegetational history, postglacial climatic changes were clear enough, no doubt, in regions where forests rapidly reoccupied moraine. But in more and more places where animals and plants had survived glaciation, *any* picture of their habitats, whether or not they changed after 9000 B.C., was clouded by the ambiguity of pollen percentages. Neither that ambiguity nor the idea of human disturbance was particularly new in 1961, as I have also tried to show. In fact the prevalence of disturbance had struck me with special force in 1953, when Richard West found weed pollen in the interglacial forest at Hoxne,

Suffolk, a famous site of Acheulian (middle Paleolithic) man. But I think
it was no accident that the problem of the causes (i.e. the circumstances)
of extinction began to seem acute to Paul Martin only after he had
worked in savanna country. There, on the whole, there was (and for the
moment still is) more game, more kinds of game, and more hunting
pressure. There, too, the confrontation of tree pollen with NAP could
have led to a serious misreading of Pleistocene events.

THE MANAGEMENT OF EMPTY NICHES

In the savanna and desert country between the Nile Valley and the
Caspian, where both the fact and the idea of human influence have been
familiar longest, scholarly uncertainty about climate also set in very early.
Few features of a desert landscape are more impressive, to a traveler in a
Land Rover, than a dry wadi; and few features suggest more about
changing climate while proving less. An urban, north-European ambi-
valence about deserts, so evident in the works of D. H. Lawrence, was
one ingredient, along with a dash of Marxism, in a sweeping generalization that gradually took shape. As the *oasis theory* of civilization, formulated in particular by V. Gordon Childe, it gained wide currency.
Stimulated by the crowded conditions around shrinking waterholes, so it
was thought, animals, plants, and starving hunters eventually domesticated each other. The underlying assumption, post-Pleistocene desiccation, was so obvious as not to need much discussion. Later excavations,
especially around Jarmo in Iraq, produced so much skepticism about
oases that the theory emerged in inverted form: post-Pleistocene desiccation is clear enough, but it is probably the consequence, not the cause, of
human habitation.

   None of the experts would put the matter so flatly, and even if true in
the Fertile Crescent it need not be true in Arizona or Australia. But a
major reason for the inversion is that the circularity of mammalian
evidence had finally been noticed. If the hippopotamus no longer thrives
in Palestine, the reason may be not that the climate has changed but that
someone, perhaps Natufians, killed him.

   Leaving aside the ambiguities of Pleistocene environments, and of
radiocarbon dating (which has developed some new ones since most of
this book was written), let us consider some implications of Paul Martin's
idea. Whatever the causes of late-Pleistocene extinction, the fact is that
there was a lot of it; and although it is hard to be sure how abnormal it
was, some, at least, of the big Pleistocene mammals were lost without
leaving replacements. In the language of ecology, to say this is to say that
some niches were left vacant. (Ecologists always use *niche* to mean what

an organism *does* in nature; *habitat* is where it *lives*.) And this is interesting, for a wry reason: niche theory can be very abstruse, with much mathematical talk about overlapping and noncontiguous niches, and why (or whether) there are more niches in the tropics than in the Arctic; but when the theoreticians stop for questions they usually admit, with embarrassment, that a niche not occupied, or occupiable, by a known organism is an unhelpful construct, like a hoop snake. Until we can discipline our imaginations, and design a theory that predicts unknown niches instead of describing known organisms, some obviously vacant niches, such as those once filled by moas, dwarf mammoths, and glyptodonts, will continue to repay study.

And the value of this kind of study is not purely theoretical. Noticing that most of the extinct mammals and birds were large herbivores living in savanna country, and that some of them like ground sloth, water buffalo, Columbian mammoth, and giant wallaby were powerful converters of carbohydrate to protein, we can also infer that they had a powerful influence on the makeup, as well as on the amount, of vegetation. (These quantitative inferences are among those that most need more research.) Plainly, their niches on all continents are currently filled, however badly, by domestic cattle and sheep. Any one-for-one replacement, like that of bison by Herefords on the western plains, might be ecologically sound, though the issue has always been prejudged, up to now, without much ecological inquiry. One can state categorically, with no inquiry, that no one or two domestic species can replace *all* the lost herbivores with equal success.

The reason, of course, is to be found in the long evolution of ecological order, with a community of some twenty species of native African ungulates providing a relevant example. Shouldering each other into separate niches, over the course of the Pleistocene or longer, they evolved into short-grass grazers, long-grass grazers, browsers with short and with long necks, species with narrow or wide ranges, species more and less tolerant of thirst and of tsetse flies, species with patrilineal and matrilineal social organizations, and so on and on. Such a diverse system, naturally, is considerably more resistant to disturbance, whether by locusts or by hunters, than the one-grazer-one-carnivore system favored in the Fertile Crescent, and propagated thence around the world. It is also inherently more efficient, when considered as a converter of solar energy to protein, as some African ranchers have been finding out.

How much of this kind of order was destroyed, inadvertently, when Upper Paleolithic people finished off the game in such places as the Dordogne, and then moved on, is impossible to say. This book simply opens that inquiry; the history of vegetation, which contains even more of the same kind of order, contains the methodological clues, but they are beset by ambiguities. There is, as yet, no general agreement that people, not changing climate, were responsible for the emptying of niches. Until the inquiry has progressed a little farther, it is premature to contemplate refilling them—repopulating Arizona, say, with horses, camels, elephants, antelopes, tapirs, and peccaries—for we have one witless menagerie in New Zealand and scarcely need another. As we become more confident as to which niches were emptied artificially, better ideas for filling them artificially are bound to emerge. Meanwhile, to destroy any more ecological order, deliberately, in the interest of propagating cattle, would seem—no stronger word is helpful—simply daft.

FACETS OF THE PROBLEM

PAUL S. MARTIN

Geochronology Department
University of Arizona
Tucson, Arizona

PREHISTORIC OVERKILL[1]

Abstract

A sudden wave of large-animal extinction, involving at least 200 genera, most of them lost without phyletic replacement, characterizes the late Pleistocene. Except on islands where smaller animals disappeared, extinction struck only the large terrestrial herbivores, their ecologically dependent carnivores, and their scavengers. Although it may have occurred during times of climatic change, the event is not clearly related to climatic change. One must seek another cause. Extinction closely follows the chronology of prehistoric man's spread and his development as a big-game hunter. No continents or islands are known in which accelerated extinction definitely predates man's arrival. The phenomenon of overkill alone explains the global extinction pattern. This interpretation of the cause of late-Pleistocene extinction was advanced by Wallace in the World of Life *(1911). It finds chronological support in recent discoveries. It clarifies an otherwise incomprehensible part of the Pleistocene fossil record.*

The end of the Ice Age saw the sudden decline of an extraordinary number of large vertebrates. Unlike the relatively gradual, essentially orderly replacement of new genera seen earlier in the Pleistocene and Tertiary, extinction rates suddenly skyrocketed. New genera did not appear. There was no generic replacement either by immigration or evolution (Martin, 1958, p. 400). As a result,

> We live in a zoologically impoverished world, from which all the hugest, and fiercest, and strangest forms have recently disappeared ... yet it is surely a marvelous fact, and one that has hardly been sufficiently dwelt upon, this sudden dying out of so many large Mammalia, not in one place only but over half the land surface of the globe [Wallace, 1876, p. 150].

At the time he wrote, Wallace regarded the cause of extinction as a direct outcome of the worldwide effects of Pleistocene glaciation.

1. Contribution 128, Program in Geochronology, University of Arizona.

But in the voyage of the *Beagle*, Darwin had already shown that extinct Pleistocene fauna occurred in beds younger than the last glaciation. Wallace himself came to reject the effects of the glacial epoch as a sufficient explanation. In the *World of Life* (1911, p. 264), he wrote:

> What we are seeking for is a cause which has been in action over the whole earth during the period in question, and which was adequate to produce the observed result. When the problem is stated in this way, the answer is very obvious. It is, moreover, a solution which has often been suggested, though generally to be rejected as inadequate. It has been so with myself, but why I can hardly say.

While crediting it to Lyell, Wallace reached the view that seems to me best supported by subsequent evidence, that no known environmental defects or crises, other than those brought by prehistoric man, can adequately account for the sequence of events. I would depart from Wallace's view in only one regard—he apparently also believed, following Lyell, in certain deep-seated general causes operating to exterminate large animals at the end of each geological era.

I do not consider the intriguing question of accelerated extinction at the end of the earlier geological eras (Bramlette, 1965; Newell, 1966) relevant to the matter at hand. In the late Pleistocene one has a far more detailed stratigraphy and chronology to work with. But the main point is that one finds a totally different pattern in the Pleistocene, one affecting mainly one class of organisms, the Mammalia. There is no upturn in extinction rate among marine organisms, such as typifies the close of Permian, Triassic, and Cretaceous. The phenomenon of accelerated extinction is unknown in the marine Pleistocene. If it had occurred, Lyell's method of dating marine Cenozoic beds would not have been so simple or successful.

THE PATTERN OF PLEISTOCENE EXTINCTION

I shall attempt to sketch salient features of late-Pleistocene generic extinction, with emphasis on North America, revising some interpretations presented in an earlier effort (Martin, 1958, p. 394–413). The reason for concentrating on genera is pragmatic. The generalized ecologic, chronologic, and phylogenetic interpretations for discussing an extinct genus are likely to be speculative enough without entering a taxonomic level in which more than a dozen valid specific names may be available in a group that could not possibly have evolved into as many good biological species. Are we to infer a dozen allopatric species in a genus that is seldom or never known to be represented by two distinct morphological forms in a single fossil horizon? Even in the case of a thoroughly and very skillfully revised group, with much of its synonymy resolved, the critical identification of a species from a carefully dated outcrop of considerable archaeological or paleoecological significance may require presence of diagnostic parts such as horn cores or complete jaws. Ecologists, long subjected to various pressures to study modern communities at the species level whenever possible, may not fully anticipate or appreciate the hazards of trying to study Pleistocene mammals at the species level.

To turn now to the matter of Pleistocene chronology. Although many large extinct animals have yet to be dated by C^{14} and although the method itself continues to present discordant results, especially when applied incautiously, it seems possible to conclude on the basis of both relative and absolute dating that throughout the Americas, in Australia, and on the islands of Madagascar and New Zealand a major wave of generic extinction occurred once only, and at a time within the last 15,000 years. This was not the case in Africa and Southeast Asia, where most generic extinction occurred some tens of thousands of years earlier, essentially beyond the reach of the C^{14} dating method.

Apart from small oceanic islands, the animals lost were mainly "big-game" mammalian and avian herbivores of over 50 kg adult body weight (see Table 1). Doomed by the collapse of the herbivores was a retinue of ecologically dependent carnivores, scavengers, commensals, and, presumably, various unknown parasites. One need not assume any narrow predator–prey relationship. In fact, most mammalian predators seek a variety of prey species. One can assume that the loss of thirty-one genera of large herbivores at the end of the last glaciation of North America (Table 1) reduced the variety of carnivores. In other words, while one cannot say that saber-tooth cats disappeared *because* of the extinction of their supposed prey (such as mammoths), one can say that there had to be some feedback, some extinction of carnivores, when various herbivores disappeared. It happened that the saber-tooth was among those lost and the jaguar among those surviving. The fact that prehistoric man would not have hunted and killed saber-tooth cats or other large carnivores is not a valid criticism of the hypothesis of over-kill. The question is whether or not he triggered extinction of the herbivores.

Generic extinction did not occur only at the end of the Pleistocene. *Nannippus, Plesippus, Stegomastodon, Titanotylopus, Canimartes, Trigonictis*, and other genera listed in Table 1 disappear from the United States at the end of the Blancan, over a million years ago. But the adaptive niches for horses, mastodons, camels, and large mustelids continued to be occupied (see Hibbard et. al., 1965, p. 520). In contrast, in the late Pleistocene, the life forms lost were not replaced or maintained by related species. Possibly the browsing and grazing niches so suddenly abandoned by large animals in the late Pleistocene were partly refilled by an increase in biomass of small mammals and insects. But in the strict sense, the record is one of extinction without replacement.

Continental extinction of late-Pleistocene age also differs from that earlier in the Cenozoic in the lack of appreciable change among the small vertebrates. On the North American continent there is no terminal Pleistocene loss of small mammalian genera comparable to the loss of *Prodipodomys*, *Pliophenacomys*, *Pliopotamys*, *Pliolemmus*, and *Bensonomys* in the early-Pleistocene (Table 1). As in the case of the large herbivores mentioned above, most of the lost Blancan genera of small mammals have clear-cut phyletic replacements in the younger Pleistocene faunas. In contrast, there is very little difference between the Wisconsin glacial-age small vertebrates and the modern fauna, even at the species level. Of seventy small mammals—shrews, bats, and rodents—all but six are assigned to living species (Hibbard, 1958, with minor additions). Of sixty-nine amphibians and reptiles (excluding turtles and tortoises) of the same age, only three are considered extinct species (Gehlbach, 1965). Of seventy-three freshwater mollusks found in late-Pleistocene and Recent faunas of southwestern Kansas and northwestern Oklahoma in Jinglebob or younger faunas, only one is extinct (Hibbard and Taylor, 1960). Very few extinct species ·of mollusks are found in local faunas younger than the Sanders ("Aftonian"; see Taylor, 1965, p. 605).

Survival of the small includes survival of the pelagic. Unlike the Pliocene–Pleistocene boundary, which is marked in certain deep-sea cores by the loss of Discoasters, a reduction in variation of the *Globorotalia menardii* complex, disappearance of *Globigerinoides sacculifera fistulosa*, and appearance of *Globorotalia truncatulinoides* (Ericson et al., 1963), the Pleistocene–Recent boundary cannot be recognized by marine guide fossils. Around 11,000 years ago, the planktonic Foraminifera of the Atlantic changed as cool-water faunas gave way to warm. But the climatic shift was not accompanied by biotic extinction. At no time in the Pleistocene was there massive marine extinction comparable to the loss of belemnites, ammonites, and nanno-plankton that marks the Upper Maestrichtian–Danian boundary, commonly correlated with extinction of the dinosaurs (Bramlette, 1965). Nor were the largest mammals of the world, the cetaceans, affected by late-Pleistocene extinction.

Finally, late-Pleistocene extinction is not evident in the plant kingdom. While a major depression 20,000 years ago, with a worldwide drop in

vegetation zones of roughly 1,000 meters, is evident in pollen profiles of
the time taken in unglaciated, mountainous areas of most continents,
there are no extinct late-Pleistocene genera among the diatoms or
vascular plants, two groups of organisms with extraordinarily rich fossil
records. Only in the early Pleistocene, best known in western Europe,
is the local extirpation of warm-temperate plants well known (Leopold,
1967). The lost European genera, such as *Liquidambar, Nyssa, Sequoia,
Sciadopitys, Magnolia, Tsuga, Juglans, Eucommia,* and *Pterocarya,*
survive in parts of eastern Asia or North America. It is notable, and
I believe highly relevant to my interpretation, that despite all the
theoretical reasons why glaciation should have made Western Europe a
geographic trap for temperate biota, there is less, not more, evidence of
generic extinction of mammals there than on other continents.

For these reasons, late-Pleistocene extinction must be regarded as
imbalanced. It left empty niches in the terrestrial ecosystem, niches
previously occupied by a succession of large herbivores through the
Neogene. *Only on oceanic islands* were numerous small vertebrate genera
obliterated. Among the animals lost were giant marsupials in Australia,
moas in New Zealand, giant lemurs and struthious birds in Madagascar,
about twenty-eight genera of mammals and one genus of tortoise in
North America (Table 1), and a still poorly known but probably larger
number of mammalian genera in South America. Late-Pleistocene
generic extinction is less well marked in northern Eurasia. There
Mammuthus (mammoth), *Coelodonta* (woolly rhino), and *Megaceros*
(Irish elk) were the only common late-Pleistocene genera to disappear.
Thanks to a refuge in the unglaciated eastern Canadian Arctic, *Ovibos*
(musk-ox) survived in the New World. In both Africa and Southeast
Asia, a major episode of Pleistocene extinction antedates the late Würm
and apparently coincides with the end of the Acheulean cultural stage,
ca. 40,000–50,000 B.P. Can the cause of this peculiar pattern be found in
its chronology?

THE PLEISTOCENE OF O. P. HAY

With the recognition of multiple glaciation and the evidence from the
mid-continent of four major drift deposits separated by well-developed
soils, those in search of a climatic explanation for extinction recognized
the logic of seeking some chronological order. If all of the four classic
midcontinental glaciations—Nebraskan, Kansan, Illinoian, and Wisconsin—were of roughly equivalent magnitude, they should represent
roughly equivalent climatic changes that would have imposed a roughly
equivalent stress on the fauna. Extinction would be progressive, and its

chronology should show that animals climatically more sensitive were lost first. With this expectation, Hay proposed the following chronology in 1919 and again in 1923. I have attempted to indicate current correct generic names in certain cases.

A. Extinct by the end of the Kansan

Megatherium (= *Eremotherium*) *Eschatius* (= *Camelops*)
Glyptodon *Camelops*
Stegomastodon *Camelus*[2]
Anancus (= *Stegomastodon* *Hydrochoerus*
 or *Cuvieronius*) *Aftonius* (= *Euceratherium*)
Gomphotherium (= *Stegomastodon* *Leptochoerus*[3]
 or *Cuvieronius*) *Trucifelis* (= *Smilodon*)
Elephas (= *Mammuthus*) *imperator*

B. Extinct by the end of the Sangamon

Mylodon (= *Paramylodon*) *B. antiquus*
Tapirus *Aenocyon* (= *Canis dirus*)
Equus *Dinobastis*
Taurotragus[2] *Smilodon*
Sangamona *Smilodontopsis* (= *Smilodon*)
Bison latifrons

C. Extinct by the end of the Wisconsin

Megalonyx *Bootherium*
Elephas (= *Mammuthus*) *Mylohyus*
Mammut *Platygonus*
Cervalces *Bison occidentalis*
Symbos *Castoroides*

Hay's chronology was soon challenged. Bryan and Gidley (1926) pointed out that bones of extinct camels, a group Hay considered extinct by mid-Pleistocene, were found in playa-lake deposits of unmistakably late-Pleistocene age. Stratigraphy and radiocarbon dates show decisively that extinct *Tapirus*, *Smilodon*, *Paramylodon*, *Mammuthus imperator*, *Camelops*, and *Euceratherium* survived to, and *through*, the last glaciation. Thus most of the genera in list B plus some of those from list A must be added to list C. Others in list A are now regarded as having been

2. Old World genus, misidentified.
3. Extinct before the end of the Tertiary.

misidentified or as extinct before the end of the Tertiary. As Romer (1933), Colbert (1942), and others have noted, Hay's chronology for the large mammals is without substance. If Hay's Pleistocene extinction chronology survives, it is in the small mammal pattern (Table 1). Unlike the megafauna there is much more extinction in the early Pleistocene (Blancan) than later, as one might expect if climatic change initiated the main faunal changes. What about the megafauna?

Among large genera (50 kg or more in adult body weight), the recent list of Pleistocene distributions for the United States in Hibbard et al. (1965, p. 573, copied in Table 1), indicates four large mammals lost by the end of the Kansan, three at the end of the Yarmouth, none at the end of the Illinoian, and one at the end of the Sangamon. The spectacular upset comes at the end of the Wisconsin. By the end of the late-glacial, thirty-three genera are going or gone, far more than disappear in the rest of the Pleistocene put together (Table 1).

If the genera are arranged in the form of land-mammal ages, which avoids the assumption of correct glacial and interglacial age assignment, the results are similar. Extinction of eleven genera of large size occurs in the roughly two million years represented by the Blancan, six in the Irvingtonian, and thirty-five at the end of the Rancholabrean.

As matters stand, it is far easier to date post-Blancan Pleistocene local faunas by the arrival of new Eurasian large mammals than by the loss of old ones. It is unfortunate for Pleistocene geologists that the pattern of extinction is so disharmonic. If the extinct genera were more evenly spaced either in a random pattern or in one closely related to multiple glaciation as Hay proposed, relative faunal dating in the Pleistocene would not be so difficult.

For example, how can one tell a Sangamon interglacial fauna from a relatively warm interstadial fauna of the Wisconsin or even the Illinoian? Slaughter and Ritchie (1963) compared the Clear Creek fauna, from near Dallas, Texas, with the Jinglebob, Cragin Quarry, and Good Creek faunas, all referred to the Sangamon. The Clear Creek Formation includes *Geochelone* and probably *Bison latifrons*, both commonly considered pre-Wisconsin guide fossils. But a radiocarbon date of 28,840 ± 4,740 (SM 534) on shells from the Clear Creek Formation indicates interstadial age in the standard Wisconsin glacial chronologies of most geologists.

Unless the Clear Creek radiocarbon date is shown to be seriously in error or improperly associated with the fauna, it establishes *Geochelone* and *Bison latifrons* in a post-Sangamon horizon. For *Geochelone* there is no doubt; a small species, *G. wilsoni*, is reported from late-Wisconsin or post-Wisconsin deposits in Friesenhahn Cave and at Blackwater

TABLE 1. Pleistocene Extinct Mammals, Continental North America (Hibbard et al., 1965)

| | Pliocene | Blancan | | | Irving-tonian | Rancholabrean | | | Direct assn. with man | <15,000 B.P. by C¹⁴ | In beds <15,000 B.P. |
		Nebraskan	Aftonion	Kansan	Yarmouth	Illinoian	Sangamon	Wisconsin			
LARGE (> 50 kg)											
Machairodus, saber-tooth cats											
Ceratomeryx, extinct pronghorn											
Rhynchotherium, mastodons											
Pliauchenia, extinct camels											
Borophagus, bone-eating dogs											
Ischyrosmilus, saber-tooth cat											
Chasmaporthetes, extinct hyena											
Glyptotherium, glyptodons											
Nannippus, three-toed horses											
Plesippus, zebrine horses											
Stegomastodon, mastodons											
Titanotylopus, giant camel											
Hayoceros, extinct pronghorn											
Glyptodon, glyptodons											
Platycerabos, extinct bovid											
Stockoceros, extinct pronghorns											
Mammut, American mastodons									×	×	×

TABLE 1.—continued

| | Blancan | | Irving-tonian | | Ranchola-brean | | | | | |
	Pliocene	Nebraskan	Aftonion	Kansan	Yarmouth	Illinoian	Sangamon	Wisconsin	Direct assn. with man	<15,000 B.P. by C^{14}	In beds <15,000 B.P.
Megalonyx, ground sloths										?	×
Tanupolama, extinct llamas									×	?	×
Cuvieronius, extinct mastodons									×		×
Platygonus, extinct peccaries										×	×
Camelops, extinct camels									×	×	×
Equus, horses									×	×	×
Paramylodon, ground sloths										?	×
Capromeryx, extinct pronghorns									?		×
Castoroides, giant beavers										×	×
Arctodus, giant short-faced bears									?		×
Nothrotherium, small ground sloths									?	×	×
Chlamytherium, giant armadillos											×
Dinobastis, saber-tooth cat									?		×
Smilodon, saber-tooth cats										×	×
Hydrochoerus, capybaras											×
Mammuthus, mammoths									×	×	×
Mylohyus, woodland peccaries										×	×
Euceratherium, shrub-oxen									×		×
Preptoceras, shrub-oxen									×		×

TABLE 1.—continued

| | Pliocene | Blancan | | Irvingtonian | | Rancholabrean | | | Direct assn. with man | <15,000 B.P. by C¹⁴ | Inbeds <15,000 B.P. |
		Nebraskan	Aftonion	Kansan	Yarmouth	Illinoian	Sangamon	Wisconsin			
Tetrameryx, extinct pronghorns									×	×	×
									×	×	×
*Tapirus, tapirs											
*Tremarctos, spectacled bears											×
Bootherium, extinct bovid										×	×
Cervalces, extinct moose											×
Brachyostracon, glyptodon											
Boreostracon, glyptodon											
Eremotherium, giant ground sloth											×
Neochoerus, extinct capybara											×
*Saiga, Asian antelope											
*Bos, yak											
Sangamona, caribou?											×
Symbos, woodland musk-ox											×

SMALL (< 50 kg)

	Pliocene	Nebraskan	Aftonion	Kansan	Yarmouth	Illinoian	Sangamon	Wisconsin	Direct assn. with man	<15,000 B.P. by C¹⁴	Inbeds <15,000 B.P.
Buisnictis, extinct mustelid											
Notolagus, extinct rabbits											
Ogmodontomys, extinct voles											
Paracryptotis, extinct shrew											
Brachyopsigale, extinct mustelid											

TABLE 1.—continued

| | Pliocene | Blancan | | | Irving-tonian | Ranchola-brean | | | Direct assn. with man | <15,000 B.P. by C14 | In beds <15,000 B.P. |
		Nebraskan	Aftonion	Kansan	Yarmouth	Illinoian	Sangamon	Wisconsin			
Symmetrodontomys, extinct mouse	—										
Nekrolagus, extinct rabbit	—										
Dipoides, extinct beavers	—										
Hesperoscalops, extinct moles	—										
Heterogeomys extinct gophers	—										
Pratilepus, extinct rabbits	—										
Cosomys, extinct vole	——										
Prodipodomys, extinct kangaroo rats	————										
Bensonomys, extinct mice	————										
Paenemarmota, giant woodchuck	————										
Trigonictis, extinct grison	————										
Procastoroides, extinct beaver		———									
Nebraskomys, extinct voles		———									
Canimartes, extinct mustelid		———									
Pliopotamys, extinct voles		———									
Pliolemmus, extinct vole		———									
Pliophenacomys, extinct voles	————————										
Osmotherium, extinct skunk					—						

TABLE 1.—continued

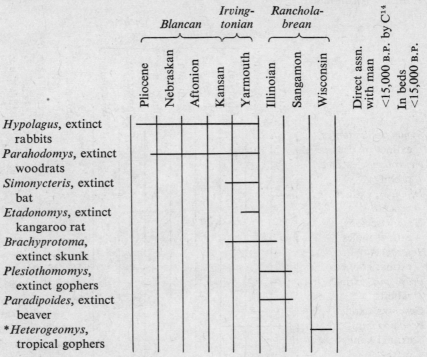

* Living in Asia or tropical America.

Draw (Slaughter, 1966). Thus the genus, if not the giant species, can be regarded as lasting to the end with the mammoth, horse, sloth, camel, and other typical members of the terminal Pleistocene megafauna. The point to stress again is the lack of a "logical" or progressive pattern of extinction in the Pleistocene of North America and the hazards of using extinct megafauna as guide fossils. Archaeologists especially need to guard against assigning a pre-Wisconsin age to artifacts associated with "interglacial" glyptodonts, extinct armadillos, or *Geochelone*.

Because virtually all the extinct Pleistocene megafauna of North America survived the stress of multiple glaciation, either some unique natural catastrophe must have precipitated extinction or else natural environmental changes had nothing to do with the event. The former interpretation is presented by Slaughter (1967). The environment in southwestern United States during the time of extinction is reviewed by Mehringer (1967). While the climate had been changing prior to the time of extinction, the fossil pollen record shows that conditions by

11,000 years ago were almost back to "normal"—back to the type of vegetation and climate supposedly typical of the earlier interglacials, an environment then congenial to mammoth, sloth, horse, and other extinct species. Why desert herbivores in particular, such as *Camelops* and *Nothrotherium*, should suddenly disappear at the moment of postglacial climatic recovery, with its expansion of arid habitats, is hard to explain. The ecology of extinction in the humid East is equally mysterious. Why should *Mammut* and *Symbos*, which from their associated fossil pollen record inhabited a boreal environment of spruce during the late-Wisconsin (Semken et al., 1964; Ogden, 1965, p. 494–95), suddenly disappear just when the boreal vegetation zone was expanding into central Canada from its shrunken full-glacial position outside the Wisconsin ice margin?

The point is pithily put by Hibben (1946, p. 176):

> Horses, camels, sloths, antelopes, all found slim pickings in their former habitat. But what was to prevent these animals from simply following the retreating ice to find just the type of vegetation and just the climate they desired? If Newport is cold in the winter, go to Florida. If Washington becomes too hot in the summer, go to Maine.

All extinction explanations must accord to the extinct fauna at least some reasonable biotic adaptability comparable to that of the living large mammals. In a year or two, any of the latter might migrate several times back and forth over the 1,000-km distance through which Wisconsin-age vegetation zones slowly moved.

RADIOCARBON CHRONOLOGIES

Although C^{14} dates more or less associated with extinct fauna promise a great deal of chronological refinement that should reveal much about the pattern and, it is hoped, the cause of extinction, certain hazards of "push-button" dating and even the overly enthusiastic extraction of poorly documented dates from date lists need be acknowledged. Before summarizing the more reliable radiocarbon dating results as they bear on Pleistocene extinction, I will cite some of the vexing cases. Hester (1960, p. 58) lists in detail the reasons why date and fauna may not go together; he also summarizes various questionable dates.

A good example of a misleading date on a very important extinct fauna is L-211, 2,040 ± 90, from St. Petersburg, Florida. Partly on the basis of L-211, I reached the conclusion (Martin, 1958), repeated in Pearson (1964), that the extinct megafauna of Florida survived long beyond the time of continental extinction elsewhere in North America.

The date was questioned by Hester (1960), and the dubious stratigraphic association is reviewed by Bullen (1964), who discounts the age on cultural grounds as well. Finally, one of the strongest points against late-postglacial survival of extinct fauna in Florida, as elsewhere, is failure to find such animal remains in postglacial archaeological sites. While late survival of extinct fauna in Florida remains a dim possibility, there is as yet no reliable evidence. At present it seems best to reject L-211.

Stratigraphic chaos, apparently the result of intense solifluction, has thus far defeated attempts at direct radiocarbon dating of the abundant remains of extinct horse, bison, and mammoth in Alaskan muck (W-891, W-1106, W-1108, W-1111, W-1113, W-937). The dates, on wood associated with the bones, appear much too young to represent the true age of the fauna and are thought to be intrusive. A more reliable and possibly terminal date is on hair and hide of *Bison (Superbison) crassicornis*, St-1663, 12,000 B.P. (in Péwé et al., 1965, p. 33). The muck deposits of unglaciated central Alaska are mainly of Wisconsin age.

In the case of Big Bone Lick, Kentucky, one of the richest late-Pleistocene deposits in eastern North America, two samples of wood initially thought to be associated with bones of extinct species proved modern (W-908, W-1357). W-1358, 10,600 ± 250, was recovered from wood associated with a proboscidean tusk and may be close to the terminal age of *Equus* cf. *complicatus*, *Mylodon*, *Mammut*, and *Mammuthus* in the area.

An anomalous set of dates, perhaps illustrating the hazards of attempting to use modern methods on a site repeatedly excavated in the past, is the Mother Grundy's Parlor series (Q-511, 552, 553/4). Charcoal of postglacial age definitely is not to be associated with late-glacial mammoth, hyena, and reindeer from this Upper Paleolithic cave (see Garrod, 1926, p. 135–45).

Dates on bone and teeth or tusk would seem to promise an easy solution to direct dating of the fauna. But even when inorganic carbon is removed and organic residues alone are dated, the results can be glaringly discordant with stratigraphic expectation. The noncarbonate fraction of bones of the first mammoth to be dated from Finland yielded an age of about 9,000 years (Tx-127), decidedly too young to represent the late-glacial pollen associated with the mammoth bones and several thousands of years younger than current geological estimates for mammoth extinction in Europe (Butzer, 1964, p. 410). A bone-collagen age of 8,000 B.P. (UCLA 705) from an ilium of dwarf mammoth from Santa Rosa Island is much younger than charcoal dates (of cultural origin?) associated with mammoth in the area. Even skeptics of the hypothesis making man the fundamental cause of extinction would probably agree

CORPORATION OF THE CITY OF HOBART

TOWN HALL, ·
HOBART, TAS.
Phone: 34 6111

11/8/75

PARKING METER INFRINGEMENT — FINAL NOTICE

Mr. Wayne Richard Sigle**n**,

23D Digney Street,

<u>DYNNRYNE</u>, Tas.

VEHICLE REG. NUMBER	DATE OF INFRINGEMENT	PARKING METER REPORT No.	AMOUNT DUE
EA-8090	3/7/75	48350	**$1.50**

No payment has been made to the Corporation in respect of the parking meter infringement detailed above.

If the amount shown above is paid to the Corporation forthwith, no further action will be taken on the matter.

If it becomes necessary for the Corporation to institute legal proceedings a Magistrate may order a forfeit up to a maximum of $25.00.

<u>PLEASE PRESENT THIS NOTICE INTACT WHEN MAKING PAYMENT.</u>
ADDRESS FOR REMITTANCE BY MAIL: CITY TREASURER,
G.P.O. BOX 503E,
HOBART, 7001.

J. S. LEWIS, Town Clerk.

RECEIVED PAYMENT AS PER CASH REGISTER FIGURES

that the dwarf mammoth on this island could not have survived man's initial occupation by several thousand years. Either UCLA-705 must be too young or UCLA-106 and L-209-T (ca. 4,000 years older) must predate man's arrival (see also Meighan, 1965, p. 711). In the case of M-1516, a 4,000-year-old date was obtained on peccary (*Platygonus compressus*) bone from within a sand ridge of the shore of Glacial Lake Warren, which was drained and abandoned over 11,000 years ago. The absence of *Platygonus* from any postglacial Indian middens and from other postglacial vertebrate deposits is a further reason for regarding a 4,000-year record as quite improbable. Organic bone dates from Bonfire Cave, Texas, were several thousand years younger than the age of associated 10,000-year-old charcoal (Pearson et al., 1966).

NORTH AMERICAN EXTINCTION

Jelinek (1957) and Martin (1958) were among the first to propose radiocarbon chronologies for the late-Pleistocene megafauna of North America. The most thorough and most often cited effort was that of Hester (1960), in which dates associated with sixteen extinct genera were presented. Martin and Hester regarded 8,000 B.P. as the terminal date for many genera. Many issues of *Radiocarbon* have appeared since Hester's compilation; the new dates are listed in Table 2, and the more important ones are plotted in Figure 1. While they almost double the number of dated fossil records for the extinct fauna, they add only two genera, *Mylohyus* and *Tetrameryx*, to Hester's list. None of the new dates applies to the following sparsely and questionably dated genera: *Tanupolama*, *Smilodon*, *Sangamona*, *Arctodus*, *Euceratherium*, *Preptoceras*, and *Stockoceros*. The majority of the extinct late-Pleistocene genera and species have not been critically dated by radiocarbon (Table 1).

A much greater yield of radiocarbon dates of deposits apparently postdating the time of extinction is now at hand. Negative evidence is accumulating that makes questionable some of the younger dates on extinct fauna that Martin and Hester accepted. If native elephants, camels, horses, and sloths were still present in continental North America less than 10,000 years ago, their remains are unaccountably absent from a number of carefully excavated and carefully dated archaic and Paleo-Indian sites (Haynes, 1967). It is theoretically possible that the main wave of extinction, which now appears to have occurred around 11,000 B.P., left lingering enclaves of survivors that lasted well into the mid-postglacial. Such enclaves would be difficult to locate and even more difficult to date in the narrow stratigraphic units where they might

TABLE 2. North American Radiocarbon Dates Associated with Extinct Genera, Less than 20,000 B.P. (Radiocarbon, Vols. 1–8)

Site	Extinct genus	Sample	Years B.P.	Comment
1. Aden Crater, N.M.	Nothrotherium	*Y-1163a	9,840 ± 160	Body tissue affected by preservatives
		Y-1163b	11,080 ± 200	
2. Berrien Springs, Mich.	Mammuthus jeffersoni	*M-1400	8,260 ± 300	Tusk
3. Big Bone Lick, Ky.	Equus	W-1358	10,600 ± 250	Wood with tusk
	Mylodon = Paramylodon	*W-1357	< 200	Wood
	Mammut			
	Mammuthus	*W-908	< 250	Wood
4. Bonfire Shelter, Tex.	Mammuthus	Tx-153	10,230 ± 160	Sample overlies extinct fauna, with extinct Bison
	Camelops			
	Equus			
5. Byron, N.Y.	Mammut	W-1038	10,450 ± 400	Plant material beneath bone
6. Domebo, Okla.	Mammuthus	*TBN-311	4,952 ± 304	Untreated tusk
		SM-610	10,123 ± 280	Lignitic wood
		SM-695	11,045 ± 647	Wood
		SI-172	11,200 ± 500	Bone organics
		SI-175	11,200 ± 600	Humic acids
7. Ciudad de los Deportes, Mexico D.F., Mexico	Equus, Mammuthus, other fauna of Upper Becerra Formation	UCLA-111	18,700 ± 450	Wood from stump of Cupressus
8. Clovis, N.M.	Mammuthus	A-481	11,170 ± 360	Silty clay around skull
9. Elkhart, Ind.	Mammut	*M-694	9,320 ± 400	Bone
10. Fairbanks, Alaska	Mammuthus	*L-601	21,300 ± 1,300	Sample impregnated with preservatives

Site	Genus	Lab No.	Date	Notes
11. Gratiot County, Mich.	Mammut	*M-1254	10,700 ± 400	Molar; late-glacial pollen types
12. Gypsum Cave, Nev.	Nothrotherium	LJ-452	11,690 ± 250	Sloth dung
13. Jaguar Cave, Idaho	Camelops Equus	Geochron. Lab. Isotopes Inc.	11,580 ± 250 10,370 ± 350	Fauna associated with older date
14. Kalamazoo County, Mich.	Symbos	*M-639	13,200 ± 600	Bone
15. Kings Ferry, N.Y.	Mammut	Y-460	11,410 ± 410	Spruce wood from bone layer
16. Lehner Ranch, Ariz.	Mammuthus Equus Tapirus	M-811 K-554 A-42 A-378	11,290 ± 500 11,170 ± 140 11,240 ± 190 10,940 ± 100	Haynes gives average date for the bone bed and Clovis level as 11,260
17. Lindenmeier Site, Colo.	Camelops Mammuthus	I-141 *I-473 *I-632	10,780 ± 135 7,200 ± 200 11,200 ± 500	Folsom level / Bone and tusk fragments / Bone and tusk fragments
18. Lloyd Rock Hole, Pa.	Mylohyus	Y-727	11,300 ± 1,000	Fossil pollen record indicated mixing
19. Lubbock Reservoir, Tex.	Equus Mammuthus Camelops	I-246	12,650 ± 250	Shells
20. McCullum Ranch, N.M.	Mammuthus Equus Camelops, sloth	A-375	15,750 ± 760	No occupation
21. Monterey Bay, Calif.	Hydrodamalis steller	SI-115	18,940 ± 1,100	Historic distribution on Bering Island only
22. Murray Springs, Ariz.	Mammuthus	*A-69 bis.	8,270 ± 260	Overlies elephant bones
23. Northern Lights, Ohio	Castoroides	Y-526	11,480 ± 160	Wood associated with skull

Table 2 (continued).

Site	Extinct genus	Sample	Years B.P.	Comment
24. Novelty Mastodon, Ohio	Mammut	OWU-126	10,654 ± 188	Wood, boreal pollen types
25. Pontiac, Mich.	Mammut	No lab. designation	11,900 ± 350	Organic material, spruce pollen
26. Powder Mill Cave, Mo.	Aenocyon	*GX-145	13,170 ± 600	Bone fragments
27. Rampart Cave, Ariz.	Nothrotherium	L-473A	9,900 ± 400	Sloth dung, surface
		L-473C	11,900 ± 500	Sloth dung, 18 inches
		L-473D	>38,300	Sloth dung, 54 inches
28. Rancho la Brea, Calif.	Extinct fauna in Pit 3	Y-354b Y-354a Y-355a Y-355b Y-355A bis LJ-55	Range between 13,900 and 15,400	Wood not necessarily associated with any extinct species
	Extinct fauna in Pit 9	UCLA-773D	13,300 ± 160	
29. Rawhide Butte, Wyo.	Mammuthus	*A-366	10,550 ± 350	Date younger than animal
30. Rochester, Ind.	Mammut	I-586	12,000 ± 450	Wood
31. Rodney, Ontario	Mammut	S-29	11,400 ± 450	Wood
		S-30	12,000 ± 500	Muck
32. Russell Farm, Mich.	Mammut	*M-347	5,950 ± 300	Tusk
33. San Bartolo Atepehuacan, Mexico	Mammuthus, plus Upper Beecrra Formation	M-776	9,670 ± 400	Charcoal from lacustrine deposit; stone implements
34. Sandusky County, Ohio	Platygonus	*M-1516	4,290 ± 150	Bone; much too young to date Lake Warren beach containing bones

Site	Species	Lab no.	Date	Remarks
35. Santa Isabel Iztapan, Mexico	*Mammuthus* and Upper Becerra Formation	*M-774	2,640 ± 200	Bone; associated with stone implements; date much too young
36. Santa Rosa Island, Calif.	*Mammuthus exilis*	UCLA-106 UCLA-705	11,800 ± 800 8,000 ± 250	Charcoal Bone collagen; discrepancy with L-290-T, 12,500; charcoal
37. Scotts, Mich.	*Symbos*	*M-1402	11,100 ± 400	Bone
38. Seattle, Wash.	Fossil sloth	UW-8	12,300 ± 200	Peat, from pelvis
39. Sheridan, N.Y.	*Mammut*	*M-490	9,200 ± 500	Bone
40. Sullivan Creek, Alaska	*Mammuthus* *Equus*, other extinct species	*W-891 *W-937 *W-1106 *W-1108 *W-1111	2,520 ± 200 200 ± 200 < 200 6,730 ± 260 < 200	Wood, muck, etc. should be reliable dates; association with extinct fauna questionable
41. Sulphur River Formation, Tex.	*Mylohyus* *Equus*	SM-532 SM-533	9,550 ± 375 11,135 ± 450	*Mylohyus* slightly above younger date, *Amer. Antiq.*, v. 30, p. 351
42. Tule Springs, Nev.	*Teratornis* Sloth *Mammuthus* *Equus* *Camelops*	UCLA-503, 507, 512, 514, 518, 521, 522, 543, 604, 636, *549		Ten of eleven dates range between 11,500 and 14,000. UCLA-549 is slightly younger (9,520 ± 300) but not properly associated with fauna (C. V. Haynes, pers. corres.)
43. Tunica Bayou, La.	*Mammut*	W-944	12,740 ± 300	Wood, with bones
44. Tupperville, Ontario (Ferguson Farm)	*Mammut*	*S-16	6,250 ± 250	Gyttja immediately overlying skull; may be younger

Table 2 (continued).

Site	Extinct genus	Sample	Years B.P.	Comment
45. Ventana Cave, Ariz.*	Nothrotherium, Tetrameryx, Tapirus, Equus, other species of the volcanic debris unit	A-203	11,300 ± 1,200	Charcoal
46. Genessee Co., Mich.	Mammuthus	M-1361	11,400 ± 400	Tusk, but agrees with pollen analysis of Oltz and Kapp
47. Kendall Co., Tex.	Extinct vertebrate fauna	Tx-250	>10,600	Cave deposit, species not listed
48. Tupperville, Ontario (Perry Farm)	Mammut	GSC-211 S-172	11,860 ± 170 12,000 ± 200	Plant remains Plant remains
49. U.P. Mammoth, Wyo.	Mammuthus	I-449	*11,280 ± 350	Tusk, with associated artifacts
50. Wilson Butte Cave, Idaho	Horse, camel, sloth	M-1409 M-1410	14,500 ± 500 15,000 ± 800	Stratum C, with artifacts Stratum E, with artifacts

* Doubtful or unacceptable date, including dates on bone

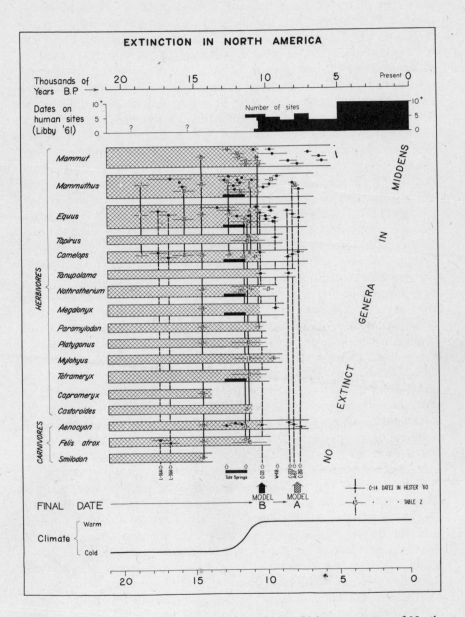

FIG. 1. Carbon-14 dates and extinction of certain late-Pleistocene genera of North America.

be sought. But even in the case of mastodon, to be discussed below, the claims for survival into the postglacial period (the last 10,000 years) can be questioned on grounds of ambiguous stratigraphic association, C^{14} age determination, or other uncertainty. In attempting to approach the terminal data for an extinct species, a certain number of "overshots" is to be expected. I am prepared to depart from my earlier view that a provocatively young date is likely to be accurate unless proved otherwise. The late- and postglacial sites in which dated beds of 10,000 to 8,000 B.P. fail to show a trace of extinct fauna include Frightful Cave; Coahuila (Taylor, 1956); cave deposits in the valley of Tehuacan, 9,000 years and younger in age (MacNeish, 1964); Danger Cave, Utah (Jennings, 1957, and Tx-85–89); Russell Cave, Alabama (M-766, M-590); Stanfield-Worley Bluff Shelter, Alabama (M-1152-3, M-1346-8); Kincaid Shelter, Texas (Tx-17-20); Eagle Cave, Texas (Tx-153); Raddatz Rockshelter, Wisconsin (Wittry, 1959, M-812, 813); and the pre-ceramic shell middens of Florida (Bullen, 1964) and California.

What do radiocarbon dates show concerning the extinct fauna itself? It is possible that extinct animals survived into the archaic of the early or mid-postglacial to coexist with human cultures that had abandoned their custom of hunting big game and were now hunting smaller prey and unknowingly making the first experiments with plant domestication? If all radiocarbon dates that Hester listed as questionable are discounted, the genera *Arctodus*, *Castoroides*, *Paramylodon*, *Platygonus*, *Smilodon*, and *Symbos* are associated with terminal dates of roughly 11,000 B.P. or older. None of the newer dates alters this interpretation (Table 2, Fig. 1). For the genera *Aenocyon* (*Canis dirus*), *Camelops*, *Equus*, *Mammut*, *Mammuthus*, *Megalonyx*, *Nothrotherium*, and *Tanupolama*, Hester lists terminal dates of roughly 8,000 B.P., with 6,000 B.P. for the mastodon. These form the basis of his conclusion that most of the fauna survived into the early postglacial.

For all genera except *Mammuthus* and *Mastodon* (to be considered below) there are only *three* sites, Gypsum Cave (C-222, 8,500 B.P.), Whitewater Draw (A-67, 8,200 B.P.; C-216, 7,800 B.P.), and Evansville, Indiana (W-418, 9,400 B.P.), on which the evidence for postglacial survival rests. One might expect, if they are valid terminal dates, to see them verified by newer records. But of the dates to appear since Hester's review (Table 2), only those that might easily have been contaminated, or those in uncertain stratigraphic association with the extinct fauna, are of postglacial age. Carefully controlled recent excavations at sites containing abundant bones of several genera of extinct animals in beds repeatedly and carefully dated, such as Tule Springs, Nevada, Blackwater Draw, New Mexico, and Lehner Ranch, Arizona, yielded no evidence of generic survival beyond roughly 11,000 years ago. New dates on sloth (*Nothrotherium*) dung from Aden Crater (Y-1163b) and Gypsum Cave (LJ-452) are much closer to 11,000 than 8,000 years old. And despite its ideal nature from many standpoints, even sloth dung may not be totally free from contamination. A very small amount of younger wood or fiber, easily introduced into a cave by wood rats (*Neotoma*) could contaminate the sample, a possibility I would not discount for L-473A, 10,000 B.P., from Rampart Cave.

North American Mastodons and Mammoths

Haynes (1964) has reviewed the radiocarbon dates and cultural content of some of the best known Early Man–mammoth sites in North America. There is little doubt that Clovis fluted-point hunters pursued the mammoth for a very short period of time in western North America before being replaced after 11,000 B.P. by hunters who used Folsom points and killed *Bison*. Carefully dated extinct bison sites are well known in the postglacial period. There is no question of extinct *Bison* living thousands of years after 11,000 B.P. But what about the mammoth? Haynes (p. 1412) cautiously concludes that the change from Clovis to Folsom may be related to a decline in the mammoth populations. Not only are there no well documented cases of mammoth associated with man in postglacial deposits of the last 10,000 years, there are no well-documented cases of postglacial mammoth sites without him. Of the post-10,000-year dates on mammoth, M-744, W-288, JBN-311, and O-171 are questioned either by Hester or by authors of the date lists themselves. M-1400 is on tusk. C-216, A-67, and A-69 from Whitewater Draw and Murray Springs, Arizona, are not satisfactorily documented as far as the alleged association with mammoth is concerned. UCLA 705 was discussed above. The youngest securely dated records of mammoth are those of 11,000-year vintage from the Lehner Site, Tule Springs, Domebo (Leonhardy, 1966), and Clovis (Blackwater Draw). One possible exception is the Mexico City mammoth of San Bartolo Atepehuacan, found with an obsidian flake and no fewer than fifty-nine small chips of basalt and obsidian. A date on associated carbon fragments was 9,700 B.P. (M-776 Aveleyra, 1964, p. 404). Elsewhere there is every reason to assume that New World mammoths and their hunters had disappeared before 10,000 B.P.

There remains the matter of mastodon extinction, an event that most authors have regarded as postglacial, significantly later than the extinction of other Pleistocene genera and later than the early hunters. If it can be shown that mastodons were little affected by the intrusion of the

Clovis hunters and indeed survived them by 4,000 years, as Griffin claims (1965, p. 658), the case for prehistoric man as the major cause of extinction would certainly need to be seriously modified or abandoned. In addition to Griffin, Martin (1958), Hester (1960), Skeels (1962), and most recent authors except Quimby (1960) have accepted a terminal radiocarbon date at about 6,000 B.P. Commonly cited are dates on tusk from Lapeer County, Michigan, and on the Washtenaw County, Michigan, mastodon (M-347 and M-67 respectively), both around 6,000 B.P. In addition, there is an organic date of 6,250 ± 250 (S-16) on gyttja from Ferguson Farm, Tupperville, Ontario, immediately overlying bones of a mastodon. A collagen date on the bones themselves is significantly older (GSC 614, 8,910 ± 150). While even the latter is best regarded as a minimum date only, it is considered closer to the true age of the mastodon than the gyttja which is now thought to be intruded (A. Dreimanis, personal correspondence).

Is it possible that all postglacial dates on mastodons are overshots? No skeptical archaeologist would consider accepting a radiocarbon date of 6,000 to 8,000 years on an alleged Clovis site before subjecting it to the most minute excavation and examination, without demanding an effort at replication of the date on the critical beds, without considering carefully all the possibilities of intrusion, and without a field demonstration of the evidence to equally critical colleagues.

There are three reasons why cautious second thoughts may now be needed regarding widely accepted claims of postglacial mastodon survival. The first is the radiocarbon dates published in the last few years and listed in Table 2. Except for the Ferguson Farm date discussed above, all the new dates, which are on wood, gyttja, or material other than the bones themselves are of late-glacial age, ranging between 12,700 and 10,500 B.P. (W-1358, W-1038, Y-460, OWU-126, Pontiac, Michigan; I-586, S-29, S-30, and W-944). In contrast, of four mastodon dates on bone or tusk, three (M-694, M-347, and M-490) are younger than 9,500 B.P. The discrepancy between these and the organic dates may be attributed to inorganic carbonate replacement or to humic acid contamination rather than to a real difference in age of the fossils.

The second is the fact that palynological study of beds containing mastodon and mammoth bones in the northeast indicates an association with spruce–pine pollen zones (Ogden, 1965) and presumably spruce forests. This environment disappears from the Great Lakes with the retreat of Valders ice about 10,500 years ago (Wright, 1964). Ogden quotes the 8,400-year radiocarbon date, supposedly associated with the Orleton Farms mastodon (M-66), but he does not mention that the pine–spruce pollen counts from near the mastodon bones must predate

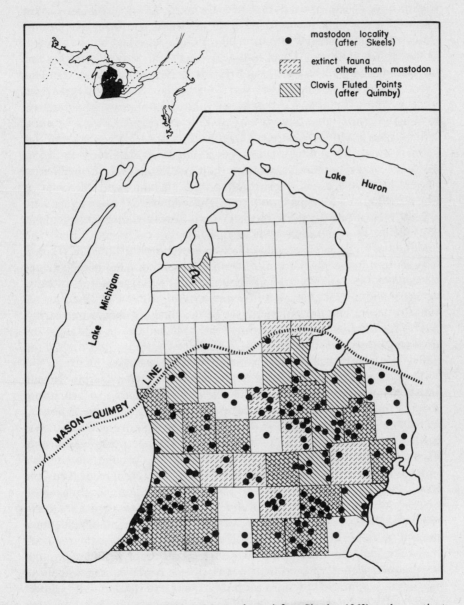

FIG. 2. Distribution of Michigan mastodons (after Skeels, 1962), other extinct Pleistocene genera, and fluted points (after Quimby, 1960).

this interval. Griffin (1965, p. 656) has rejected as not in direct association the Archaic projectile points found near the Orleton and other mastodons. The radiocarbon date (M-66) may be no more reliable.

The third difficulty in accepting a 6,000-year extinction date for mastodon emerges from study of its fossil distribution (Fig. 2). This proboscidean is the most common Pleistocene fossil in northeastern United States, so it may have once been as important in vertebrate biomass of the region as *Loxodonta* is in the game parks of central Africa. Skeels (1962) reports 163 records from Michigan alone. She proposes that its extinction there was hastened by its failure to follow the boreal forest environment across the Greak Lakes into Canada after deglaciation. But the northern limit of the mastodons in Michigan is not the Straits of Mackinac, as one might imagine if the elephants were trapped in a cul-de-sac south of the Great Lakes. The distribution of the 163 fossils is remarkable (Fig. 2). Although both mastodons and mammoths succeeded in crossing the Grand River outlet of Lake Warren, they did not range beyond Osceola and Gladwin counties at the latitude of Saginaw Bay (44° N). Equally remarkable is the distribution of fluted points from surface sites in Michigan. As Quimby (1958, 1960) noted, they are found only in the same part of the state as fossil mastodons— but south of what I have drawn as the "Mason–Quimby Line." One doubtful record, possibly an import, is that of a fluted point in Grand Traverse County (J. B. Griffin, personal correspondence).

The fluted-point hunters are regarded as occupying the lower peninsula of Michigan between the time of deglaciation and before the abandon- ment of the Main Algonquin Lake level, which was 24 ft above present Lake Huron (Griffin, 1965, p. 659). In the chronology of the Great Lakes post-Algonquin beaches were forming 10,500 to 10,000 years ago. From Skeel's map, it seems possible that a few mastodons saw the initial fall of Lake Algonquin; there are three fossil records in the Saginaw Bay region that appear to lie below Algonquin beach levels.

The mystery of why fauna and fluted points apparently terminated at the Mason–Quimby Line (Fig. 2), rather than reaching the Valders ice margin, remains unexplained. But it is a minor matter compared with that of explaining why, if the species survived until 6,000 years ago, mastodons failed to spread throughout the state, and from the beachhead in southern Ontario on into central Canada. Beyond possible inter- glacial records there are no bones of mastodon north of the fluted-point line of Mason (1962) copied on the inset of Figure 2. A more than coincidental association between fluted points and mastodons in the Southeast is mapped by Williams and Stoltman (1965, p. 677).

Griffin (1965) protests the lack of stratigraphic association between fluted points and extinct fauna in the East. But there are very few stratified fluted-point sites anywhere in mastodon country east of the Mississippi, and fewer in which suitable conditions for bone preservation exist. Unless more substantial documentation is forthcoming, present claims of postglacial mastodon survival based on radiocarbon evidence alone are insufficient. Meanwhile, the Mason–Quimby Line is evidence of the sort to be expected if overkill were the cause of mastodon extinction.

This hypothesis also makes credible some of the peculiar cultural attributes of the Paleo-Indians. Mason (1962, p. 242) concluded:

> It seems more than coincidental that the end of the Paleo-Indian cultural dominance, as measured by radiocarbon and other dating techniques, agrees closely with the demise of the fossil Pleistocene big-game animals; or to put it another way, that it was during the period characterized archaeologically by such artifact types as Folsom and Clovis that the great Pleistocene extinctions were taking place. It would push the limits of credibility to view as likewise coincidental the fact of the emergence of the generalized subsistence basis of the Archaic cultures during the disappearance of the Pleistocene fauna and fluted points. In other words, there is expressed a functional relationship between these culture types and the total ecology of which they are parts.

I have purposely avoided the question of a "pre-projectile-point stage." Bryan (1965) develops the hypothesis of a generalized leaf-shaped, percussion-flaked, stone-point tradition innovated in North America fairly early in the Wisconsin glacial stage. Müller-Beck (1966) states that "The first invasion of man in the New World for which a reliable archaeological reconstruction seems possible—there could have been earlier invasions—took place about 28,000 to 26,000 years ago." Both authors advance their conclusions on typological and paleontological grounds, recognizing that there is no indisputable radiocarbon-dated evidence for man in the New World older than that associated with the fluted-point hunters of around 12,000 years ago. If C^{14} dates from Wilson Butte Cave, Idaho, can be replicated, it may extend man's New World chronology.

The possibility that *Homo sapiens* spread into the Americas long before the late-glacial by no means eliminates the hypothesis of overkill. One may assert that the postulated users of core tools, choppers, and perhaps even bone tools were not specialized for killing big game, and thus had little effect on the megafauna, unlike the Clovis hunters of elephants or the Folsom hunters of extinct *Bison*. Possibly the easily hunted giant species of *Geochelone*, as yet unknown in beds of late-

(handwritten margin notes:)
① Extinct fauna surviving beyond hunters
②
Massive unbalanced Pleistocene extinction before man.

glacial age of the American continent, owe their extinction to a pre-projectile-point culture. What would upset the hypothesis of overkill would be clear-cut cases on the continent of many of the extinct animals surviving beyond the time of the big-game hunters, or clear-cut cases of massive unbalanced Pleistocene extinction anywhere before man.

NEW ZEALAND

That extinction of a variety of medium to large-sized herbivores can occur within a few hundred years after prehistoric man's initial appearance is shown by the extinction chronology of New Zealand (Fleming, 1962). New Zealand occupies 103,000 square miles, slightly smaller than the State of Colorado. There were no native terrestrial mammals, but twenty-seven species of extinct moas, including a 10-ft-tall *Dinornis maximus*, have been discovered in astonishing numbers in postglacial deposits, 800 to the acre in Pyramid Valley and that many in a pocket ($30 \times 20 \times 10$ ft deep) at Kapua (Duff, 1952). These giant flightless birds can be traced to the late-Miocene (*Anomalapteryx*). Regarding extinction, the orthodox theory was that many were extinct before man arrived, thus most had died out naturally (Duff, 1963a, p. 6). Partly on the basis of radiocarbon dates of moa bones, stomach contents, associated charcoal, and tussock bedding and partly on recent archaeological findings, Fleming (1962) has concluded that such was not the case. Twenty-two of the extinct moa species have now been found in association with prehistoric man (Fig. 3, Table 3). Sixteen moas have been dated by radiocarbon analysis of the bone itself, although some of the bone may have been contaminated by younger humates. In addition to the moas, a number of other birds became extinct in the same general period; half of these have been found in cultural association. Reviewing the last ten years of New Zealand archaeology, Golson and Gathercole (1962) conclude: "Nevertheless one definite result has emerged from this aspect of the decade's work. Possible climatic and genetic factors notwithstanding, man as the moa's first mammalian predator was a prime instrument in its extinction."

Duff (1963a, p. 6) has abandoned his hypothesis that moa extinction must have been due largely to natural causes, although he is still concerned with the absence of the giant moa *Dinornis maximus* in most of the moa-hunter camps; he still suggests a considerable reduction in moa numbers before man's arrival.

Fleming (1962) reports little evidence of early and mid-Holocene moa extinction, an extremely important fact in comparing the New Zealand pattern with that of Australia and the Americas. Despite

(handwritten margin notes:)
27 species of MOAS; no native terrestrial mammals
22 Associations with man
No mid-Holocene MOA extinctions

TABLE 3. *New Zealand Dates and Extinct Fauna*

Locality	Fauna	Sample	Date A.D.	Comment
1. Blenheim	(Moa hunter site)	NZ-50	1100	Charcoal
2. Paptowai, Otago	(Moa hunter site)	NZ-134	1185	Charcoal
3. Hina Hina	(Moa hunter site)	NZ-53	1210	Charcoal
4. Tahakopa River Mouth, Otago	*Dinornis maximus*	NZ-136	1320	Charcoal
	Euryapteryx gravis	NZ-137	1490	Bone of *Euryapteryx*
		NZ-138	1490	Bone of *Dinornis*
		NZ-139	1640	Bone of *Euryapteryx*
		NZ-140	1560	*Moa* sp. All bone dates may be too young.
5. Hawksburn Valley, Otago	Moa bones	NZ-62	1350	Charcoal
		NZ-61	1360	Charred wood with *Moa* bone fragments
		NZ-60	1500	Femur of *Euryapteryx*
		NZ-59	1550	Burned *Moa* bone
6. Tautuku, Otago	*Dinornis torosus*	NZ-146	1670	In occupation deposit

piedmont glaciation and widespread periglacial phenomena on South
Island and despite volcanism with the extensive blanketing of North
Island by nutrient-poor pumice and ash, no species of the giant birds
are definitely known to have disappeared before man's arrival.

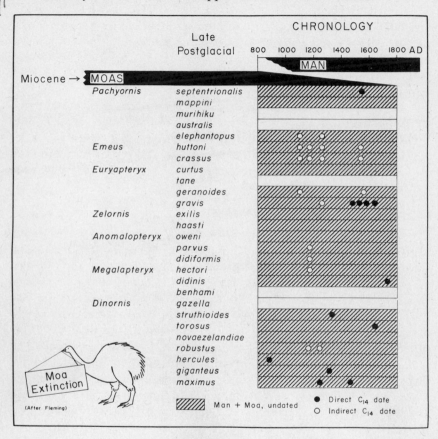

FIG. 3. Carbon-14 dates and extinction of moas in New Zealand (after Fleming, 1962).

Duff (1963b) notes that the moa hunters must have independently
developed techniques for seeking their prey; they were of East Polynesian
origin and, unlike the Upper Paleolithic cultures, had no tradition of
big-game hunting. They developed new techniques fairly rapidly. A
simple one was fire, widely used in South Island, to judge by charcoal
horizons in the soil. Golson and Gathercole claim that the moa hunters
retained their tools even after the birds were largely extinct and marine
resources were their main source of food. Retention of hunting tools

and persistence of a hunting tradition through the time of minimum numbers of his prey would establish man's role as a superpredator, still selecting big game whenever possible, even after his main food supply came from other sources.

Ultimately, when the Maori arrived in A.D. 1350, most of the giant birds were gone, so when first questioned on the subject a hundred years ago the Maori could provide no convincing accounts of the birds. Less explicable was the initial refusal of New Zealand scientists to regard prehistoric man as the cause for moa extinction. Fleming (1962, p. 116) remarks:

> It seems we are reluctant to blame our fellow men for a pre-historic offense against modern conservation ideals and would rather blame climate or the animals themselves. The simplest explanation is to attribute all late Holocene extinction to the profound ecological changes brought about by the arrival of man with fire, rats, and dogs.

AUSTRALIA

Crop contents of the extinct giant marsupial (*Diprotodon*) were dated at over 40,000 B.P. (NZ-205); dentine from a lower jaw of *Diprotodon* was 6,700 B.P. (NZ-206), apparently too young to represent the true age of the fossil (Table 4). At Lake Menindee a rich assemblage of extinct marsupials can be associated with GaK-335 (19,000 B.P.) and LJ-204 (26,000 B.P.). Although Gill (1963) concludes that aboriginal entry began at least 20,000 years ago, the evidence of man at the time has been questioned by Mulvaney (1964). More convincing dates on prehistoric man are GaK-334 (11,600 B.P.) from Nools, where Tindale recovered a flake assemblage below a microlith assemblage, and those at Kenniff Cave, where Mulvaney (1964) obtained samples dated at 13,000 and 16,000 years (NPL 33 and 68) and associated with a "Tasmanoid" industry.

Younger dates from Lake Menindee, NZ-66 and W-169 of 6,000 and 8,600, respectively, are now thought to be associated with essentially modern faunas (Hubbs et al., 1962). A terminal date for *Nototherium* (not to be confused with the southwestern United States sloth *Nothrotherium*) may be Gx-105, 14,000 B.P., on bone fragments of a jaw. Although all bone carbonate dates are suspect, the age agreed with the collectors' estimate and is also equivalent with Y-170, 13,700 B.P., according to Hubbs et al. (1962), the youngest dating definitely applicable to a varied assemblage of giant marsupials. This is in accord with absence of extinct fauna from Nansump Cave in beds dated at 12,000 B.P. (Lundelius, 1960.)

TABLE 4. Australian Dates and Extinct Fauna

(Radiocarbon, Vols. 1–7)

Locality	Fauna	Sample	Date	Comment
1. Lake Menindee, N.S.W.	*Macropus Terragus* (giant kangaroo)	LJ-204	26,300 ± 150	Charcoal from hearth
	Extinct animals	GaK-335	18,800 ± 800	Charcoal from hearth
2. Boolounda Creek, S. Australia	*Nototherium*	GX-105	14,000 ± 250	Bone fragments from jaw
3. Orroroo, S. Australia	*Diprotodon*	NZ-381	11,100 ± 130	Molar
4. Lake Callabona, S. Australia	*Diprotodon*	NZ-206	6,700 ± 250	Dentine from lower jaw, supposedly same animal as WZ-205, >40,000 (crop contents)
5. Lake Menindee, S. Australia	Extinct genera	NZ-66	6,570 ± 100	Shells associated with Tartangan artifacts

Thus it appears that both prehistoric hunters and the main wave of extinction swept through Australia slightly before these events occurred in North America. Many additional geochronological data are needed before intercontinental cultural, climatic, and extinction chronologies can be compared more critically. But there is no longer doubt that man and the extinct Australian marsupials coexisted (Gill, 1963).

TROPICAL AMERICA

Although the extinct late-Pleistocene fauna from Central and South America is less well known than that of the United States, it is obvious that the inventory of extinct genera and species exceeds that of higher latitudes. In a preliminary account of a single fauna in Bolivia, Hofstetter (1963) recovered the following extinct genera, far more than are known from any single fossil locality in North America: *Nothropus*, *Megatherium*, *Glossotherium*, *Lestodon*, *Scelidotherium*, *Glyptodon*, *Chlamydotherium*, *Neothoracophorus*, *Hoplophorus*, *Panochthus*, *Neochoerus*, *Theriodictis*, *Arctotherium*, *Smilodon*, *Macrauchenia*, *Toxodon*, *Cuvieronius*, *Notiomastodon*, *Hippidion*, *Onohippidium*, *Palaeolama*, and *Charitoceros*. It is apparent that the Pleistocene game range of South America was especially well stocked, as one would expect in a tropical ecosystem. Extinction impoverished the tropical American fauna to a greater degree than that of the temperate regions. Did it occur before, after, or coincidental with extinction in temperate North America?

In South America, sloth dung, one of the best materials for critical radiocarbon dating, indicates survival to just over 10,000 years ago of ground sloth associated with extinct horse at Mylodon Cave and Fells Cave (Sa-49, W-915, C-484) (Table 5). If C-485 (8,639 B.P.) on burned bone from Palli Aike Cave, Chile, and Sa-47 (6,500 B.P.), Ponsomby, Patagonia, are also correctly associated with sloth and horse, a remarkably late survival could be claimed. However, no extinct animal remains were found in somewhat older rock shelters from Minas Gerais, Brazil (P-521, P-519), excavated by W. B. Hurt. For this reason the use of Palli Aike, Ponsomby, and certainly the 3,000-year age on "extinct giant bear" from Minas Gerais (M-354) may be questioned as valid terminal dates for the fauna. If the date of 14,000 B.P. (M-1068) from Falcon, Venezuela, associated with extinct giant sloths, horse, and mastodon, is also associated correctly with big-game hunters in South America, it would, of course, obliterate the concept of their relatively late (12,000 B.P.) arrival in the New World. Apparently, more dates support the view that extinction in South America coincided with or slightly postdated that in North America, but those who believe in a slow,

TABLE 5. *South American Dates Associated with Extinct Fauna*

(Radiocarbon, Vols. 1–7)

Locality	Fauna	Sample	Date	Comment
1. Muaco Site, Falcon, Venezuela	Mastodon *Megatherium, Equus*	M-1068	14,300 ± 500	Association with Early Man needs verification
2. Fell's Cave, Chile	Giant sloth, horse	W-915 Sa-49	10,720 ± 30 10,200 ± 400	Charcoal from firepit Sloth dung, C-484, in same layer as 10,832 ± 400
3. Ponsomby, Patagonia	Extinct horse	Sa-47	6,500 ± 400	Date on peat deposit associated with the oldest archaeological level

steady reduction of the late-Pleistocene fauna over tens or hundreds of thousands of years have yet to be confronted with the sort of dating evidence that invalidates this interpretation elsewhere.

In the West Indies, extinct vertebrates were of too small a size to have suffered extermination on the mainland (Martin, 1958, p. 409). Rouse (1964) attributes the extinction of at least some of the twenty-two genera of mammals found in prehistoric middens and cave earth to the arrival of man two to four thousand years ago. Subfossil "giant" species of small or medium-sized terrestrial vertebrates are still being discovered (cf. Ethridge, 1964), and a giant land snail, presumably also extinct, is known from Hispaniola (Clench, 1962). Whereas some of the native West Indian fauna may have disappeared as a result of catastrophic post-Columbian ecologic changes, brought by the introduction of *Rattus*, it appears that more disappeared before the fifteenth century. Here, as elsewhere, the main circumstance pointing toward prehistoric man's role in extinction, without shedding light on details of the process, is the matter of chronology. The fauna survives until man arrives.

On the continent, there is some archaeofaunal evidence of local extirpation of medium-sized animals in certain intensely occupied areas like the Valley of Mexico, where Vaillant (1944) reported deer (*Odocoileus*) to be virtually exterminated two thousand years ago. The postglacial withdrawal of mule deer and antelope from southern parts of the Mexican Plateau has been attributed to vegetation change (Alvarez, 1964; Flannery, 1966), but overkill by expanding prehistoric populations seems at least equally probable. Peccary, marmot, and porcupine bones are notably scarce or absent in refuse from the more densely inhabited parts of the prehistoric Southwest. Local hunting may have wiped out these mammals during the late Pueblo period. In the Antilles, late prehistoric extinction of the larger lizards, rodents, and sloths probably occurred as a result of intense seasonal search for animal protein, when the relatively numerous prehistoric tribes were not cultivating manioc and maize, their mainstay. A comparable region in which the effect of prehistoric man on extinction of medium- to small-sized animals remains to be determined is the islands of the Mediterranean. The disappearance of *Myotragus* in Minorca seems much more closely timed to the earliest record of human occupation of the island than was once realized (Waldren, personal correspondence).

AFRICA

The "rose-colored glasses" view of prehistoric man in Africa is well put by Harper (1945, p. 15): "As long as the African Continent was occupied by primitive savages, without modern weapons, animal life was, in a

large sense, in a virtual state of equilibrium." I shall take this opportunity
to point out a grave error in the assumptions of various scientists writing
on the question of big game and the Pleistocene (e.g. Eiseley, 1943;
Mason, 1962, p. 243; Leopold *in* Talbot and Talbot, 1963, p. 5; and,
alas, Martin, 1958, p. 412). These authors failed to realize that Africa,
no less than the other continents, suffered its episode of accelerated
megafaunal extinction. Perhaps some of them were thinking of the last
twenty thousand years, when it is true that practically no extinction
occurred (Flint, 1957, p. 277; Butzer, 1964, p. 400). Perhaps others were
misled by Theodore Roosevelt's chapter (1910), "A Railroad Through
the Pleistocene," where he compares the game of the East African plains
with the American Pleistocene fauna. Whatever the reason, they have
assumed that the African megafauna survived the Pleistocene unscathed,
and Eiseley in particular has used this point as an argument against the
hypothesis of New World overkill.

Although its fossil fauna is far from adequately known, roughly fifty
genera disappeared during the Pleistocene (see Hopwood and Hollyfield,
1954; Cooke, 1963). Furthermore, in Africa, as in America, most of the
surviving large animals are also known as Pleistocene contemporaries
of the extinct genera. The living genera of African big game represent
only 70 per cent of the middle-Pleistocene complement (Martin, 1966).
Thus despite its extraordinary diversity, the living African fauna must
be regarded as depauperate, albeit much less so than that of America or
Australia.

The time of "middle"-Pleistocene extinction was barely within the
range of reliable dating by radiocarbon—i.e. over just forty thousand
years ago. Fortunately, the rich archaeological content of many fossil
beds aids in age interpretation. Toward the end of the Acheulian, and
often associated with the stone bifaces and other tools of these big-game
hunters in sites such as Olduvai (Bed IV) in Tanzania, Olorgesailie and
Kariandusi in Kenya, and Hopefield and the Vaal River gravels in
South Africa, the following large mammals are last recorded: *Meso-
choerus, Tapinochoerus, Stylochoerus, Libytherium, Simopithecus,
Archidiskodon (Elephas)*, and *Stylohipparion*. Eight additional extinct
genera of the period are known only from middle, or occasionally late,
Pleistocene sites in South Africa (Cooke, 1963, p. 98–101). All are absent
from Middle Stone Age sites, and thus were extinct *before* the major
depression of African montane vegetation zones of full- and late-glacial
age recently reported by pollen stratigraphers (Coetzee, 1964; Livingstone, 1962; Morrison, 1961; Van Zinderen Bakker, 1962).

On stratigraphic and faunal evidence, Leakey (1965) attributes extinction of the Olduvai Gorge genera to drought. Clark (1962) places the

evolved Acheulian at about 57,000 B.P., and, on the basis of intercontinental correlation with the Brørup Interstadial, he considers the
First Intermediate Period (after the Acheulian) of 40,000 B.P. to have
been dry.

If drought decimated the African mainland fauna at the end of the
Middle Pleistocene, it managed to leave unscathed the endemic and
ecologically vulnerable insular fauna of Madagascar. Seven genera of
extinct lemurs, the pigmy hippopotamus, two species of giant tortoise,
and two genera of struthious birds occur in very late Pleistocene beds.
No earlier episode of extinction is known there. All the animals were
contemporaries of prehistoric man, who did not reach the island until
remarkably late in the postglacial. One date on charcoal associated with
pottery and iron hooks is also a time when the roc, *Aepyornis maximus*,
was abundant (GaK-276, A.D. 1100). Unless substantial paleobotanical
evidence for a unique drought can be found in the "First Intermediate
Period" in Africa, or evidence for a major decline of the Malagasay
fauna prior to man's arrival, the evidence for a climatic cause of
extinction in Africa suffers from the same ad hoc appeal that has made
it an unacceptable explanation for the pattern elsewhere.

Late-Pleistocene extinction in Africa long precedes that in the
Americas and Australia, as would be expected in view of man's gradual
evolution in Africa. A major point for paleontologists to recognize is that
the question "Why no extinction in Africa despite man's antiquity?"
is misleading. There was a major wave of generic extinction in Africa
although not so intense as in South America. Extinction in Africa seems
to coincide with the maximum development of the most advanced
early Stone Age hunting cultures, the evolved Acheulian of abundant,
continent-wide distribution. The case of Africa neither refutes the
hypothesis of overkill nor supports the hypothesis of worldwide climatic
change as a cause of extinction (Martin, 1966).

CONCLUSION

In continental North America, the only major episode of generic
extinction in the Pleistocene occurred close to eleven thousand years ago
(Fig. 1). Provisional ages for the start of major extinction episodes
elsewhere are: South America, 10,000 B.P.; West Indies, mid-postglacial;
Australia, 13,000 B.P.; New Zealand, 900 B.P. (Fig. 3); Madagascar,
very late postglacial (800 B.P.); northern Eurasia (four genera only),
13,000 to 11,000 B.P. (Table 6); Africa and probably Southeast Asia,
before 40,000 to 50,000 B.P. Radiocarbon dates, pollen profiles
associated with extinct animal remains, and new stratigraphic and

TABLE 6. Eurasian Dates and Extinct Fauna

Locality	Fauna	Sample	Date	Comment
1. Lena River, Siberia	*Mammuthus*	Y-633	30,000	Skin
2. Lake Nojiri, Japan	*Megaceros*	GaK-269	31,000 ± 2,500	Wood, 97 cm
	Elephas sp.	GaK-268	21,600 ± 900	Wood, 61 cm
		GaK-267	16,150 ± 550	Wood, 40 cm
3. Yokoyama, Japan	*Elephas*, horse *Megaceros*	GaK-312	28,400 ± 1,800	Charred wood
4. North London, England	Mammoth, reindeer	Q-25	28,000 ± 1,500	Plant debris
5. Chekolsouka, Siberia	*Mammuthus*	MO-215	26,000 ± 1,600	Mammoth hair
6. Cambridge, England	Reindeer, *Mammuthus*	Q-590	19,500 ± 650	Plant detritus
7. Lascaux Cave, France	(Magdalenian paintings)	GrN-1632	17,190 ± 140	Charcoal
8. Lascaux Cave, France		Sa-102	16,100 ± 500	Charcoal
9. Hanaizumi Formation, Japan	*Palaeoloxodon tokunagai Megaceros, Loxodonta*	Y-594	15,850 ± 360	Wood with worked (?) bones
		N-132, N-133 N-141-3	14,900 to > 36,800	Relationship to fossils not given in date list
10. Naguno, Japan	*Megaceros*	GaK-161	15,750 ± 390	Wood with giant elk bones
11. Pont du Chateau, France	Elephant, rhino, horse, cave bear, etc.	Sa-103	13,500 ± 450	Peat, beneath the fauna

Location	Species	Lab no.	Date	Notes
12. Peggau, Austria	Cave bear	GrN-2036	13,370 ± 150	Charred wood
13. Vailly-sur-Aisne, France	Mammoth	Sa-53	11,550 ± 450	Tooth
14. Taimyr Peninsula, Taimyr Lake, Siberia	Woolly mammoth Woolly mammoth	T-297 Mo-3	11,450 ± 250 11,700 ± 300	Sinews *Salix* wood from mammoth horizon
15. Bernese Overland, Switzerland	Cave bear	B-152 B-153	10,150 ± 200 9,500 ± 150	Bones Bones
16. Calabria, Italy	"Pleistocene fauna with extinct species"	R-186	10,030 ± 90	Charred bones, Upper Paleolithic industry
17. Kunda, Estonian SSR	Mammoth	TA-12	9,780 ± 260	Tusk, does not belong to Mesolithic campsite
18. Couternon, France	*Elephas primigenius*	Gif-341	9,440 ± 350	*Pinus sylvestris* fragment; mammoth found at same level
19. Helsinki, Finland	Mammoth	Tx-127	9,030 ± 105	Bone, too young to agree with pollen content
20. Derbyshire, England	Mammoth, hyena, horse, etc.	Q-551 Q-552 Q-553/4	8,800 ± 300 7,662 ± 140 6,915 ± 140 6,705 ± 140	Charcoal, unaccountably young ages
21. La Manche, France	Mammoth tooth	Gif-342	8,720 ± 300	Total carbonate sample

archaeofaunal evidence show that, depending on the region involved, late-Pleistocene extinction occurred either after, during, or somewhat before worldwide climatic cooling of the last maximum of Würm–Weichsel–Wisconsin glaciation (Fig. 4).

While it occurred at a time of climatic change, the pattern appears to be independent of a climatic cause. Outside continental Africa and Southeast Asia, massive extinction is unknown before the earliest known arrival of prehistoric man. In the case of Africa, massive extinction coincides with the final development of Acheulean hunting cultures, which are widespread throughout the continent.

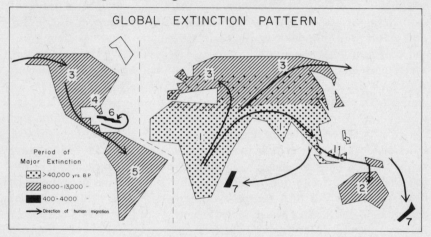

FIG. 4. The global pattern of late-Pleistocene extinction in sequence: 1, Africa and southern Eurasia; 2, New Guinea and Australia; 3, Northern Eurasia and northern North America; 4, southeastern United States; 5, South America; 6, West Indies; 7, Madagascar and New Zealand. In each case, the major wave of late-Pleistocene extinction does not occur until prehistoric hunters arrive.

Yet the notion of prehistoric overkill is commonly dismissed out of hand. In his book on extinct and vanishing birds of the last few hundred years Greenway (1958, p. 29) suggests that prehistoric men and birds "arranged a means of living together to the end that no birds were extirpated." At least one very perceptive neo-Darwinian evolutionist and humanist probably speaks for many in regarding it as "almost inconceivable that Indians alone put an end to the whole vast horse population of the late Pleistocene over so enormous an area." After considering and discounting all other possible explanations of horse extinction at the end of the Pleistocene, Simpson (1961, p. 200) held:

This seems at present one of the situations in which we must be humble and honest and admit that we simply do not know the answer. It must be remembered too that extinction of the horses in the New World is only part of a larger problem. Many other animals became extinct here at about the same time. The general cause of extinction then or at earlier times must have been the occurrence of changes to which the animal populations could not adapt themselves. But what precisely were those changes?

Indeed, it is not when horses alone but when the full complement of extinct Pleistocene animals are considered, when all major land masses are included in the analysis, and especially when the chronology of extinction is critically set against the chronology of human migrations and cultural development (as in Fig. 4) that man's arrival emerges as the only reasonable answer to Simpson's question. To be sure there is much ignorance left to admit. We must beg the question of just how and why prehistoric man obliterated his prey. We may speculate but we cannot determine how moose, elk, and caribou managed to survive while horse, ground sloth, and mastodon did not. One must acknowledge that within historic time the Bushmen and other primitive hunters at a Paleolithic level of technology have not exterminated their game resources, certainly not in any way comparable to the devastation of the late-Pleistocene. These and other valid objections to the hypothesis of overkill remain. But thus far the hypothesis has survived every critical chronological test. On a world scale the pattern of Pleistocene extinction makes no sense in terms of climatic or environmental change. During the Pleistocene, accelerated extinction occurs only on land and only after man invades or develops specialized big-game hunting weapons.

It seems to me that the chronologic evidence strongly supports the conclusion of an earlier Darwinian who took pains not to dismiss the phenomenon as trivial and who ended a lifetime of study by concluding, in a generally overlooked part of his work, that man must in some way be the destructive agent (Wallace, 1911, p. 261–67).

The thought that prehistoric hunters ten to fifteen thousand years ago, (and in Africa over forty thousand years ago) exterminated far more large animals than has modern man with modern weapons and advanced technology is certainly provocative and perhaps even deeply disturbing. With a certain inadmissible pride we may prefer to regard ourselves, not our remote predecessors, as holding uncontested claim to being the arch destroyers of native fauna. But this seems not to be the case. Have we dismissed too casually the possibility of prehistoric overkill? The late-Pleistocene extinction pattern leaves little room for any other explanation.

I have sought and received stimulating conversation and correspondence on this problem from colleagues, teachers, and students. Without attempting to evaluate the magnitude of their help, or to imply their endorsement of my use of it, my grateful thanks at least are due the following: J. B. Griffin, J. E. Mosiman, L. S. B. Leakey, D. Livingstone, J. G. Clarke, C. V. Haynes, V. M. Bryant, P. J. Mehringer, J. Schoenwetter, M. S. Stevens, J. Elson, C. W. Hibbard, Roger Duff, R. J. Mason, Ruth Gruhn, D. S. Byers, C. Ray, K. P. Koopman, A. Dreimanis, M. K. Hecht, and C. A. Reed. It goes without saying that each contributor to the chapters in this book has greatly added to my efforts at understanding the extinction problem. A final acknowledgment remains, above all, to Marian.

References

Alvarez, T., 1964, Nota sobre restos oseos de mamiferos del Reciente, encontrados cerca de Tepeapulco, Hidalgo, Mexico: *Mexico Inst. Nac. Antrop. Hist., Dept. Prehistoria,* Publ. 15, p. 1–15

Aveleyra Arroyo de Anda, L., 1964, The primitive hunters, p. 384–412, *in* West, R. C., *Editor, Handbook of Middle American Indians,* vol. *1,* Natural Environment and Early Cultures: Austin, Univ. Texas Press, 570 p.

Bramlette, M. B., 1965, Massive extinctions of biota at the end of Mesozoic time: *Science,* v. *148,* p. 1696–99

Bryan, A. G., 1965, Paleo-American prehistory: Idaho State Univ. Museum, Occ. Papers, no. 16, 247 p.

Bryan, K., and Gidley, J. W., 1926, Vertebrate fossils and their enclosing deposits from the shore of Pleistocene Lake Cochise, Arizona: *Amer. Jour. Sci.,* v. *211,* p. 477–88

Bullen, R. P., 1964, Artifacts, fossils, and a radiocarbon date from Seminole Field, Florida: *Florida Acad. Sci. Quart. Jour.* (1963), v. *26,* p. 293–303

Butzer, K. W., 1964, *Environment and archeology; An introduction to Pleistocene geography:* Chicago, Aldine, 524 p.

Clark, J. D., 1962, *Carbon-14 chronology in Africa south of the Sahara:* Tevoren, Belgium, IV⁰ Congrès Panafricain de Préhistorie et de l'Etude du Quaternaire, Actes, Sect. III, p. 303–11

Clench, W. J., 1962, New species of land mollusks from the Republica Dominica: *Brevoria.,* v. *173,* p. 1–5

Coetzee, J. A., 1964, Evidence for a considerable depression of the vegetation belts during the upper Pleistocene on the Eastern African mountains: *Nature,* v. *204,* p. 564–66

Colbert, E. H., 1942, The geologic succession of the Proboscidea, p. 1421–52, *in* Osburn, H. F., *Proboscidea,* vol. *2:* New York, American Museum of National History

Cooke, H. B. S., 1963, Pleistocene mammal faunas of Africa, with particular reference to Southern Africa, p. 65–116, *in* Howell, F. C., and Bourliere, F., *Editors, African ecology and human evolution:* Chicago, Aldine, 666 p.

Duff, R., 1952, *Pyramid valley:* Christchurch, Pegasus Press, 48 p.

—— 1963a, *The problem of* Moa *extinction:* Thomas Cawthron Mem. Lect. 38, Cawthron Inst., Nelson, N.Z., Stiles, 27 p.

—— 1963b, New Zealand archaeology: *Antiquity,* v. *37,* p. 65–8

Eiseley, L. C., 1943, Archaeological observations on the problem of post-glacial extinctions: *Amer. Antiq.,* v. *8,* p. 209–17

Ericson, D. B., Ewing, M., and Wollin, G., 1963, The Pleistocene epoch in deep-sea sediments: *Science,* v. *146,* p. 723–32

Ethridge, R., 1964, Late Pleistocene lizards from Barbuda, British West Indies: *Florida State Mus. Bull. Biol. Sci.,* v. *9,* p. 43–75

Flannery, K. V., 1966, The postglacial "readaptation" as viewed from Mesoamerica: *Amer. Antiq.,* v. *31,* p. 800–05

Fleming, C. A., 1962, The extinction of moas and other animals during the Holocene period: *Notornis,* v. *10,* p. 113–17

Flint, R. F., 1957, *Glacial and Pleistocene geology:* New York, Wiley, 553 p.

Garrod, D. A. E., 1926, *The Upper Paleolithic age in Britain:* Oxford, Clarendon Press, 211 p.

Gehlbach, F. R., 1965, Amphibians and reptiles from the Pliocene and Pleistocene of North America; A chronological summary and selected bibliography: *Texas Jour. Sci.,* v. *17,* p. 56–70

Gill, E. G., 1963, The Australian aborigines and the giant extinct marsupials: *Austr. Nat. Hist.,* v. *14,* p. 263–66

Golson, J., and Gathercole, P. W., 1962, The last decade in New Zealand archaeology: *Antiquity,* v. *36,* p. 168–74

Greenway, J. C., Jr., 1958, Extinct and vanishing birds of the world: New York Amer. Comm. Inst. Wild Life Protection, Spec. Publ., v. *13,* 518 p.

Griffin, J. B., 1965, Late Quaternary prehistory in the northeastern woodlands, p. 655–67, *in* Wright, H. E., Jr., and Frey, D. G., *Editors, The Quaternary of the United States:* Princeton Univ. Press, 922 p.

Harper, F., 1945, Extinct and vanishing mammals of the Old World: New York Zool. Park Special Publ. no. *12,* 850 p.

Hay, O. P., 1919, On the relative ages of certain Pleistocene deposits: *Amer. Jour. Sci.,* v. *47,* p. 361–75

—— 1923, The Pleistocene of North America and its vertebrated animals: Carnegie Inst. Washington, Publ. 322, 499 p.

Haynes, C. V., Jr., 1964, Fluted projectile points; Their age and dispersion: *Science,* v. *145,* p. 1408–13

—— 1967, Carbon-14 dates and Early Man in the New World: (this volume)

Hester, J. J., 1960, Pleistocene extinction and radiocarbon dating: *Amer. Antiq.,* v. *26,* p. 58–77

Hibbard, C. W., 1958, Summary of North American Pleistocene mammalian local faunas: *Michigan Acad. Sci. Papers,* v. *43,* p. 3–32

Hibbard, C. W., Ray, C. E., Savage, D. E., Taylor, D. W., and Guilday, J. E., 1965, Quaternary mammals of North America, p. 509–25, *in* Wright, H. E., Jr., and Frey, D. G., *Editors, The Quaternary of the United States:* Princeton Univ. Press, 922 p.

Hibbard, C. W., and Taylor, D. W., 1960, Two late Pleistocene faunas from southwestern Kansas: *Univ. Michigan Mus. Paleont. Contrib.*, v. *16*, p. 1–223

Hibben, F. C., 1946, *The lost Americans:* New York, Crowell, 200 p.

Hofstetter, R., 1963, La faune Pleistocene de Tarija (Bolivia); Note préliminaire: *Mus. Nat. d'Histoire Naturelle, Bull.*, v. *35*, p. 194–203

Hopwood, A. J., and Hollyfield, J. P., 1954, *An annotated bibliography of the fossil mammals of Africa* (1792–1950): London, British Museum (Natural History), p. 1–194

Hubbs, C. L., Bien, G. S., and Suess, H. E., 1962, La Jolla natural carbon measurements II: *Radiocarbon*, v. *4*, p. 204–38

Jelinek, A. J., 1957, Pleistocene faunas and early man: *Michigan Acad. Sci., Arts and Lett., Papers*, v. *42*, p. 225–37

Jennings, J. S., 1957, Danger Cave: *Univ. Utah Anthropol. Papers 27*, p. 1–328

Leakey, L. S. B., 1965, *Olduvai Gorge 1951–1961:* Cambridge Univ. Press, 118 p.

Leopold, Estella B., 1967, Late-Cenozoic patterns of plant extinction: (this volume)

Libby, W. S., 1961, Radiocarbon dating: *Science*, v. *133*, p. 621–29

Livingstone, D. A., 1962, Age of deglaciation in the Ruwenzori range, Uganda: *Nature*, v. *194*, p. 859–60

Leonhardy, F. C., 1966, Domebo; A Paleo-Indian mammoth kill in the prairie-plains: Museum of the Great Plains Contrib., no. 1, 53 p.

Lundelius, E. L., Jr., 1960, Post-Pleistocene faunal succession in western Australia and its climatic interpretation: Internat. Geol. Cong., 21st Session, Nordern, Rept., pt. 4, p. 142–53

MacNeish, R. S., 1964, Ancient Mesoamerican civilization: *Science*, v. *143*, p. 531–37

Martin, P. S., 1958, Pleistocene ecology and biogeography of North America, p. 375–420, *in* Hubbs, C. L., *Editor, Zoogeography:* Publ. 51, Amer. Assoc. Adv. Sci., 509 p.

———— 1966, Africa and Pleistocene overkill, *Nature*, v. *212*, p. 339–42

Mason, R. J., 1962, The Paleo-Indian tradition in eastern North America: *Current Anthropol.*, v. *3*, p. 227–78

Mehringer, P. J., Jr., 1967, The environment of extinction of the late-Pleistocene megafauna in the arid southwestern United States: (this volume)

Meighan, C. W., 1965, *Pacific Coast archeology*, p. 709–20, *in* Wright, H. E. Jr., and Frey, D. G., *Editors,* The Quaternary in the United States: Princeton Univ. Press, 922 p.

Morrison, M. E., 1961, Pollen analysis in Uganda: *Nature*, v. *190*, p. 483–86

Müller-Beck, H., 1966, Paleohunters in America; Origins and diffusion: *Science*, v. *152*, p. 1191–210

Mulvaney, D. J., 1964, Tasmanoid industry (premicrolith), Kenniff Cave: *Antiquity*, v. *38*, p. 263–67

Ogden, J. G., 1965, Pleistocene pollen records from eastern North America: *Botan. Rev.*, v. *31*, p. 481–504

Newell, N. D., 1966, Problems of geochronology: Acad. Nat. Sci. Philadelphia, *Proc.*, v. *118*, p. 63–89

Pearson, F. J., Davis, E. M., and Tamers, H. A., 1966, University of Texas radiocarbon dates IV: *Radiocarbon*, v. *8*, p. 453–66

Pearson, R., 1964, *Animals and plants of the Cenozoic era:* London, Butterworth's, 236 p.

Péwé, T. L., Ferrians, O. J., Nichols, D. R., and Karlstrom, T. W. V., 1965, INQUA VII Congress Guidebook, Field Conference F, *Central and South Central Alaska:* Lincoln, Nebraska Acad. Sci., 141 p.

Quimby, G. I., 1958, Fluted points and geochronology of the Lake Michigan Basin: *Amer. Antiq.*, v. *23*, p. 247–54

―――― 1960, *Indian life in the Upper Great Lakes*, 11,000 B.C. to A.D. 1800: Univ. Chicago Press, 182 p.

Romer, A. S., 1933, Pleistocene vertebrates and their bearing on the problem of human antiquity in North America, *in* Jenness, D., *Editor*, *The American aborigines, their origin and antiquity:* Univ. Toronto Press

Roosevelt, Theodore, 1910, *African game trails:* New York, Scribner's, 529 p.

Rouse, I., 1964, Prehistory of the West Indies: *Science*, v. *144*, p. 499–513

Semken, H. A., Miller, B. B., and Stevens, J. B., 1964, Late Wisconsin woodland musk oxen in association with pollen and invertebrates from Michigan: *Jour. Paleont.*, v. *38*, 823–35

Simpson, G. G., 1961, *Horses:* The Natural History Library Edition: Garden City, Doubleday, 323 p.

Skeels, M. A., 1962, The mastodons and mammoths of Michigan: Michigan Acad. Sci., *Arts Lett.*, *Papers*, v. *47*, p. 101–33

Slaughter, B. H., 1966, An ecological interpretation of the Brown Sand Wedge local fauna, Blackwater Draw, New Mexico, and a hypothesis concerning late-Pleistocene extinction, *in* Wendorf, F., and Hester, J. J., *Editors*, *Paleoecology of the Llano Estacado*, Vol. 2: Santa Fe, Ft. Burgwin Res. Center

―――― 1967, Animal ranges as a clue to late-Pleistocene extinction: (this volume)

Slaughter, B. H., and Ritchie, R., 1963, Pleistocene mammals of the Clear Creek local fauna, Denton County, Texas: Dallas, *Sou. Meth. Univ. Grad. Res. Center Jour.*, v. *31*, p. 117–31

Talbot, L. M., and Talbot, M., 1963, The wildebeest in western Masailand, East Africa: Wildlife Monog. no. *12*, 88 p.

Taylor, D. W., 1965, The study of Pleistocene nonmarine mollusks in North America, p. 597–611, *in* Wright, H. E., Jr., and Frey, D. G., *Editors*, *The Quaternary of the United States:* Princeton Univ. Press, 922 p.

Taylor, W. W., 1956, Some implications of the carbon-14 dates from a cave in Coahuila, Mexico: *Texas Archaeol. Soc. Bull.*, v. *27*, p. 215–34

Vaillant, G. T., 1944, *The Aztecs of Mexico:* Middlesex, Penguin Books, 333 p.

Van Zinderen Bakker, E. M., 1962, A late-glacial and post-glacial climatic correlation between East Africa and Europe: *Nature*, v. *194*, p. 201–03

Wallace, A. L., 1876, *The geographical distribution of animals*, Vol. 1: London, MacMillan, 503 p.

———— 1911, *The world of life:* New York, Moffat, Yard, 441 p.

Williams, S., and Stoltman, J. B., 1965, An outline of southeastern United States prehistory with particular emphasis on the Paleo-Indian era, p. 669–83, *in* Wright, H. E., Jr., and Frey, D. G., *Editors, The Quaternary of the United States:* Princeton Univ. Press. 922 p.

Wittry, W. L., 1959, The Raddatz rockshelter, SK5, Wisconsin: Wisconsin Archaeol., n.s., v. *40*, p. 33–69

Wright, H. E., Jr., 1964, Aspects of the early postglacial forest succession in the Great Lakes region: *Ecology*, v. *45*, p. 439–48

[Handwritten annotations at top: "Animal husbandry = game scarcity in any particular areas" with arrow to "human predation" and "environmental change"; "a) climatic b) geologic c) organic d) other (?)"; "Stratigraphic record"; "— except disease caused by microbes."; "— ?"]

JOHN E. GUILDAY

Carnegie Museum
Pittsburgh, Pennsylvania

DIFFERENTIAL EXTINCTION DURING LATE-PLEISTOCENE AND RECENT TIMES

Abstract

In any deteriorating ecosystem, large herbivores are more drastically affected than are small herbivores, by virtue of their greater demands upon the system for space, food, and cover. This produces a situation of differential extinction; the large "big-game" forms are eliminated. This is a natural consequence of interspecific ecological competition operative throughout the history of terrestrial vertebrate evolution.

Our inability to account for individual extinctions, such as that of the ground sloths, and for continental extinction patterns is not due to lack of adequate reasons but merely to lack of grounds for singling out one or several of a great variety of possible causes. The great prime mover was the unprecedented (at least in mammalian history) harshening of the environment during the Pleistocene, with sudden and great changes in the distribution of temperature and moisture throughout the world. To single out a particular predator or a set of circumstances is fun but futile.

Attempts to sharpen the focus of the extinction picture by singling out possible causes may not be possible. It is inconceivable that the same extinction pattern prevailed throughout the globe without being affected by local conditions —accidents of geography, local climates, different faunal and floral associations. The fact that these late-Pleistocene extinctions were so widespread and geographically almost simultaneous does call for a major overlying cause, however. I suggest that the prime mover was post-Pleistocene desiccation. Evidence for such an episode is present on all continents, and its effects would have been both swift and lethal. It may have been the spur to turn man from hunting to a life centered around animal husbandry and agriculture.

For at least a century naturalists have noted and speculated upon the cause of worldwide extinction of many large terrestrial mammals in the wake of the Pleistocene glaciations. Also at this time man, an increasingly efficient predator, began to increase in numbers and to colonize new areas such as the Western Hemisphere. It was inevitable that these two events would be associated. Was the ascendance of man responsible for extinction? If not, then how explain the curious fact that the animals involved were primarily "big game" species? A new and highly efficient predator had suddenly appeared; big game seemed to vanish just as suddenly. The phenomenon was virtually universal.

It appears to me that in the absence of man much the same pattern of extinction would have occurred. This is not to imply that man did not have a hand in it. As an integral species in a changing ecological picture, he most surely played some part. His role was decisive on islands up to the size of Madagascar, where man's initial arrival coincided with major destruction of both habitat and fauna, and in areas like the Near East, where animal husbandry and agriculture gave man the ability to destroy major habitats at a very early date. But in the Western Hemisphere, he may have delivered no more than the final coup de grace to isolated remnants already doomed by rapid postglacial environmental changes.

dietary and ecological Tolerances ?

Extinction results from the inability of an organism to cope with the problems brought on by changing environmental pressures. The number of species involved depends upon the speed and the intensity of such changes. Such major disasters as volcanic eruptions and tidal waves may wipe out an entire fauna overnight, whereas other slow but inexorable changes may take centuries to bring about extinction.

True extinction (end of a phyletic lineage without phyletic replacement) has occurred throughout the history of life on earth. Among the terrestrial vertebrates, the fossil evidence suggests two striking episodes of extinction: one at the Mesozoic–Tertiary transition saw the extinction of the last of the dinosaurs, the other at the Pleistocene–Recent transition saw the sudden dramatic disappearance of large mammals in most but not all parts of the world.

COMPETITIVE CONSTRICTION

I shall concentrate on the fate of the grassland herbivores. Pleistocene extinction of large carnivores was just as dramatic but can be directly attributed to the demise of their herbivorous prey—the large gregarious herbivores, primarily such browsing and grazing animals as horses, camels, elephants, and ground sloths.

Evolution involves the constant genetic alteration of populations to a continuously changing environment. Species of animals can flourish

Modern populations are in many cases limited by human activities

side by side only as long as they are not in serious competition. This is well demonstrated by the modern fauna of the African savanna, where the habitat is sufficiently diversified to include a variety of subhabitats (some of them merely seasonal in nature and seemingly rather subtle), allowing many species of savanna herbivores to live together. In such situations the habitat may seem so bountiful that there appears to be no appreciable competition within the mixed herbivore fauna. This may give a false impression of lack of interspecific competition in a community that may be highly competitive during times of environmental crisis. The times of crisis, often cyclic and of a seasonal nature, determine the numbers of species in a fauna. Stress may be completely lifted between crisis, when species may not compete, but during a crisis species must resort to the ecological niche to which they have become adapted. They will survive or perish with the fortunes of their habitat.

Vesey-FitzGerald (1960) gives a dynamic picture of the grazing succession throughout the year in the Rukwa Valley of Tanzania. Eighteen species of large herbivores occupy the same range but are not in serious competition, by virtue of a variety of habitat subtypes created by annual floodings, drought, grass fires, and even by the physical presence of other animals. Grazing by large herds tends to retard normal plant succession. Prior utilization of certain swampy grassland by big animals such as the elephant or the buffalo seems necessary before it can be used by smaller grazers. Different species exhibit varying degrees of dependence upon water or shade during the dry season and varying degrees of adaptation to flooded pasturages at periods of high water. Browsing species (giraffe, kudu) do not participate in the annual grazing cycle. Even such a small subhabitat as the sprouts of trampled vegetation in hippo paths is utilized by the smaller antelope.

In the event of a major climatic change (see Fig. 1) two things will occur. (1) Some subhabitats will be destroyed; others, perhaps once marginal under the old climatic regime, will expand. (2) Many once noncompeting herbivores will be thrown into intense competition; their numbers will wax and wane with the fortunes of the subhabitat to which they have become bound by selective evolution. ——

Competition will be felt first by the larger species in the community, which by virtue of their size require larger areas of primary habitat to maintain themselves. If a particular habitat starts to break up, the effects of such a change will disturb the elephant, whose range must be reckoned in square miles, sooner than it will the equally herbivorous mouse that can maintain high numbers in a much smaller area. Eventually both might be affected, but initially it is the large mammal that must make adjustments. If it cannot, the road to extinction lies before it.

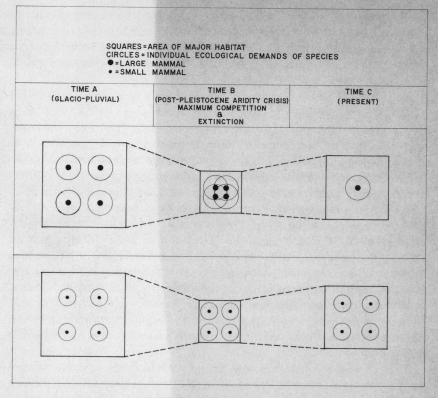

FIG. 1. Differential extinction model.

Extinction rarely occurs overnight, but once the train of events leading to it are set out in motion they are inexorable unless the environment changes back again before numbers have fallen so low that the birth rate will no longer exceed the toll of heightened environmental pressure, or unless the animal undergoes an evolutionary adjustment. If it is too specialized, if it is thrown into competition with superior species, if it cannot migrate into suitable habitat, if the environmental change occurred too fast (making evolutionary change impossible), then extinction will result. All these problems become acute as the size of the habitat dwindles, again weeding out large forms first.

Small mammals, because of their size and smaller home range, can survive in microhabitats barred to larger forms. They are not so drastically affected by a given climatic change as a large mammal. An environmental change, say from park-savanna to steppe, would eliminate the large browsing forms from the fauna but would have less effect upon the herbivorous rodents. I am not implying that small mammals are not sensitive to ecological change; some are adapted to extremely narrow ecological niches. What I am saying is that they are capable of living in smaller areas than are larger species and are therefore less vulnerable to a deteriorating climate—as long as small relict areas of primary habitat remain open to them.

THE GLOBAL GRASSLAND

Let us examine the major grassland areas of the world today. Nowhere on earth are big-game species more numerous than in the grasslands of Africa. Among the herbivores, the savanna fauna of Northern Rhodesia includes nineteen species of large mammals (those a hunting culture would classify as big game) and eighteen small mammals (Ansell, 1960), about a 1 to 1 ratio. In striking contrast to the African situation and in contrast to the situation in the Pleistocene, as shown by their respective fossil records, all other grassland faunas today are deficient in large mammals. The Great Plains (Nebraska: Jones, 1964) has, or had in pre-settlement days, seventeen small herbivores but only five large ones. The Pampas (Argentina: Grassé, 1955) has thirteen small and three large herbivores, and the Eurasian steppes (Ellerman and Morrison-Scott, 1951) approximately twenty-seven small and seven large herbivores. In all three areas the ratio is now approximately 4 to 1 in favor of the small herbivores.

After the recession of the last ice sheet (Wisconsin times) no small genera were lost from the North American fauna (Hibbard et al., 1965, p. 513). The fossil record does show extinction of elephants, horses, camels, ground sloths, all but one pronghorn, several ovibovids, peccaries, and the giant beaver. Three orders and seven families of large mammals were eliminated; the grazers were reduced to two species, the bison and the pronghorn.

The situation in the Pampas was, if anything, more dramatic. The large endemic herbivores, litopterns, notoungulates, and typotheres vanished. The large herbivores which had invaded South America via the reconstituted Panamanian land bridge became extinct—elephants, horses, and some camels, as well as the endemic sloths and glyptodonts. As in North America, one genus of giant rodent, in this case the rhinoceros-sized *Eumegamys*, became extinct.

The Eurasian continent lost elephants, rhinoceroses, hippopotami, many bovids, equids, and the large beaver *Trogontherium*. Again, the smaller herbivores survived.

The situation in Australia was much the same, the extinction of all herbivorous marsupials larger than a kangaroo occurred in late-Pleistocene times.

Africa—A "Living Pleistocene" Fauna

How did the savanna fauna of Africa escape the extinction effect prevalent elsewhere? What occurred that affected most of the large grazing faunas of the world but hardly affected the savanna fauna of Africa? Before attempting to answer this, we must examine two more areas, both in Africa.

In the Sahara Desert proper, there are approximately thirteen genera of small herbivores, primarily desert-adapted rodents, but only four large herbivores: gazelle, addax, oryx, and a wild ass. The relative percentage of large versus small herbivores in the fauna resembles that of the modern temperate grassland communities. During the late Pleistocene under more pluvial conditions, the area supported large herbivores in abundance. There is evidence that at least some of the Sahara enjoyed a higher rainfall in early postglacial times—desert-locked fish and crocodiles of Congolese affinity still survive in the waterholes of the mid-Saharan massifs (McBurney, 1960); abandoned stream channels, and hippopotamus, elephant, giraffe, various antelopes, warthogs, and crocodilians are found as fossils in the southern Sahara. Some of these faunas have been dated archaeologically as late as 4,000 years ago. The mid-Saharan rock carvings of buffaloes, giraffes, elephants, and crocodilians testify to the presence of man, the hunter of big game, in what is now a sterile desert.

The once rich large-mammal fauna is now gone from the Sahara. The segment of it that retreated north, including zebra, elephant, gnu, lion, and others, has become extinct in relatively recent times. It was wiped out not by man the hunter, but by man the herdsman, exterminating the habitat with flocks of domestic herbivores, primarily goats. The segment of the original Saharan fauna that migrated south and east, following the climatically dictated migration of the grassland habitat, is essentially intact today at the generic level. Extinctions have occurred in the African fauna throughout the Pleistocene, but the late-Pleistocene die-off did not occur (see Fig. 2).

The effect of aridity on the African Pleistocene megaherbivores has been inferred by Leakey at Olduvai Gorge (1965, p. 76). There the major faunal break in the Pleistocene sequence was a period of aridity:

> The upper part of Bed IV is mainly a brown aeolian tuff which appears to be unfossiliferous. It is regarded as indicating renewed dry conditions, possibly even sub-desertic. It seems likely that it was this dry period which was responsible for the extinction of so many of the animals which are characteristic of the upper part of the Upper Pleistocene in East Africa. We know from the evidence of

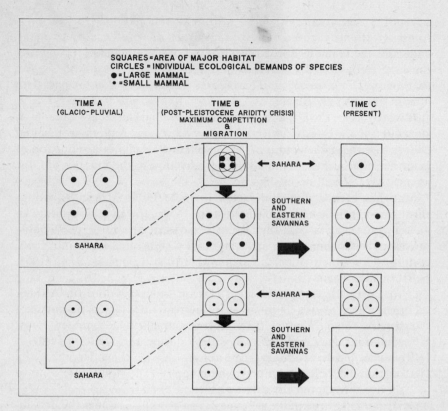

FIG. 2. Pleistocene survival of African megafauna.

sites in Kenya that the fauna of the Upper Pleistocene, in Gamblian pluvial times, is with few exceptions the same as that of the present day.

Man was apparently present throughout the episode; indeed, much of the bone recovered has been from human occupation sites.

At the opposite end of the aridity scale, the rainforest of the Congo supports a large number of small herbivores, but only two large endemic herbivores (excluding the great apes), namely the okapi (a primitive giraffe at an essentially Miocene level of giraffe evolution) and the bongo. Scarcity of big game in dense forest is readily related to the inaccessibility of food—within reach only of smaller arboreal forms. Those savanna ungulates living in the Congo basin are largely confined to grassland and forest-edge situations and do not occur in true rain-forest. The hippopotamus occurs throughout, but because of its amphi-bious habits it is not so tied to any major biome as long as there is streamside vegetation. The fossil history of the mammals of the rain-forest is extremely poor, but to judge from the presence of so many endemics, the fauna has probably survived essentially intact throughout the late Tertiary and Quaternary, and there is no reason to suspect that the ratio of large to small herbivores was any larger in the past than it is at the present time.

It would seem, then, that savanna areas and parklands are capable of sustaining the greatest variety of large herbivores. Any environmental change favoring either more or less vegetation results in a reduction of environmental niches, hence a relative reduction in the number of large as opposed to small herbivores.

Man has been a common factor in these three African situations throughout their late-Pleistocene history. Man, the hunter, appears to have fled the Sahara with the advent of postglacial aridity, presumably pursuing the savanna fauna as it migrated with its grasslands into more easterly and southerly highland areas, where it still survives. Bond (1963, p. 310) points out:

> Whatever the changes in local climate and vegetation (in Africa), there have always been areas of retreat into which mammalian species could migrate. Thus those animals which prefer forest conditions may have spread widely during the wetter phases of climate, but have always been able to survive the drier phases by migration to higher ground where forest conditions were able to survive throughout the driest phases.

This is, naturally, true of the savanna fauna as well. Consider the immensity of the area involved. The African biota is not and was not restricted by any geographic peculiarity (such as the Central American corridor). It migrated freely with minimal adjustment.

In brief, the Sahara gives us a picture of a rich post-Pleistocene large-mammal fauna, including man as a hunter, and a sudden relative decrease of big game from the fauna. If that fauna were not intact today in eastern Africa thanks to the accidents of geography, one might be tempted to say that here, too, was a case of man killing off the big game, when in reality he had no lasting effect on big-game populations in that era before animal husbandry or extensive agriculture enabled him to build up his numbers and destroy habitat.

North America

At the close of the Pleistocene the Great Plains of North America were relatively rich in large mammals. In Recent times this vast area was occupied only by the bison, the pronghorn, and marginally by three deer and a sheep. As stated previously the drastic elimination of large mammals suggests not increased predation but something far more basic, the extinction of subhabitats or ecological niches. To increase interspecific competition among the large herbivores, post-Pleistocene desiccation and habitat destruction need not have been nearly so drastic as it was in the Sahara. Unlike the African grasslands, the North American savanna had no suitable area to which it could migrate and still hold its identity. It could only deteriorate in place.

Full-glacial times in higher latitudes were associated with pluvial periods in lower latitudes. And while it is true that vertebrate populations were forced south as glaciation proceeded to its climax, the increase in precipitation in lower latitudes must have transformed large areas of what is currently desert in the Southwest and Mexico into suitable habitat for megaherbivores. As recession of the ice commenced, lower-latitude habitats began to dry up. This, not the glacial maximum, was the time of *major* adjustment for megaherbivores. Increasing desiccation and warmth, as reflected by the receding ice, did not draw biotic zones north in the way that they were able to expand south—deserts expanded, habitats deteriorated, and the time of trial for the megaherbivores was at hand.

We have evidence of some midwestern forms in eastern North America during post-Pleistocene times (New Paris No. 4, Pennsylvania; Natural Chimneys, Virginia; Florida), but the East had its own forest fauna and flora blocking wholesale invasion, and, being more humid to begin with, the effects of a drier climatic episode or episodes were less strongly felt. Mexico is a geographic cul-de-sac. Any attempted migration in that direction would end in compression and increased interspecific competition. Any eastward expansion of the southwestern desert would encroach still farther upon the Great Plains. If the Great Plains were to deteriorate by partial desiccation for even a relatively short space of time, biological winnowing would occur by increasing competition for decreasing pasturage associated with the disappearance of heretofore available ecological niches. Otherwise it is hard to explain the fact that in Recent times three species of deer (primarily woodland browsers) competed with true grazers—the bison and the pronghorn—in marginal plains habitats, while many large savanna herbivores (elephants, sloths, camels, horses) met with extinction. Because of their low-crowned molars, deer are less suited to a parkland environment than are some of the many species of large herbivores with higher-crowned teeth which did not survive. Deer seem to have invaded the Great Plains from their primary forest habitat, utilizing the corridor woodlands of the Great Plains river valleys. These animals occupied vacant ecological niches that had returned after deterioration and reconstruction but were now minus the original large grazing mammals. This does not imply that prior to Altithermal times there were no deer on the Great Plains. It does mean that in time of climatic stress deer could retire to their primary (forest) habitat, whereas parkland grazers could not, and perished with the habitat.

As we have seen from the Recent distribution of large mammals in Africa, savanna rather than true prairie is capable of supporting a greater variety of megaherbivores. There is evidence for post-Pleistocene savanna in what is now prairie and semidesert country. Pollen studies of sediments from Hackberry Lake, Nebraska, suggest that 5,000 years ago the area contained many more trees, primarily pines, than it does today (Sears, 1961). Trees occur today throughout the area on isolated ridges and scarps and are viewed as relict stands of a former widespread woodland reminiscent in broad outline of the African savannas. Citing pollen studies by Hafsten that indicate that the Llano Estacado of eastern New Mexico and Texas was at least partially forested during the late Pleistocene, Wells (1965) states:

> These data suggest that, during Wisconsin time, pine forest or woodland grew extensively on the southern part of the Llano Estacado, which borders on the Chihuahuan Desert province and is at present the most arid sector of the entire Great Plains region. If this area was not arid enough to harbor extensive treeless grassland during a pluvial of the Pleistocene, it is unlikely that any other area of the plains region was.

Mehringer (1965), working specifically with fossil pollen from southern Nevada, concludes: "It seems apparent that many areas of the arid western United States were not arid during parts of the late Pleistocene ... and many areas of the modern deserts and desert grassland were covered by woodlands or parklands."

In the Mohave Desert area of southern Nevada, fossil woodrat nests produced well-preserved juniper (*Juniperus osteosperma*) twigs and leaves that represent a lowering of the woodland zone from its present position in that area by about 600 m as recently as 10,000 years ago (Wells and Jorgensen, 1964). Observing that the present lower limits of woodland lie at about 1,700 m on peaks higher than 1,900 m, and that woodland is absent from 1,700 m to 1,900 m on peaks that do not exceed this altitude, Wells and Jorgensen conclude, "The present lowermost limits of piñon–juniper woodland were formerly not only *lower*, but also (more recently) *greater* in elevation than at present, probably during the warm Hypsithermal climatic phase of post-Wisconsin time, beginning about 8000 years ago." This appears to be a very tangible case of deterioration of habitat during the Climatic Optimum followed by reconstitution—

juniper woodland moving down from refuge areas afforded by higher peaks but unable to recolonize lesser peaks from which they were eliminated.

The situation in the Great Plains appears to be not equivalent but parallel to that in the Sahara, with one important difference: the Great Plains savanna biota could not migrate as a continuing ecosystem to areas rendered suitable for colonization as a result of those same climatic changes. Martin et al. (1961), on the basis of the pollen content of dung of the Shasta ground sloth *Nothrotherium* from Rampart Cave, Arizona, concluded that the ecological niche to which *Nothrotherium* was apparently adapted still exists today and is unoccupied. They assume that the niche has been continuously available since the time of *Nothrotherium* and invoke Early Man as the logical agent of extinction. It appears equally possible that the *Nothrotherium* habitat was extinguished through progressive desiccation and has since returned but without *Nothrotherium*.

Just as man fled the Sahara with the migration of the savanna big game in postpluvial times, the Paleo-Indian tradition came to an end with the extinction of the large herbivores in the American grasslands—victims of environmental change. Great Plains habitat under desiccation could only deteriorate; it could not migrate. Under our present climatic regime the Great Plains is a harsh environment. It is doubtful if many of the late Pleistocene megaherbivores could survive there now.

It is noteworthy that the primary Great Plains megaherbivore today, *Bison bison*, is relatively dwarfed, the smallest known species of the genus in the New World. It dates no farther back than Paleo-Indian times as a species. Differentiated from other bison solely on small stature and minor cranial and horn characters, it most probably represents the dwarfed modern form of *B. occidentalis*. This dwarfing, which Kurtén refers to the "population-density factor," is a common phenomenon in insular situations or during periods of heightened environmental stress and is "most likely to represent an adaptation for keeping population numbers high in spite of a severe limitation of habitat and food supply. An adequate population level is necessary for the maintaining of genetic variability and to ensure survival despite accidents" (Kurtén, 1965a, p. 62).

Post-Pleistocene dwarfing of *Bison* might logically be related to a time of heightened environmental stress, as aridity caused the deterioration of the Great Plains biome. The bison survived this time of stress and had no trouble recovering and flourishing despite continued human predation, as demonstrated by the unnatural spectacle of millions of individuals of but a single species.

Although temperate forest faunas were not so drastically affected as grassland faunas, they experienced late-Pleistocene climatic changes.

The pre-extinction fauna of the eastern United States included mammoth, several musk-oxen, caribou, mastodon, several peccaries, moose, deer, ground sloth, horse, giant beaver, and *Cervalces* (an extinct moose). Of those that have either become extinct or do not occur today in the temperate deciduous forest south of the Canadian Life Zone, all except the mastodon and the peccaries are believed to be either grazing forms or swamp dwellers. This suggests a former habitat no longer in existence. The browsing deer and elk and the omnivorous bear alone survived among the large mammals in a closed deciduous temperate forest. As suggested by the 11,300-year-old cave biota of New Paris No. 4 in central Pennsylvania (Guilday et al., 1964), a parkland environment, with its multiplicity of environmental niches, could support a greater variety of herbivorous big game.

As to why the mastodon and the peccary died out, all that one can point to is the existence of an environment that has since changed and forced species into closer competition. In a situation of a progressively closing forest canopy, primary grazing types would be most affected. The fossil record seems to bear this out. I suggest that the mastodon could not survive in a closed-canopy forest in close competition with more efficient browsers and was not confined to but flourished best in the same habitat as that of the equally defunct giant beaver—extensive lake margins and swamps, the habitat occupied today by the moose. The late-Pleistocene picture in the eastern United States is that of parkland. The presence of midwestern (parkland) forms is established by cave faunas. As the forest canopy closed it is conceivable, for instance, that the omnivorous peccary was forced into closer and closer competition with other omnivores such as the black bear. This implies a reduction of former ecological niches. The habitat need be suppressed just enough to decrease the numbers of one competing species while increasing those of the other. The bear not only competed for food, but in all probability it preyed on peccary whenever the occasion forced it to co-exist in the same ecological niche.

I do not point to the bear as being the cause of the extinction of the peccary or suggest that the moose usurped the mastodon's habitat, but I do suggest that the ecology of late-Pleistocene eastern North America contained a variety of habitats—extensive lakes and swamps, parklands now generally absent—all replaced by closed deciduous forest with its shade-starved understory and long winter-resting period. Elimination or creation of habitats through climatic change, overspecialization, and heightened competition between species was ultimately the cause of the Pleistocene extinctions. The die-off of larger specialized forms is a natural consequence of their more urgent requirements and slower reproductive rates. The fact that they were attractive prey to Early Man by virtue of their size is understandable, but this is unrelated to their disappearance. Much the same extinction pattern would have developed in any case.

South America—Habitat Shrinkage

Differential extinction on the South American continent was more intense than in North America, resulting in the extinction of all of the large Pampean mammals. The savanna area of the South American continent is much smaller than that of North America and more circumscribed by physical barriers—the sea, the Andes, the lowlands of the Amazon basin. In addition, the native fauna, endemic since the Tertiary, was faced with invasion from North America in late-Pliocene times. As a result, the only large herbivores left on the Pampas after the Pleistocene were temperate forms of North American origin. Again, desiccation, destroying a habitat that unlike the African savanna had no room to migrate, may have favored the ascendance of the new North American immigrants. As in North America, primarily large mammals were affected. In each case the primary grazing species was not a native animal but an immigrant—the bison from the Palearctic, the guanaco from North America. Gone was a host of endemics in both continents. In the inevitable course of selective evolution they had become too specialized to adjust to the rapid changes characteristic of the Pleistocene and were eliminated. These archaic forms may well have survived in marginal habitats in the absence of superior competitors, to recolonize their primary habitat should it be reconstituted. The fossil record of the modern fauna shows that they did not. I suggest the near elimination of the Pampas as a major habitat, again through desiccation in early postglacial times, with fragmentation into small upland areas along the fringes of the Amazon basin or the Andean foothills, areas not extensive enough to support the larger herbivores. Unlike the African savanna, which could persist in the uplands of East Africa, the Pampas was restricted by the Andean massif and could only deteriorate. It is interesting to note that the only successful large herbivore on the Pampas, the Guanaco, was derived from a genus otherwise restricted to a mountain habitat. Pampas deer, like the cervids of the Great Plains, are browsing animals, able to flourish in a marginal habitat (grasslands) only in the absence of primary grazers. The reintroduction of the horse into the Pampas, as noted by Martin (1958), was eminently successful. It apparently reoccupied an unexploited habitat from which it disappeared in the late Pleistocene.

It is almost axiomatic that no carnivore or combination of predators can exterminate its prey, or even control its numbers, unless the prey is extremely localized. On the contrary, the numbers of the predator are governed by the numbers of prey. It is unreasonable to consider the primitive hunter as an exception. The *complete* destruction of the endemic large herbivores and the survival of one true grazing form on the Pampas (apparently a montane derivative) and several browsers, again would seem to indicate a destruction of primary habitat, a drastic reduction of large mammals, and subsequent reconstruction of the grassland habitat whose large herbivore fauna remains depauperate, not in numbers, but in variety.

It is highly probable that in the absence of man many large herbivores might have been able to stage a comeback. Caught between the Andean massif, the sea, and the Amazon basin to the north, this grassland area, as in North America, was not free to migrate in the African sense, and it is entirely possible that the ground sloths or the elephants may have been reduced (like the whooping crane) to such scattered remnants that the slow breeding rates characteristic of large animals could no longer compensate for normal environmental attrition, predation, accidents, and disease. In such cases man may well have acted as the coup de grace. This implies only that man may have been responsible as one predator out of many for isolated cases of extinction, but only of those species which were so "ecologically sick" that the last push over the edge was all that was needed. Given a healthy ecological picture, predation has no lasting effect upon numbers—if it had, we should have devoured ourselves eons ago. Displacement is another matter. We may never know what effect agriculture had upon remnant populations of mammals in Central America and Mexico, with the resultant destruction of habitat again acting upon remnant forms.

Australia—"Mid-Recent Aridity"

The situation in Australia seems to have been the same as that of the Sahara or the North and South American grasslands. Eliminated were all the larger herbivorous marsupials—*Palorchestes*, a ten-foot-tall kangaroo, the rhinoceros-sized *Diprotodon*, a giant wombat, and a koala (Keast, 1959). On the basis of carbon-14 dating of fresh-water mussel shells associated with *Diprotodon* at Menindee, western New South Wales, Gill (1963) believes the giant fauna survived at least until 6,570 years ago. On the basis of C^{14} dates man is believed to have been in Australia for about 13,000 years, and there appears to be no serious doubt that early aborigines did prey upon giant marsupials.

I suggest that man and the giant marsupials coexisted in a natural predator–prey relationship as long as the interior of the continent could

support a large herbivore population in a savanna habitat, but that post-
Pleistocene desiccation, affecting all but the continental fringes, destroyed
most of a former savanna habitat that could not migrate as did the African grasslands. Wood (1959) and Crocker (1959) both invoke "mid-
Recent" aridity to explain modern plant distributions in Australia.
Tindale (1959) states that the beginning of Recent times in Australia was
marked by the onset of widespread aridity, which Gill (1955) equates
with the Climatic Optimum or Hypsithermal of the Northern Hemisphere. Keast (1959) attributes the demise of the giant herbivores to a
combination of increasing aridity and the introduction of Man and the
dingo. Tindale discounts any effects of aridity and advocates the combination of man/fire/dingo. He asks those who stress an aridity crisis during
the Climatic Optimum to account for the die-off of large forms only. In
this as in other cases, differential extinction is easily explained by heightened interspecific stresses affecting large mammals to a greater extent
than small ones by virtue of their greater demands upon the ecosystem.

[handwritten marginalia: NO Evidence To support Holocene "Aridity", in fact, mid-Holocene was wetter from pollen in Victoria Perhaps drier in Tasmania]

The familiar pattern was set up, involving lessening range, heightened
interspecific competition, and increased vulnerability of large as opposed
to small forms. Again, in the absence of man, some relic populations may
have hung on. Presumably these were slow breeders, almost immune to
adult predation other than by man. But the present large herbivore
fauna of the savanna—kangaroos of the genus *Macropus*—suggests, as
do the qualitatively depauperate large herbivore faunas of the North and
South American grasslands, a drastic reduction of habitat, resulting in
intense competition for limited range. Because of its equatorial situation,
Australia would seem to be one area where the Slaughter theory (1967)
of extinction through heightened mortality of young born in unfavorably
cold seasons under a worsening climatic regime could not apply. Yet the
same pattern of differential extinction affecting primarily the large
mammals took place in Australia at almost the same time as in America.

LATE-PLEISTOCENE EXTINCTIONS

The magnitude of the late-Pleistocene die-off is exaggerated by factors
inherent in the preservation of fossil faunas in North America. They are
from a very narrow time zone, either associated with and collected by the
Paleo-Indian, or preserved in post-Wisconsin cave and bog desposits.
As a result we have a relatively rich fauna from a limited horizon, with
two sharply artificial cutoff points associated with the appearance and
disappearance of the Paleo-Indian hunting tradition. Extinction took
place throughout the Pleistocene, but the late-Pleistocene die-off is
impressed upon us by our relatively poor knowledge of earlier Pleistocene faunas. If we had comparable faunal samples from successively

earlier glaciations the picture would appear much less dramatic. The order Proboscidea, for example, contained fifteen genera in the Pliocene (Simpson, 1945). Only four (*Mammut, Mammuthus, Elephas,* and *Loxodonta*) survived into late-Pleistocene times. Only two, *Elephas* and *Loxodonta,* survive today. The little that we do know suggests that extinction was progressive throughout the Pleistocene, although not always at the same rate. Of the North American large mammal genera, 18 % of those then living did not survive the late Pliocene, 23 % appeared for the last time in Blancan times, 18 % in Rancholabrean, and 30 % in Wisconsin times (Hibbard et al., 1965).

Kurtén, working with Pleistocene carnivores from Palestine and Florida, noted that those species becoming extinct in the late Pleistocene had a long history of size deterioration that began before the Wisconsin and proceeded to extinction. Since increase in size is one of the commonest evolutionary phenomena, Kurtén rightly recognized progressive diminution as a sign of mounting ecological stress reflecting itself in gross size. He concludes, "It seems that many of the 'sudden' extinctions at the end of the Ice Age are in fact the result of population declines dating back well into the last glaciation or possibly even the last interglacial" (Kurtén, 1965b, p. 248).

The Pleistocene mammalian faunas in Europe are much better known than those of North America. Of twenty-two genera of large mammals known from the earliest Pleistocene (Günz), 50 % did not survive beyond that point; of twenty-seven genera known from the early and middle Pleistocene (Riss-Mindel), 25 % became extinct at the close of that interval. Of the thirty-two genera known from the latest Pleistocene (Würm), only three are extinct: an elephant (the last of the fauna), one rhinoceros (again, the last in the fauna), and the Irish elk (Thenius, 1962). In both areas, the small mammal faunas of the Pleistocene (Hibbard et al., 1965; Jánossy, 1961), as far as is known, had no extinctions above the level of species.

EARLY MAN

Early Man on the Great Plains was a big-game hunter, as abundant archaeological evidence now attests. He has often been incriminated as the *cause magnifique* of the extinction of the "big game" from the area, if not directly, then indirectly (Prufer, 1963).

Indirectly, Early Man without domestic animals, without agriculture, leading the nomadic life forced by migrating game, could have had no lasting effect upon the environment. Using fire, he could have altered large areas drastically. But, as game-management research in Africa has brought out, fire is not necessarily detrimental, and may be beneficial, in reconstituting pasturage. Apparently it is necessary to maintain a grass-land in Kruger National Park, where Brynard (1964, p. 371) states,

Bush encroachment in the Park has increased considerably . . . and it is largely due to the fact that fires occurred less frequently after the establishment of the Reserve when it was uninhabited by natives and hunters.

The carrying capacity for grazing animals of the area involved would tend to decrease in proportion to the intensity of bush encroachment and the opposite would be true for the browsers.

After a trial period of the elimination of fire, it was discovered that, whereas some few long-grass grazers were benefited and increased steadily in numbers, the majority of grazers were unable to adapt themselves and steadily declined in numbers. They were forced to move to areas where grass was still burned regularly and, incidentally, where they were slaughtered by the hundreds. Controlled burning has been reinstituted at Kruger Park as a necessary game-management operation. Again, I point to the African savanna fauna as being a living example of the effects of man upon large game populations, both direct and indirect, for thousands of years.

Directly, the primitive hardware and the low numbers of early nomadic hunting cultures could have little effect upon the elimination of any one species from a fauna unless that animal was confined by extreme specialization and restricted both in range and numbers. For example, why was the musk-ox, whose only defense is to form a circle and glower, able to survive and prosper despite both Indian and Eskimo predation until the introduction of firearms? Obviously the production of offspring exceeded predation pressure.

Although we have abundant evidence that the Paleo-Indian hunted elephants, what evidence do we have that he specialized in big game? An elephant carcass is not easily scattered, and it is no surprise that the carcass lay where it fell, with irretrievable projectile points and skinning flakes about, to be buried in aggrading flood plain silts for us to recover. A smaller animal might easily have been shouldered and taken to some upland camp or simply scattered upon butchering. Small animals may have been eaten more often than large ones. Concentrated killing of mammoths by Paleolithic Man occurred in the Balkans, and some northern hunting cultures were almost entirely caribou predators, as the historic plains tribes were buffalo hunters. But these dependent situations arose only where no other species was as readily available; either the site was situated at a bottleneck along a migratory route where the animal became vulnerable to predation, or the habitat was too severe to support a variety of game.

The role of man in initiating, or at least influencing, the extinction of the large herbivores may have been truly decisive in the Near East and Indian savanna grassland areas where man competed directly for water and forage by virtue of his rule as an animal husbandman. Here, through the early acquisitions of both agriculture and livestock herds, the "balance of power" shifted. In this area of the world and in this area alone, man may well have been the precipitating and deciding factor in the extinction of many of the large late-Pleistocene mammals. In all other areas his influence is really an unknown. To judge from the African picture, however, the general pattern of mammalian extinction, and the lack of effect of predation by any carnivore (including Early Man) in a well-adjusted ecosystem, man was not the precipitating factor of the late-Pleistocene die-off, even though he may have been a local contributing factor.

I should like to thank the many persons who were kind enough to read this manuscript and share their views both pro and con with me. It was greatly appreciated. The diversity of opinion was both welcome and instructive. Special thanks to Mrs. Elizabeth Myers and Mr. Harry K. Clench for their help and suggestions.

References

Ansell, W. S. H., 1960, *Mammals of Northern Rhodesia:* Lusaka, Northern Rhodesia Government Printer, 155 p., plus supplement, 24 p.

Bond, Geoffrey, 1963, Pleistocene environments in Southern Africa, p. 308–34, *in* Howell, F. C., and Bourliere, F., *Editors, African ecology and human evolution:* Viking Fund Publication 36, 666 p.

Brynard, A. M., 1964, The influence of veldt burning on the vegetation and game of the Kruger National Park, p. 371–93, *in* Davis, D. H. S., *Editor., Ecological studies in Southern Africa:* The Hague, Junk, 415 p.

Crocker, R. L., 1959, Past climatic fluctuations and their influence upon Australian vegetation, p. 283–90, *in* Keast, A., Crocker, R. L., and Christian, C. S., *Editors, Biogeography and ecology in Australia:* The Hague, Junk, 640 p.

Ellerman, J. R., and Morrison-Scott, T. C. S., 1951, *Checklist of Palaeartic and Indian mammals:* London, British Museum (Natural History), 810 p.

Ewer, R. F., and Cooke, H. B. S., 1964, The Pleistocene mammals of Southern Africa, p. 35–48, *in* Davis, D. H. S., *Editor, Ecological studies in Southern Africa:* The Hague, Junk, 415 p.

Gill, E. D., 1955, The problem of extinction with special references to Australian marsupials: *Evolution,* v. 9, p. 87–92

—— 1963, The Australian aborigines and the giant extinct marsupials: *Austr. Nat. Hist.*, v. *14*, p. 263–66

Grassé, Pierre-P., 1955, *Traité de Zoologie:* Tome XVII, vols. *1, 2*, Paris, Masson, 2300 p.

Guilday, J. E., Martin, P. S., and McCrady, A. D., 1964, New Paris No. 4: A Pleistocene cave deposit in Bedford County, Pennsylvania: *Natl. Speleol. Soc. Bull.*, v. *26*, p. 121–94

Hibbard, C. W., Ray, C. E., Savage, D. W., Taylor, D. W., and Guilday, J. E., 1965, Quaternary mammals of North America, p. 509–25, *in* Wright, H. E., Jr., and Frey, D. G., *Editors, The Quaternary of the United States:* Princeton, Princeton Univ. Press, 922 p.

Jánossy, D. von, 1961, Die Entwicklung der Kleinsäugerfauna Europas im Pleistozän (Insectivora, Rodentia, Lagomorpha): *Zeit. Säugetierkunde*, v. *26*, 64 p.

Jones, J. K., Jr., 1964, Distribution and taxonomy of mammals of Nebraska: *Univ. Kansas Publ., Mus. Nat. Hist.*, v. *16*, 356 p.

Keast, Allen, 1959, The Australian environment, p. 15–35, *in* Keast, A., Crocker, R. L., and Christian, C. S., *Editors, Biogeography and ecology in Australia:* The Hague, Junk, 640 p.

Kurtén, Björn, 1965a, The Carnivora of the Palestine caves: Helsinki, *Acta Zool. Fennica*, v. *107*, 74 p.

—— 1965b, The Pleistocene Felidae of Florida: Florida State Bull., *Biol. Sciences*, v. *9*, p. 215–73

Leakey, L. S. B., 1965, Olduvai Gorge 1951–1961: Cambridge Univ. Press

Martin, P. S., 1958, Pleistocene ecology and biogeography of North America, p. 375–420, *in* Hubbs, C. L., *Editor, Zoogeography:* Amer. Assoc. Adv. Sci. Publ., v. *51*, 509 p.

Martin, P. S., Sabels, B. E., and Shutler, Dick, Jr., 1961, Rampart Cave coprolite and ecology of the shasta ground sloth: *Amer. Jour. Sci.*, v. *259*, p. 102–27

McBurney, C. B. M., 1960, *The Stone Age of Northern Africa:* Middlesex, England, Penguin Books, Pelican Book A342, 288 p.

Mehringer, P. J., Jr., 1965, Late Pleistocene vegetation in the Mohave Desert of southern Nevada: *Arizona Acad. Sci. Jour.*, v. *3*, p. 172–88

Prufer, O. H., 1963, Ice Age overkill: *Explorer*, v. *5*, p. 7–13

Sears, P. B., 1961, Pollen profile from the grassland province: *Science*, v. *134*, p. 2038–40

Simpson, G. G., 1945, The principles of classification and a classification of mammals: *Amer. Mus. Nat. Hist. Bull.*, v. *85*, 350 p.

Slaughter, B. H., 1967, Animal ranges as a clue to late-Pleistocene extinction: (this volume)

Thenius, Erich von, 1962, Die Grossäugetiere des Pleistozäns von Mitteleuropa: *Zeit. Säugetierkunde*, v. *27*, p. 65–83

Tindale, N. B., 1959, Ecology of primitive aboriginal man in Australia, p. 36–51, *in* Keast, A., Crocker, R. L., and Christian, C. S., *Editors, Biogeography and ecology in Australia:* The Hague, Junk, 640 p.

Vesey-FitzGerald, D. F., 1960, Grazing succession among East African game animals: *Jour. Mamm.*, v. *41*, p. 161–72

Wells, P. V., 1965, Scarp woodlands, transported grassland soils, and concept of grassland climate in the Great Plains region: *Science*, v. *148*, p. 246–49

Wells, P. V., and Jorgensen, C. D., 1964, Pleistocene wood rat middens and climatic changes in Mohave Desert; A record of juniper woodlands: *Science*, v. *143*, p. 1171–74

Wood, J. G., 1959, The phytogeography of Australia (in relation to radiation of Eucalyptus, Acacia, etc.), p. 291–302, *in* Keast, A., Crocker, R. L., and Christian, C. S., *Editors, Biogeography and ecology in Australia:* The Hague, Junk, 640 p.

WILLIAM ELLIS EDWARDS

Institute of Archaeology and Anthropology
University of South Carolina
Columbia, South Carolina

THE LATE-PLEISTOCENE

EXTINCTION AND DIMINUTION IN

SIZE OF MANY MAMMALIAN SPECIES[1]

Abstract

Extinction is commonplace. But the late-Pleistocene interval was virtually unique in intensity, in intrafamilial selection of larger, more gregarious forms, in the lack of replacement by ecological equivalents, and likely in rapidity.

Extinction resulted not from "racial senility" or "orthogenesis" but from unsuccessful adaptation to physical, biotic, or sociocultural changes. Extinction seems closely correlated with climatic change. But absolute survival ability is generally far broader than relative survival ability; changes in climatic zones are usually gradual, enabling faunas to shift comparably instead of undergoing extinction; and continental elimination of many mammalian genera is unknown for preceding deglaciations. The apparent rejoining, after isolation by ice sheets, of Old World-Alaskan and American terrestrial faunas immediately before extinctions suggests communicable disease, after diffusion of differentiated parasites through nonadapted herbivore populations, with greater devastation of gregarious and larger (more slowly reproducing) forms. But in comparable historical cases new parasite–host equilibria simply formed.

Especially in America, biological evolution of prey defense could not keep pace with rapid, accelerating cultural evolution of hunting techniques by humans. Uniquely efficient predation likely effected many extinctions, especially among larger, relatively non-adaptable, competing predators. But most subsequently extinguished herbivores probably survived through predator–prey equilibrium, in which the prey density is just enough to sustain the predators. Eventually, human hunters exceeded this density by cultural reversion to plant and animal collecting. With antiadaptive, traditional hunting emphasis, they exterminated all

1. This paper summarizes some of the discussions in Edwards (1966a). It was made possible in part by National Science Foundation Grant GS-690.

but the smaller, the relatively solitary, the excessively formidable, the indigenes of inaccessible terrain, and the occupants of those areas not permitting economic diversification.

Lack of climatic correlates and similar poikilothermic and homoiothermic changes invalidate the climatic interpretation of postglacial invertebrate and vertebrate diminution. Hunting technology reduces defensive advantages of larger prey and emphasizes concealment and speed. Increased pressure of human predation results in decline in average prey age and thus size. Furthermore, faster-maturing and therefore smaller adults are genetically selected. Finally, humans may bypass smaller individuals for efficiency or conservation, so dwarfs are favored in the resulting selection.

Enlargement, likely accelerated by early hominids, was probably reversed within the Lower Paleolithic by more effective hunting (until some recent recovery). Predators and carrion eaters also decreased. Man diminished as well, although primarily through shifting to plant foods and through developing food preservation and storage.

More than a million species of animals and several hundred thousand species of plants at present known to science were preceded by a much greater number of organisms (Simpson, 1952). Obviously extinction is commonplace, and at least three types can be recognized: (1) extinction of one or more variant forms in a polymorphic population through intraspecies replacement; (2) extinction of a species with simultaneous replacement by an ecologically equivalent competing species; (3) extinction without competition or replacement by an ecologically equivalent form. The first two types represent normal evolutionary changes based on "relative survival ability", in competition with "superior" forms. It is much more difficult to account for nonreplacement extinction, which seems to typify the late Pleistocene. Where not otherwise indicated, the term extinction will here be employed in the sense of worldwide or at least continental disappearance of entire species.

Striking intervals of terrestial quadruped extinction occurred at the ends of the Paleozoic, Mesozoic, and Cenozoic geological eras. By far the most drastic changes among terrestial vertebrates occurred at the close of the Mesozoic. But the latest interval of extinction, beginning in America only about 15,000 years ago, is unique or nearly so in several respects. First, the larger-bodied and more gregarious members of the orders and families were affected most and were not replaced (type 3 extinction). Second, the intensity of extinction was quite variable; for example, relatively more occurred in the Americas than in northern Eurasia. Finally, there is reason to believe that the change was much more rapid at the end of the Cenozoic, perhaps several hundred times faster than in the earlier intervals of extinction. These unique features suggest a unique cause.

The entire Quaternary comprises only about 0.1 % of the history of life on earth; the postglacial is less than 1 % of the Quaternary. Because culture-bearing man is a recently developed and large primate, disproportionately more attention has been accorded to the immediate geologic past and to mammals, especially the larger forms. Furthermore, the fossil record tends to be much more abundantly preserved for the immediate past and biased in favor of the larger forms, which leave disproportionately thick and more conspicuous skeletal remains. Our present view of the late-Pleistocene extinction rate is therefore exaggerated, especially for mammals and forms of large size. But a fairly comprehensive survey of extinguished forms in the remarkably brief interval involved reveals that the extreme and unique nature of late- and postglacial extinctions (reptilian and avian as well as mammalian) is far more real than apparent (Brodkorb, 1963; Charlesworth, 1957; Edwards, 1950, 1954; Hibbard et al., 1965; Martin and Mehringer, 1965; Reed et al., 1965; Rouse, 1952; Schultz and Frankforter, 1946; Selander, 1965; Simpson, 1943, 1944, 1953, p. 22, 27).

The causes of the continental extinction of many late-Pleistocene species has frequently received the facile explanation that many factors were involved. At times it has also been argued that no real answer can be given until all the facts are in—or that likely there will never be sufficient data upon which to reconstruct the complex relationships and processes long past.

Somewhat mystical concepts, such as those of "racial senility" and the related principle of "orthogenesis" (straight-line evolution), may immediately be excluded from consideration. Extinction was quite surely caused by unsuccessful adaptation to environmental changes related to physical, biotic, or sociocultural phenomena. Climatic change, parasitic disease, and human hunting activities seem to represent the most likely mechanisms.

THE CLIMATE HYPOTHESIS

In Eurasia and in America the highest frequency of rapid population decline and extinction seems to have coincided fairly well with extreme late- and postglacial changes in climate. But climatic causation of extinction requires more critical evaluation. The range of potential environmental tolerance (absolute survival ability) is generally far broader than the actual range (relative survival ability) of a species. This is demonstrated by consideration of such relatively recent contractions in range as that of lions, known in protohistoric Greece, and, more strikingly, by a variety of laboratory experiments (e.g. Cole, 1957). The changed climate probably fell well within the environmental tolerance range of most extinguished forms, even without any genetic adjustment. Extinction in response to climatic change probably required a preceding replacement by an equivalent form (an immigrant from an adjacent region or an indigenous form undergoing population growth) which was relatively more fit under the altered environmental conditions (Beaufort, 1951, p. 28; Darlington, 1957, p. 134, 180, 251, and 560). However, direct competition with others in the same ecologic niche cannot easily explain the majority of extinctions here considered. What more efficient competitors replaced the large herds of Plains herbivores—camel, horse, mammoth, and extinct bison?

Furthermore, climatic change was not rapid. Changing climatic zones generally undergo a gradual geographic *shift*. Only in a minority of cases, such as on certain islands and peninsulas and possibly in Western Europe, has a major climatic zone not afforded a suitable "exit" to the indigenous flora and fauna. The often expressed concept of wholesale faunal migrations into more suitable areas is invalid. Rather, the climatic and faunal shift experienced by each generation is normally almost imperceptible. The newly invaded area is gradually filled by population increase along a relatively narrow frontier "zone of fluxion," while along the given region's opposite border of contraction the populations do not suddenly and totally emigrate but simply become gradually less until local extinction eventuates.

While I recognize the possibility of other postulates, there is no evidence that climatic change occurred in America any more rapidly at the time of extinction than during the waning of earlier Pleistocene glaciations. Beyond the low, fairly steady frequency of extinctions that likely represent gradual replacement by ecological equivalents, continental collapse of many mammalian genera is not known during any preceding interval of deglaciation.

THE DISEASE HYPOTHESIS

Epidemics have sudden and in some cases drastic effects and, because larger mammals generally have a low reproductive rate, recovery from devastating epidemics would be more difficult for larger forms. During much or virtually all of the last glaciation, ice sheets separated the terrestial fauna of the Old World and Alaska from that in the rest of America (Karlstrom, 1961; Haynes, 1964). Parasites undoubtedly continued to differentiate in each hemisphere. With the re-establishment of contact between eastern and western species, parasites diffused through herbivore populations not specifically adapted to them. Thus, as in the case of formerly isolated human populations in the late fifteenth and sixteenth centuries, epidemics might have swept both hemispheres. The effect of communicable disease would be far greater upon gregarious than solitary animals or sea mammals. The late-glacial survival of virtually all African land mammals might be attributed to the eventual amelioration of lethality of the immigrating parasites.

But it is virtually impossible for infectious disease to cause extinction, for the disease declines with the population until only scattered, resistant individuals remain. In time, through the selective evolution of both the hosts and the then endemic parasites, a more nearly symbiotic condition of mutual tolerance becomes established. No cases of extinction by disease are known for modern times. A near exception is the chestnut of eastern North America, enormously reduced in numbers in recent decades (Elton, 1958, p. 21). A few individuals, by chance "preadaptively" more resistant to the fungus (*Endothia parasitica*), do persist. Thus the chestnut may be pulled from the consuming fire of extinction. Finally, the disease hypothesis can hardly account for the lack of evidence of widespread extinctions at the start of the other interglacials. Infectious disease might in some cases have constituted a preconditioning factor to the multiple extinctions, but probably not so significant a factor as that of climatic change.

Therefore, not climate, disease, or any other "natural" phenomenon could account for more than a few of the observed extinctions. Most other major natural events were periodic during the Pleistocene, so the decisive factor in extinction must have been a novel one. The only relevant factor that was rapidly changing geographically and quantitatively during the Pleistocene was human culture. What role did cultural change play in effecting extinction?

THE CULTURE HYPOTHESIS

Extinction of both predator and prey species clearly can result from an extreme population decline. Herbivores usually maintain stability by tending to approximate the minimum density that is necessary to sustain predators, as determined by the relative efficiencies of prey defenses versus predator offenses. Denser prey makes predation unnecessarily easy, causing an increase in the population of predators. This increase in turn leads to a reduction of both prey and, eventually, predator numbers. A fairly stable dynamic equilibrium tends to be maintained, although under certain conditions oscillating prey and predator cycles may develop (Lack, 1954; Skellam, 1955). The general effect of the development or introduction of an appreciably more efficient predator is a rapid decline in game populations, soon followed by the extinction of less efficient, competing predators.

The foregoing principle is used to explain the fact that all native South American marsupial predators became extinct soon after that continent was rejoined to North America at about the beginning of the Pleistocene. Another example is the rapid extinction of thylacine wolves in competition with dingos in Australia (Gill, 1955). Thus efficient human hunters arriving in a new continent may be expected to bring about the extinction first of the most directly competing predator species, even though the predator is not itself hunted. The extinction of predators, which has perplexed many previous investigators (e.g. Butzer, 1964, p. 400), is thereby explained.

In theory, at least, if human hunters directly reduce the population density of other predators by killing them and also institute effective conservation measures, the herbivore population can attain greater abundance than it would reach in the absence of humans—the fate of the Kaibab deer, for example. But, despite certain totemic, seasonal, and other taboos that may accomplish conservation to some degree, such measures rarely seem to be very effective within any human group dependent entirely or even in small part on hunting for survival. When there is the threat of starvation—virtually inevitable through expansion of human population—abundant available animal food will generally be utilized despite taboos.

Limits on human population growth include infectious disease and warfare. Such population-reducing factors might prevent faunal reductions that could otherwise effect extinctions in time.

For hundreds of thousands of years, the advent of culture did not bring about many marked faunal changes. Hominids were merely progressively substituted for somewhat less efficient scavenger–predators. But, like a population growth curve, the curve of cultural evolution rises exponentially. It describes an overall trend of increase by geometrical progression in the efficiency of hunting weapons, including the thrusting spear, the missile spear, the spear thrower, and the bow, and of special techniques such as the fire drive. The rate of faunal adaptation to human predation through biological evolution was low compared to that of offensive improvement through cultural evolution. In the New World the effect was virtually instantaneous. The contact of American prey with highly efficient human predators—associated with the gradual dwindling to relative insignificance of predation by competing carnivores—was too brief for effective biological evolution but occurred precisely at the time when human cultural evolution was very rapid. Even the diagnostic fluted point of the Early Hunters seems to be a New

World invention. Thus for many large game species, the only uncertainty was whether cultural evolution would enable a given species to be domesticated before being exterminated. Among the mammals only the llama (guanaco) survived in this way in the New World, in contrast to the Old (Clark, 1952, p. 50; Gimbutas, 1956, p. 14, 29, 35, 37).

One final matter must be considered. Theoretically, despite wide fluctuations and lags in readjustment of predator populations, dynamic equilibrium is achieved between the populations of predators and prey within a few years or at most decades, although oscillations may continue. If, as in the case of human hunters, the predator eventually becomes disproportionately more efficient, the equilibrium will simply become established at a much lower population density for both predator and prey. Nevertheless, man was progressively becoming a uniquely efficient predator in the late Pleistocene, as climatic change and disease perhaps accelerated the decline of herbivore population in many areas. Efficient human predation might alone have caused some forms, especially among multiple prey species, to decline below the minimum population density point of no return.

Most species now extinct would likely have survived through the operation of the equilibrium principle, however, except for the effect of the nature of human culture. Man had quite surely started as a collector of plant and animal foods. Then, apparently some two million years ago, as increasingly complex social tradition finally crossed the arbitrarily drawn borderline to true culture, man increasingly relied on hunting, with development of requisite weapons and tools. It became decreasingly feasible to collect most foods while the hunters, generally increasing in individual and group size, were moving about extensively, locating ever larger game. Furthermore, Old World collecting activities almost certainly were de-emphasized or forgotten during man's spread across the arctic filter to America. But in the late Pleistocene, cultural adaptability enabled at least some of these human populations to revert to plant and animal collecting, presumably not motivated significantly by the generally very gradual and somewhat irregular decline of game but by the almost constant tendency for human overpopulation and the resulting frequent shortage of food. Plant collecting redeveloped in most of the western United States and in the higher and more arid areas of Mexico and South America; and in coastal areas (where almost all earlier sites are now submerged) and along rivers and lakes of the more humid regions of the New World, mollusk collecting early became important.

At this critical point, as the fauna declined sharply, hunting must have become a secondary and variable activity. But it continued, with high prestige value among the men, although the lower-status food-collecting activities of women now became all-important for survival value. Hunting was now often insufficiently productive to justify the expenditure of time and energy, as may possibly be exemplified by the hunting activities of certain recent groups, such as some of the Paiutes. Regardless of whether hunting was continued as an antiadaptive tradition or for functionally useful food supplementation, the shift to food collecting enabled man at this stage to maintain his population despite game decreases. The maintenance of predation pressure, for cultural reasons if no other, sealed the fate of the New World game mammals. Exceptions were the smaller animals (generally reproducing faster and less significant economically), the relatively solitary (less easily hunted), the occupants of areas not well suited to plant and animal collecting (such as boreal-forest moose, wapiti, caribou, and especially the arctic-tundra caribou and musk-ox), the indigenes of inaccessible terrain (like the tapir and mountain sheep), and the excessively formidable (such as the grizzly and black bears).

Even some local forms of sea mammals, like the Florida seal (Rouse, 1951) and the California Steller's sea cow (which survived in a refuge in the Bering Sea), were eventually exterminated. Such examples are significant, for the only sea mammals to become extinct in prehistoric times were unusually vulnerable to human hunters because of occasional sojourns on land or at least littoral habits.

The decline and extinction of mammalian populations undoubtedly hastened economic changes that culminated in the Neolithic revolution. The high proportion of extinctions in the Western Hemisphere helped to shape the course of subsequent American Indian cultures, above all limiting their potential development in animal husbandry and efficient plow agriculture; this limitation affected the rate of evolution of horticulture and the types of plants utilized.

CULTURE AND THE SIZE OF PLEISTOCENE ANIMALS

The explanation most frequently offered for changes in body size of the Pleistocene fauna (Newell, 1949; Rensch, 1954)—as illustrated by postglacial chronoclines (Huxley, 1939) in Florida (Rouse, 1951), Sout Carolina (Edwards, 1965), North Carolina (Haag, 1958), and New Yor¹ (Ritchie, 1965, p. 59, 76, 165)—is adaptation to warmer condition. (e.g. Simpson, 1949, p. 136; Rouse, 1951; Charlesworth, 1957, p. 799; Ritchie, 1965, p. 76), in conformity with Bergmann's rule (Allee et al., 1949, p. 119; Edwards, 1966b). Although diminishing size is often ascribed to climatic change (Auffenberg and Milstead, 1965, p. 560, 563), it cannot explain most of the postglacial cases. The inferred climatic correlates are scarcely observed when their chronological sequence is examined closely. Furthermore, certain cold-blooded vertebrates and invertebrates reveal size decreases comparable to those of associated homoiothermic animals, despite the fact that poikilothermic organisms generally manifest larger size in areas of warmer climate. By the process of elimination, it may be tentatively concluded that human hunting and collecting activities were responsible for the late-Pleistocene extinction of both the larger species and the larger polymorphic variants within species.

Increase in predation pressure by human food gatherers generally results in a decline in the average age and therefore in a nongenetic decline in the average size of prey, especially since the relative rate of growth typically declines but does not terminate with the achievement of reproductive maturity in cold-blooded animals such as reptiles and mollusks. With a downward shift in the age distribution, adults that mature faster and reproduce faster are genetically selected, and these tend to be smaller. Human technology, including use of missile weapons, greatly reduces the counterattacking defensive advantages of larger size and emphasizes concealment and speed of flight. At this point of increased pressure of human predation, the genetically selected optimum body size of many forms declines sharply. Also, humans may intentionally bypass smaller prey individuals for more efficient collecting or for game conservation; thus the genetically determined, mature dwarfs may escape predation by being considered less productive of food or by being mistaken for immature animals.

The African evidence (Leakey, 1965) suggests that at first the early hominids may have accelerated evolution of larger body size by their selection of smaller prey individuals (Leakey, 1960, p. 433; cf. Flint, 1957, p. 450). Apparent selection of younger game occasionally persisted into later periods (Soergel, 1912). Also much later, but probably still well within the Lower Paleolithic, hunting weapons and techniques had evolved to the point at which defensive advantages of larger body size were neutralized. Size reduction of most game species began, and presumably accelerated like interrelated cultural evolution. With the end of native hunting cultures in recent centuries, some size recovery may have occurred, as in bison added to the Eastern Woodland fauna (Griffin and Wray, 1946; Swanton, 1946), or in alligators of reportedly extreme size (Bartram, 1791, p. 128; Harper, 1958, p. 356), or in northern Florida deer. The matter deserves close attention from zoologists who are studying archaeological material.

As human activities brought about a decrease in prey size, predators and carrion eaters competing with prehistoric man must have undergone a comparable decline, for saber-toothed tigers, dire wolves, giant hyenas, and, intraspecifically, the larger variants of Asian tigers (Hooijer, 1947, 1952) all expired. Furthermore, human body size underwent a simultaneous postglacial reduction, no doubt primarily because of the shift to plant foods and the development of food preservation and storage associated with a more sedentary way of life, but in part also because of the declining size of prey species and individuals during the Upper Paleolithic and Mesolithic (Edwards, 1966b).

There is no need to invoke metaphysical concepts, such as senescent decline in species (Simpson, 1949, p. 187–210) or readjustment after orthogenetic increase (Metcalf, 1922, p. 117; Jepsen, 1949) or superiority-imparting chance mutations (Kurtén, 1957). Strong selective pressures generally maintained a close approximation to optimum body size, the resultant of various selective forces of differing magnitude and direction. I conclude that human hunting and collecting almost exclusively account for the late-Pleistocene decline in average body size of prey animals.

CONCLUSIONS

Climatic change apparently cannot account for most individual size reductions or extinctions. Similar cyclic alterations during the first 99 % of the Pleistocene did not result in the concentration of extinctions that characterized the end of the last glaciation and the early and middle postglacial, especially in the New World. Furthermore, the fact that proportionately many more land mammals became extinct than plants and marine invertebrates, which apparently have much narrower climatic tolerance (Flint, 1957, p. 444), can be reconciled with the climate hypothesis only with marked difficulty.

Epidemic parasitic disease is likewise improbable as the crucial factor, for few extinctions occurred at the ends of pre-Wisconsin glaciations. In any case, adaptive natural disease is not likely to result in complete extermination, so at best it shares with climate only the role of a preconditioning agent.

Culture, in contrast, was not only a uniquely changing factor but also, through increasing efficiency of predation and the addition of other food sources, powerful enough to bring the observed extinctions. The Old World faunal succession exhibits marked changes throughout the Pleistocene, wherever hominids were present (e.g. Leakey, 1965, who nevertheless attributes extinction in Africa to climatic change). But in the New World, where man apparently did not arrive until the last glaciation, fewer changes of comparable nature occurred until the rather abrupt events of the late Pleistocene. Furthermore, despite fluctuations, a general acceleration in the rate of extinctions seems manifested in the Old World faunas, as would have been anticipated by the culture hypothesis, because of the marked acceleration of cultural evolution. An acceleration in extinction rates has even been evident during recent centuries (e.g. Charlesworth, 1957, p. 1398).

The causes of Pleistocene extinctions are admittedly complex. Factors are at several levels of significance. First, all agents producing or even influencing mortality may be considered. At a second level of higher significance are all factors without which the specific case of extinction could not have occurred; climate, and likely in some cases disease, locally produced extinctions and brought regional population decline for many species. Apparently culture alone is properly allocated to a third, highest level of significance. Culture was importantly involved in virtually all continental extinctions of the Pleistocene except some of those involving replacement by closely related, equivalent species. Likely many or even most of the extinctions that did occur would have occurred, however delayed, through the operation of culture alone, aided only by the omnipresent factors of the first level, as seems to have happened in much of eastern North America.

I conclude that, through his omnivorous habits and through his capacity for rapid cultural evolution, man exterminated many species. During the critical interval, the population of hunters did not decline even approximately in direct proportion to that of the herbivores upon which they *in part* depended. Thus the most crucial factor in the extinctions here considered was apparently neither change in the physical nor in the biotic environment but in the sociocultural environment.

References

Allee, W. C., Emerson, A. E., Park, O., Park, T., and Schmidt, K. P., 1949, *Principles of animal ecology:* Philadelphia, Saunders, 837 p.

Auffenberg, W., and Milstead, W. W., 1965, Reptiles in the Quaternary of North America, p. 557–68, *in* Wright, H. E., Jr., and Frey, D. G., *Editors, The Quaternary of the United States:* Princeton Univ. Press, 922 p.

Bartram, William, 1791, *Travels through North and South Carolina, Georgia, East and West Florida:* Philadelphia, James and Johnson, 520 p.

Beaufort, L. F. de, 1951, *Zoogeography of the land and inland waters:* London, Sidgwick and Jackson, 208 p.

Brodkorb, P., 1963, A giant flightless bird from the Pleistocene of Florida: *Auk*, v. *80*, p. 111–15

Butzer, K. W., 1964, *Environment and archeology; An introduction to Pleistocene geography:* Chicago, Aldine, 524 p.

Charlesworth, J. K., 1957, *The Quaternary Era:* London, Edward Arnold, vols. *1* and *2*, 1700 p.

Clark, J. G. D., 1952, *Prehistoric Europe; The economic basis:* New York, Philosophical Library, 349 p.

Cole, L. C., 1957, A surprising case of survival: *Ecology*, v. *38*, p. 357

Darlington, P. J., 1957, *Zoogeography:* New York, Wiley, 675 p.

Edwards, W. E., 1950, An interpretation of the geological and cultural history of the Helen Blazes site and of the Melbourne and Vero bone beds of central-eastern Florida (unpubl. ms.)

———1954, *The Helen Blazes site of central-eastern Florida; A study in method, utilizing the disciplines of archaeology, geology, and pedology:* Ann Arbor, University Microfilms, 145 p.

——— 1965, *The Sewee Mound Shell Midden, Charleston County, South Carolina:* Columbia, United States Forest Service, 68 p.

——— 1966a, Pleistocene extinction; A series of ten papers: *Carolina Jour. Anthropol.*, v. *1*

———1966b, Determinants of racial and subracial variations in human body-size: Amer. Assoc. Physical Anthropologists, Abstracts (Berkeley, 1966)

Elton, C. S., 1958, *The ecology of invasions by animals and plants:* London, Methuen, 181 p.

Flint, R. F., 1957, *Glacial and Pleistocene geology:* New York, Wiley, 553 p.

Gill, E. D., 1955, The problem of extinction with special reference to Australian marsupials: *Evolution*, v. *9*, p. 87–92

Gimbutas, M., 1956, The prehistory of eastern Europe, part I: *Harvard Univ., Peabody Museum, Amer. School Prehistoric Research, Bull.* v. *20*, 241 p.

Griffin, J. W., and Wray, D. E., 1946, Bison in Illinois archaeology: *Illinois State Acad. Sci., Trans.*, v. *38*, p. 21–26.

Haag, W. G., 1958, *The archaeology of coastal North Carolina:* Louisiana State Univ. Studies, Coastal Studies Series 2, 136 p.

Harper, Francis, 1958, *The travels of William Bartram:* New Haven, Yale Univ. Press, 727 p.

Haynes, C. V., Jr., 1964, Fluted projectile points; Their age and dispersion: *Science*, v. *145*, p. 1408–13

Hibbard, C. W., Ray, D. E., Savage, D. E., Taylor, D. W., and Guilday, J. E., 1965, Quaternary mammals of North America, p. 509–25, *in* Wright, H. E., Jr., and Frey, D. G., *Editors, The Quaternary of the United States:* Princeton Univ. Press, 922 p.

Hooijer, D. A., 1947, Pleistocene remains of *Panthera tigris* (Linnaeus) sub-species from Wanhsien, Szechwan, China, compared with fossil and recent tigers from other localities: *Amer. Mus. Nat. Hist. Novitates*, no. 1346, 17 p.

———1952, Austromelanesian once more: *Southwest. Jour. Anthropol.*, v. *8*, p. 472–77

Huxley, J. S., 1939, Clines; An auxiliary method in taxonomy: *Bijdr. Dierk.*, v. *27*, p. 491–520

Jepsen, G. L., 1949, Selection, "orthogenesis," and the fossil record: *Amer. Phil. Soc. Proc.*, v. *93*, p. 479–500

Karlstrom, T. N. V., 1961, The glacial history of Alaska; Its bearing on paleo-climatic history: *N.Y. Acad. Sci. Ann.*, v. *95*, p. 290–340

Kurtén, B., 1957, Mammal migrations, Cenozoic stratigraphy, and the age of Peking Man and the australopithecines: *Jour. Paleont.*, v. *31*, p. 215–27

Lack, D. L., 1954, *The natural regulation of animal numbers:* London, Clarendon Press, 343 p.

Leakey, L. S. B., 1960, Finding the world's earliest man: *Nat. Geographic*, v. *118*, p. 420–35

———1965, *Olduvai Gorge 1951–61*, Vol. I: Cambridge Univ. Press, 117 p.

Martin, P. S., and Mehringer, P. J. Jr., 1965, Pleistocene pollen analysis and biogeography of the Southwest, p. 433–51, *in* Wright, H. E. Jr., and Frey, D. G., *Editors, The Quaternary of the United States:* Princeton Univ. Press, 922 p.

Metcalf, M. M., 1922, Lectures upon evolution and animal distribution: *Univ. Buffalo Studies*, v. *2*, no. 4, p. 107–83

Newell, N. D., 1949, Phyletic size increase—An important trend illustrated by fossil invertebrates: *Evolution*, v. *3*, p. 103–24

Reed, E. C., Dreeszen, V. H., Bayne, C. K., and Schultz, C. B., 1965, The Pleistocene of Nebraska and northern Kansas, p. 187–202, *in* Wright, H. E., Jr., and Frey, D. G., *Editors, The Quaternary of the United States:* Princeton Univ. Press, 922 p.

Rensch, Bernhard, 1954, The relation between the evolution of central nervous functions and the body size of animals, p. 181–200, *in* Huxley, J., Hardy, A. C., and Ford, E. B., *Editors, Evolution as a process:* London, Allen and Unwin, 367 p.

Ritchie, W. A., 1965, *The archaeology of New York State:* Garden City, Natural History Press, 357 p.

Rouse, I., 1951, *A survey of Indian River archaeology, Florida:* Yale Univ. Publ. Anthropol., no. 44, 206 p.

——— 1952, The age of the Melbourne Interval: *Texas Archeol. Paleont. Soc. Bull.*, v. *23*, p. 293–99

Schultz, C. B., and Frankforter, W. D., 1946, The geologic history of the bison in the Great Plains: *Univ. Nebraska State Mus. Bull.*, v. *3*, p. 1–10

Selander, R. K., 1965, Avian speciation in the Quaternary, p. 527–42, *in* Wright, H. E., Jr., and Frey, D. G., *Editors, The Quaternary of the United States:* Princeton Univ. Press, 922 p.

Simpson, G. G., 1943, The Tertiary of South America: *Amer. Phil. Soc. Proc.*, v. *83*, p. 649–709

———1944, *Tempo and mode in evolution:* New York, Columbia Univ. Press, 237 p.

———1949, *The meaning of evolution:* New Haven, Yale Univ. Press, 364 p.

———1952, How many species?: *Evolution*, v. *6*, p. 342

———1953, Evolution and geography: Eugene, Oregon State System of Higher Education, 64 p.

Skellam, J. G., 1955, The mathematical approach to population dynamics, p. 31–46, *in* Cragg, J. B., and Pirie, N. W., *Editors, The numbers of man and animals:* Edinburgh, Oliver and Boyd, 152 p.

Soergel, W., 1912, *Das Aussterben diluvialer Säugetiere und die Jagd des diluvialen Menschen:* Festschr. 43, allg. Vers. d. D. Anthrop. Ges. H. 2, 81 p.

Swanton, J. R., 1946, The Indians of the southeastern United States: *Bur. Amer. Ethnology, Bull.* 137, 943 p.

Extinction dependent (selective) towards animals with relatively long gestation periods

BOB H. SLAUGHTER

Shuler Museum of Paleontology
Southern Methodist University
Dallas, Texas

ANIMAL RANGES AS A CLUE
TO LATE-PLEISTOCENE EXTINCTION

Abstract

There is growing evidence that extant animals whose geologic ranges extend back into the Pleistocene have modern ranges established at about the same time as the last major period of extinction (9,000–8,000 B.P.). If we conclude that climate is the primary range-restricting element, we must also conclude that essentially modern climate formed modern (or at least historic) ranges. I believe it to be too coincidental that the formation of essentially modern climate took place during this major period of extinction; the two events are probably related.

I suggest that one of the pressures on many animals was their inability to adapt their mating habits to the changing climate. Extinction seems to have been selective toward animals with relatively long periods of gestation—not dependent on the size of the animals, as is often stated.

There are different ranks of causes of death. For example, a bison may be felled primarily by a spear. Successively higher ranks of causes include the man who launched the spear, the hunger that provoked the attack, and the circumstances that brought together these two creatures. The same is true of extinction of a species or genus, and no single direct cause was ever responsible for the extinction of any taxon. There are far too many habits and habitats to be considered. Even so, when faced with the staggering fact that about 95% of the North American megafauna became extinct during a short period some 8,000 years ago, we must first seek causes of relatively high rank.

PLEISTOCENE CLIMATE

Hibbard (1960) and MacGinitie (1958) have warned that past climate has too often been reconstructed in terms of the present. It is possible but not necessarily true that any specific climate of the past can be exactly duplicated somewhere today. Zoological studies demonstrating rather convincingly that the waters of the Gulf of Mexico are warmer now on the average than during the last major glacial advance (Frey, 1965) cannot be construed to show that the distribution of terrestrial animals was also controlled by the same warming trend. Mean annual temperatures in Texas tell us little about the effect on terrestrial animals and plants, for they do not reveal extremes. A maritime climate with an average winter temperature of 70°F, accompanied by a summer average of 80°F, would have the same annual mean temperature as a continental climate with summer mean of 95° and winter mean of 55°. It is therefore possible that Gulf water temperatures may rise while Texas winters become more severe, if summers become correspondingly hotter.

Southern animals sensitive to extreme cold may have their ranges restricted by the occasional cold fronts that drop air temperatures as much as 50° within a few hours, although the low temperature may be maintained for only a day or so. In Texas, winter climate between these cold fronts is mild. Occasional abnormally severe winters may be even more important than the more normal weather in the range restriction of certain animals. Just how far a thermally sensitive, expanding population extends its range northward may depend upon the frequency of abnormally severe winters. The northern range of such animals would extend farther if extremely cold winters occurred every twenty years than if they occurred every five years, regardless of the normal winter means. It seems doubtful that the occasional abnormally severe winters have any great effect upon the temperature of the waters of the Gulf of Mexico. Periods of desiccation recorded in pollen diagrams of the Southwest do not necessarily mean that the desiccation was continental in scope. Martin and Mehringer (1965), Mehringer (1965), and Mehringer and Haynes (1965) record the arrival of essentially modern moisture conditions in Arizona and Nevada by 11,000 B.P. Almost the same period is represented by the Brown Sand Wedge vertebrate local fauna in New Mexico, but somewhat moister conditions are indicated by the presence of tree squirrel, muskrat, meadow and prairie voles, masked shrews, etc. Meteorologists have suggested that many able paleontologists and palynologists too often draw knowledge of present climate from oversimplified maps showing temperature and rainfall means but no other important aspects of climate. For example, Baerreis and Bryson (1965) point out that desiccation trends in Nebraska are most often accompanied by increased rainfall to the south in Oklahoma. The tendency is to misinterpret changes in moisture conditions as mere east–west pulsations of the current rainfall belts.

Many workers (Dalquest, 1965; Hibbard, 1960; Hibbard and Taylor, 1960; Slaughter and Hoover, 1963; Taylor, 1965) have pointed out that the presence of heat-intolerant species in fossil faunas far south of their present ranges is not evidence of "colder climate," and likewise that cold-intolerant species in northern fossil faunas do not indicate higher summer temperatures. Instead, northern species represented as fossils in the south demonstrate cooler summers, and southern species farther north suggest milder winters. Even then, only ground-level temperatures at the locality are reflected. Slaughter and Hoover (1963) have suggested that more or less constant cloud cover accompanying pluvial stages would tend to equalize seasonal temperatures through retardation of heat loss in the winter and insulation of the ground from the summer sun, without major changes in wind direction or other factors. Most of the present discussion of paleoecology and paleoclimate will therefore be largely restricted to those places and periods represented by the data.

SOUTHWESTERN FAUNAS IN THE LATE PLEISTOCENE

Most of the work of the Shuler Museum, greatly augmented by that of Dr. Walter Dalquest, has been between the 30th and 35th parallels in eastern New Mexico, southern Oklahoma, and Texas. A series of vertebrate and molluskan faunas provides a basis for climatic inference through Wisconsin glacial advances and retreats.

As Slaughter (1966a) reported, the oldest of the series is the Easley Ranch local fauna from north-central Texas (Fig. 1; 1), which originated early in the last major interstadial. The deposit records the oldest occurrence of the armadillo *Dasypus bellus*, which apparently preceded the arrival of the cotton rat *Sigmodon hispidus*. That the summers were cool is shown by such mammalian species as *Sorex cinereus* (masked shrew), *Microtus pennsylvanicus* (meadow vole), and *Synaptomys cooperi* (southern bog lemming), as well as by numerous molluskan species that currently are restricted to areas several hundred miles farther north. In some cases, physiological changes not reflected in the skeleton may have changed the tolerance to physical environment; but when several species representing different groups, even classes, point to the same conclusions, this objection seems less formidable. That winters during Easley Ranch time were not correspondingly colder is indicated by the presence of species intolerant of cold. Common in the Easley Ranch local fauna is *Oryzomys palustris* (rice rat), a southeastern species whose current range extends northward only to the Carolinas. Of even greater significance is the giant tortoise *Geochelone*, whose intolerance of cold winters is well known. Hibbard (1960) demonstrated that these tortoises cannot withstand even light frost unless the temperatures of the following day rises

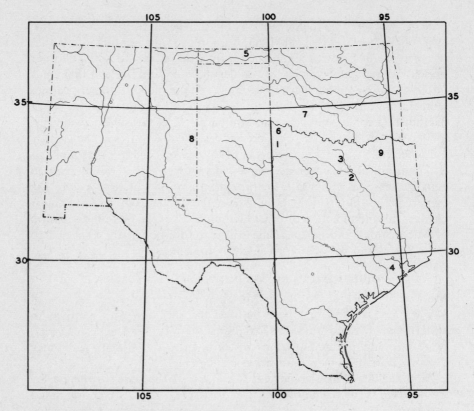

FIG. 1. Localities discussed in the text.
1. Easley Ranch local fauna (Dalquest, 1962)
2. Moore Pit local fauna (Slaughter et al., 1962; Slaughter, 1966)
3. Clear Creek local fauna (Slaughter and Ritchie, 1963)
4. Sims Bayou local fauna (Slaughter and McClure, 1965)
5. Bar M local fauna (Hibbard and Taylor, 1960)
6. Grosebeck Creek Formation (Dalquest, 1965)
7. Domebo local fauna (Slaughter, 1966c)
8. Brown Sand Wedge local fauna (Slaughter, 1966b)
9. Ben Franklin local fauna (Slaughter and Hoover, 1963)

into the sixties. Kalmbach (1943) reported the total annihilation of a
large and expanding population of *Dasypus novemcinctus* (nine-banded
armadillo) during the severe winter of 1932–33 in western Texas. The
late-Pleistocene counterpart, *D. bellus*, seems to be a replica of the modern
form but about three times larger. The armadillo has very poor temperature control and can withstand neither extreme cold nor prolonged
periods of even moderately low temperatures. It shivers continuously at
72°F (21°C) and will perish overnight in temperatures no lower than

45°F (7°C) if kept from its burrow (Talmage and Buchanan, 1954). The larger size of *D. bellus* would certainly restrict its burrowing, and therefore it may have been less tolerant of prolonged low temperatures. Even so, these cold-intolerant animals were sympatric with heat-intolerant mammals and mollusks at Easley Ranch. Apparently, both summer and winter temperatures were moderate. Although climatic conclusions are generally less easily drawn from the presence of extinct species than from fossil remains of extant species, in the case of armadillos and giant tortoises such inference is perfectly valid. The gross morphology of these forms precludes a physiological change that would noticeably affect their intolerance of extreme or prolonged low temperatures.

The next local fauna in the series is the Moore Pit (Slaughter et al., 1962; Slaughter, 1966a) from Dallas County, Texas (Fig. 1; 2), with the same indications of mild winters (giant *Geochelone*, *Dasypus bellus*, and the cotton rat *Sigmodon hispidus*). This is the earliest occurrence of this cotton rat in the Southwest. Summers had apparently warmed, in view of the loss of the typically northern mammalian species that occurred at Easley Ranch. A few heat-intolerant aquatic molluskan species still remained, but they may have been influenced by a high water table producing numerous springs, which in turn kept small bodies of water on the floodplain well cooled and oxygenated. The occurrence of southeastern species in both Easley Ranch and Moore Pit local faunas suggests more efficient moisture because of greater annual rainfall, more evenly distributed rainfall, or a decrease in evaporation. The Moore Pit assemblage lies stratigraphically lower than the same terrace that produced the Clear Creek local fauna, which has a radiocarbon date between 33,000 and 25,000 B.P. (SM-534). Other dates (Humble O-235 and H.O-248), on samples geographically and stratigraphically from the same terrace between Clear Creek and the Moore Pit, indicated an age greater than 37,000 B.P. Because the Moore Pit belongs to the same sequence of deposition, its age is believed to be slightly greater than 40,000 or 45,000 B.P.

At Clear Creek in Denton County, Texas (Fig. 1; 3), Slaughter and Ritchie (1963) found no indication of major change in the mild winter temperatures. The same giant tortoises, armadillos, cotton rats, and so on, are present. The absence of typically northern mammals indicates that summers were still relatively warm. The main difference was increased desiccation. Southwestern species such as *Cynomys ludovicianus* (black-tailed prairie dog), *Onychomys leucogaster* (grasshopper mouse), and *Notiosorex crawfordi* (desert shrew) are currently restricted to areas about 200 miles (320 km) to the west and southwest, where annual rainfall is five to ten inches (12–25 cm) less than at the fossil locality. Several southern molluskan species are abundant for the first time in our series, but there are still a few northern aquatic forms perhaps because of a sustained high water table related to the continuing aggradation of the valleys in spite of the reduction of effective moisture.

The same period of desiccation is probably recorded in the Sims Bayou local fauna of Harris County, Texas (Slaughter and McClure, 1965) (Fig. 1; 4), where such mammals as the grasshopper mouse, gray wood rat, and prairie dog are present some distance east of their current range. Giant armadillos and *Geochelone* are also present. The radiocarbon date (Shell BFP SB-1) merely indicates an age in excess of 23,000 B.P.; but the cotton rat *Sigmodon hispidus* is abundant. The earliest recorded occurrence of this form in the Southwest is at the Moore Pit, so the Clear Creek is the local fauna closest in age that suggests similar climate.

That the winters had not worsened by about 21,360 B.P. (SM-763) is demonstrated by the presence of the armadillo *Dasypus bellus*, associated with a radiocarbon date from the Bar M local fauna in northern Oklahoma (Fig. 1; 5). This fauna also includes several northern gastropod species, however, that indicate that summers were cooling.

Two local faunas of the Grosebeck Creek formation (Dalquest, 1965) bracket what is generally considered as the last glacial maximum (Fig. 1; 6). Radiocarbon dates of these sites are 19,098 ± 1,074 and 16,775 ± 565 respectively (SM-620). The typically northern mammalian and molluskan species had returned to Texas by this time, indicating summers more like the first of our series, the Easley Ranch local fauna. The giant tortoise apparently had become extinct, but armadillos were merely restricted farther southward. Winters were probably more severe than they had been for many thousands of years, with occasional hard freezes. That they were not more severe than they are today is indicated by the presence of the cotton rat *Sigmodon hispidus* and the rice rat *Oryzomys palustris*. The mean winter temperature may have been similar to that of the present, or even lower, but the cold fronts that characterize the area today may not have occurred (Dalquest, 1965).

These same mild summers remained until at least 9,550 ± 375 years ago (SM-532), as indicated by the presence of many northern mammals and mollusks among them (*Microtus pennsylvanicus*, *Sorex cinereus*, *Synaptomys cooperi*, and *Valvata tricarinata*), at the Domebo locality (Fig. 1; 7) in southern Oklahoma (Slaughter, 1966c); in the Brown Sand Wedge local fauna (Slaughter, 1966b) in eastern New Mexico (Fig. 1; 8); and in the Ben Franklin local fauna (Fig. 1; 9) in Texas (Slaughter and Hoover, 1963). Very mild winters returned, however, by about 11,000 B.P., as is demonstrated by the occurrence of (1) a southern race of box turtle in the Brown Sand Wedge; (2) a small species of *Geochelone* in the Brown

Sand Wedge and at Domebo; (3) the large armadillo *Dasypus bellus* in the Brown Sand Wedge and Ben Franklin local faunas; (4) the huge armadillo *Holmesina* at Ben Franklin; and (5) the cotton rat *Sigmodon hispidus* at all three localities. *Holmesina* weighed several hundred pounds and almost certainly could not have had extensive burrows into which to retreat from periodically inclement weather. This huge specimen probably was even less able to cope with low temperatures than the smaller armadillos.

That the major period of extinction took place about 9,000 to 7,000 B.P. (Hester, 1960) is supported by the fact that thus far not a single extinct species has been recovered from assemblages dating from after about 8,000 B.P. in Texas. Equally important is the fact that the local faunas later than 8,000 B.P. lack mammalian species that have not occurred in the same area in the very recent past. A few cold-loving aquatic molluskan species lasted a bit longer in the area (Cheatum and Slaughter, 1966), perhaps because cool springs can maintain necessary microhabitats after gross optimal conditions have passed.

CLIMATE AS THE CAUSE OF EXTINCTION

Because interglacial and interstadial conditions occurred earlier and lasted longer in the southern United States than in areas farther north, and because glacial conditions began earlier and lasted longer to the north, I do not find attempts to correlate events in the south with names of events farther north advantageous. Nevertheless, it is quite evident that there was a major interglacial or interstadial interval in southern United States roughly from 50,000 to 20,000 B.P. From the faunal evidence, I infer the following climatic changes in the Southwest during this period of glacial retreat and the Classical Wisconsin that followed (Fig. 2):

1. Early in this last major interstadial, summers were cool, there was more effective moisture than today, and winters were mild with only light frost.

2. As interglacial conditions were asserted in the south, summers became warmer, although still without the heat waves and occasional droughts that characterize the area today, and effective moisture dropped slightly below that of the present. Winters were still very mild, with only light frosts.

3. With the glacial advance of about 20,000 B.P., effective moisture increased, and summer temperatures dropped. By the time of the glacial maximum about 18,000 years ago, winters were more severe than they had been for many thousands of years, perhaps throughout the

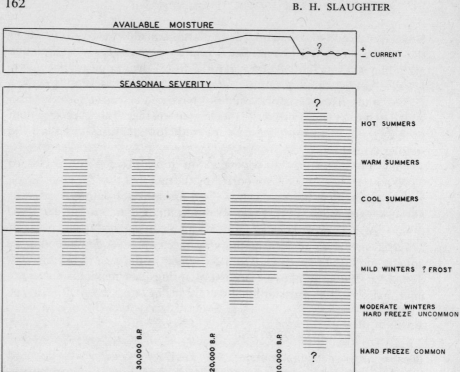

FIG. 2. Suggested moisture efficiency and severity of seasonal fluctuations during the last 50,000 years along the 35th parallel.

Pleistocene. There is no evidence, however, that winters were more severe along the 35th parallel than they are today; they were probably milder.

4. The return of armadillos to the area by about 12,000 B.P. indicated a return to very mild winter weather. However, until about 9,500 B.P. the summers remained as cool as those of the glacial maximum. The giant armadillo *Holmesina* occurred as far north as the Red River of Texas by 10,000 B.P.

If we concede that seasonal extremes are the primary modern range-restricting barriers for animals, it is reasonable to infer that the advent of essentially modern climate determined these ranges. It is hardly a coincidence that essentially modern ranges of extant animals were established during the most intense period of megafaunal extinction. The advent of essentially modern continental climate qualifies as a fundamental cause of these extinctions.

MATING SEASON AND CLIMATE

Animals of arctic and temperate regions have short mating seasons, timed for the birth of their young during spring months when temperatures are mild and when food for parent and young alike is in good supply. Farther south these seasons are longer and less well defined. For example, the modern peccary has about a six-month mating period, but its range does not extend north of the southern United States. Only a hundred miles or so north of their present range, too many of their young would arrive during severe weather. In equatorial regions most animals either have an unrestricted mating season or one that is independent of temperature. The tapir, for example, mates with the coming of the rainy season.

Animals with inflexible mating habits often are restricted in range by the season in which their young are born. The ocelot is about the same size as the bobcat; both have essentially the same diet; but the range of the bobcat extends to the Canadian border, while the northern limit of the ocelot is in Texas. Bobcat kittens are born in the spring, but ocelot young arrive in September and October—doubtless because of a retention of the mating habits of the southern hemisphere, their area of origin. Newborn ocelot kittens would not survive winters two hundred miles north of their present range.

Animals like the puma and the white-tailed deer are more flexible and have the greatest geographic ranges. The historic range of the puma extended from the Atlantic to the Pacific and from Alaska to the southern tip of South America. In the far north, puma breeding is adjusted for spring delivery; in temperate regions puma mates the year round; and in southern South America the typical time of delivery, as for ocelot, is September and October. The flexibility of mating habits gives the puma the chance to survive climatic change.

BODY SIZE AND LENGTH OF GESTATION PERIOD

The generalization that late-Pleistocene extinction was selective to size—large forms becoming extinct and smaller ones surviving—should be qualified. The small antelope *Breameryx*, weighing perhaps no more than 20 pounds (10 kg), became extinct, whereas the gray wolf, weighing up to 120 pounds (60 kg) survived. What, then, do the extinct animals have in common if not size? They all have relatively long periods of gestation. The more complex animals, with gestation periods of several months, became extinct regardless of size. *Breameryx*, as a ruminant, doubtless had a period of gestation similar to that of its modern relative, the pronghorn. The gray wolf has a gestation period of about six weeks and must be grouped with most small animals in this respect.

Present evidence indicates that winters became more severe and summers somewhat warmer about the time of the greatest extinction

(9,000–7,500 B.P.). One may visualize what would happen to a population with gestation periods of several months if winters were prolonged by two months. These animals would have mated during the preceding fall and early winter, and the young would arrive during late-winter weather when green vegetation was not available. On the other hand, animals with relatively short gestation periods (usually but not universally smaller) await clear signals for optimal weather before mating. Unusually long and severe winters will often delay mating a month or more. Such animals, awaiting the warming trend before mating, would be less affected by a sudden lengthening of the winter season and would therefore have a better chance of survival. The populations may thus be decimated by the loss of the young, even though the climate may not become severe enough to kill adults.

In areas as far south as southern Mexico, where there have probably been no severe winters for millenia, the summers could have been affected. Rising summer temperatures would have the net effect of moving biotas southward—an effect that may be as detrimental to reproduction as the lengthening of the winter season in the north. Agricultural studies have shown that when man moves animals of northern origin farther south, milk production is usually lowered and activity of the male spermatozoa is retarded. In some cases sterility follows, and reproductive organs degenerate.

CONDITIONS ON OTHER CONTINENTS

With the exception of such specialized animals as the armadillos, who have replaced insulation with armor, each group of warm-blooded animals has members adapted to almost every possible environment. Perhaps most notable are the cold-adapted elephants and rhinoceros; but horses, large carnivores, and even tapirs have on occasion developed long, thick coats to allow them to fit into ecological niches quite different from those we usually associate with them. In Eurasia, where a large range of elevation and latitude existed during the Pleistocene as well as later, extinction at the generic level largely involved a reduction of range. Lions, hyenas, elephants, horses, etc. ranged over most of Europe and Asia during the Pleistocene. Doubtless some of each group lived in and had adapted to areas where conditions were in effect very similar to those in adjacent areas after the close of the Pleistocene. These surviving forms would doubtless re-extend their ranges if and when Pleistocene climatic conditions returned. In Eurasia the number of species of horse (*sensu stricto*) was reduced as were those of other widespread groups. The most successful survivor, *Equus caballus*, doubtless was better adapted to an environment similar to the present. The success of this species in North America, where other members of the genus had failed, vouches for its adaptations, which are compatible with many different environments. Unlike some domesticated animals, the majority of their young are born in the spring, and, when returned to the wild state, as in America, they mate almost entirely seasonally, as do the surviving native ungulates.

In South America, the vicuña, a living relative of the extinct North American *Camelops* and *Tanupolama*, has survived in a selected habitat high in the mountains of Peru near the equator, protected from heat by elevation and from cold by the equatorial position (Koford, 1957). No such refuge was available to its kin in North America.

CONCLUSIONS

1. A change in the extremes in temperatures, both winter and summer, can have profound effects on animals, even though the average temperatures change only slightly. The advent of wider temperature fluctuations probably caused the extinction of much of the megafauna in the interval of about 9,000 to 7,000 B.P.

2. A breakdown in reproduction, caused by inflexible mating habits established under conditions no longer in effect, seems a most logical reason for some of the extinction that took place 8,000 years ago in North America, because it does not require changes of catastrophic magnitude, and because it allows for the survival of the smaller animals, most of which have short periods of gestation.

3. The North American continent, especially in the Southwest, does not offer a sufficient range of climates to allow many species to escape climatic change by migration.

References

Baerreis, D. A., and Bryson, R. A., 1965, Historical climatology and the Southern Plains (Abst.): VII INQUA, Abstracts, p. 12

Cheatum, E. P., and Slaughter, B. H., 1966, Notes on the alluvial history of the Lampasas River, Texas: *Southern Methodist Univ. Grad. Res. Center Jour.*, v. *35*, p. 48–54

Dalquest, W. W., 1962, The Good Creek Formation, Pleistocene of Texas, and its fauna: *Jour. Paleont.*, v. *36*, p. 568–82

———— 1965, New Pleistocene formation and local fauna from Hardeman County, Texas: *Jour. Paleont.*, v. *39*, p. 63–79

Frey, D. G., 1965, Other invertebrates—An essay in biogeography, p. 613–31, *in* Wright, H. E. Jr., and Frey, D. G., *Editors, The Quaternary of the United States:* Princeton Univ. Press, 922 p.

Hester, J. J., 1960, Late Pleistocene extinction and radiocarbon dating: *Amer. Antiq.,* v. *26,* p. 58–77

Hibbard, C. W., 1960, An interpretation of Pliocene and Pleistocene climates in North America: Michigan Acad. Sci .Arts Lett., 62nd Ann. Rept., p. 5–30

Hibbard, C. W., and Taylor, D. W., 1960, Two late Pleistocene faunas from southwestern Kansas: Univ. Michigan Mus. Paleont. Contrib., v. *16,* p. 1–223

Kalmbach, E. R., 1943, The Armadillo; Its relation to agriculture and game: Texas Game, Fish, and Oyster Comm., p. 1–61

Koford, C. B., 1957, The vicuña and the puma: *Ecological Monog.,* v. *27,* p. 153–219

MacGinitie, H. E., 1958, Climate since the late Cretaceous, p. 61–79, *in* Hubbs, C. L., *Editor, Zoogeography:* Amer. Assoc. Adv. Sci., Publ. 51, 509 p.

Martin, P. S., and Mehringer, P. J., Jr., 1965, Pleistocene pollen analysis and biogeography of the southwest, p. 433–51, *in* Wright, H. E., Jr., and Frey, D. G., *Editors, The Quaternary of the United States:* Princeton Univ. Press, 922.

Mehringer, P. J., Jr., 1965, Late Pleistocene vegetation in the Mohave Desert of southern Nevada: *Arizona Acad. Sci. Jour.,* v. *3,* p. 172–88

Mehringer, P. J., Jr., and Haynes, C. V., 1965, The pollen evidence for the environment of Early Man and extinct animals at the Lehner Mammoth Site, southeastern Arizona: *Amer. Antiq.,* v. *31,* p. 17–23

Slaughter, B. H., 1966a, The Moore Pit local fauna, Pleistocene of Texas: *Jour. Paleont.,* v. *40,* p. 70–91

———— 1966b, An ecological interpretation of the Brown Sand Wedge local fauna, Blackwater Draw, New Mexico; and a hypothesis concerning late Pleistocene extinction, *in* Wendorf, Fred, and Hester, J. J., *Editors, Paleoecology of the Llano Estacado,* v. *2*: Santa Fe, Fort Burgwin Res. Center

———— 1966c, The vertebrates of the Domebo local fauna, Pleistocene of Oklahoma, p. 31–35, *in* Leonardy, F. C., *Editor,* Domebo, a Paleo-Indian mammoth kill site in the prairie-plains: Lawton, Oklahoma, Museum of the Great Plains, Contrib. no. 1, 53 p.

Slaughter, B. H., Crook, W. W., Jr., Harris, R. K., Allen, D. C., and Seifert, Martin, 1962, The Hill-Shuler local faunas of the Upper Trinity River, Dallas and Denton Counties, Texas: Univ. Texas Bureau Econ. Geology, *Report of Investigations, 48,* p. 1–75

Slaughter, B. H., and Hoover, B. R., 1963, Sulphur River formation and the Pleistocene mammals of the Ben Franklin local fauna: *Southern Methodist Univ. Grad. Res. Center Jour.,* v. *31,* p. 132–48

Slaughter, B. H., and McClure, W. L., 1965, The Sims Bayou local fauna; Pleistocene of Houston, Texas: *Texas Acad. Sci. Jour.,* v. *17,* p. 404–17

Slaughter, B. H., and Ritchie, Ronald, 1963, Pleistocene mammals of the Clear Creek local fauna, Denton County, Texas: *Southern Methodist*

Univ. Grad. Res. Center Jour., v. *31*, p. 117–31

Talmage, R. V., and Buchanan, G. D., 1954, The Armadillo: *Rice Institute Monog. Biology*, v. *41*, p. 1–135

Taylor, D. W., 1965, A study of Pleistocene nonmarine mollusks in North America, p. 597–611, *in* Wright, H. E., Jr., and Frey, D. G., *Editors, The Quaternary of the United States:* Princeton Univ. Press, 922 p.

JAMES J. HESTER

Department of Anthropology
University of Colorado
Boulder, Colorado

THE AGENCY OF MAN
IN ANIMAL EXTINCTIONS

Abstract

Historical records of extinctions of birds and mammals of North America indicate that contributory causes include human activities. Although incomplete, the records indicate that a common pattern of these causes affected most of the mammals and birds of North America during the period of European colonization, culminating in numerous extinctions at the end of the nineteenth century. Major causes listed in order of prevalence are: hunting by primitive man, hunting by civilized men with firearms, hunting for food, isolation of populations, destruction of habitats, small size of population, hunting for a commercial market, concentration in herds or flocks, and the introduction of predators such as rats and cats. The number of causes contributing to the demise of a species ranged from one to ten. As a species was reduced in numbers, the adverse effects of other causes such as disease and low reproduction rate increased in magnitude. The native American Indians demonstrated little concern for conservation of food resources, frequently killing off local herds. In no case, however, did they cause the extinction of a species, even with the use of firearms. Presumably this is because the Indians did not have as high a population density as the colonists, nor did they kill in mass for commercial reasons.

By analogy, this information helps us understand the widespread extinction that occurred at the end of the Pleistocene age in North America. None of the causes seems to be relevant for Pleistocene man except hunting for food and the destruction of habitats by fire.

Examination of the faunal material present at early human sites indicates that only two genera, mammoth and bison, were hunted in quantity by early man. In addition, the proportion of the diet represented by extinct species varies among different sites: apparently Early Man ate whatever he could get. A final point is that several species that became extinct at the end of the Pleistocene epoch were either already extinct or greatly reduced in numbers before the known advent of man in the New World.

In summary, I take the view that Pleistocene man could not have caused the extinction of the North American megafauna until after natural causes had greatly reduced the population of each species. An alternative explanation is that change in ecologic niches resulting from the disappearance of the late-Wisconsin ice sheet subjected animal species to severe selective pressures.

The widespread extinction of the North American megafauna at the close of the Pleistocene epoch has long been of concern to researchers (Morris, 1895; Osborn, 1906; Tolmachoff, 1930; Eiseley, 1946; Peters, 1950). Basic data documenting this extinction have been accumulated in quantity by Cope (1899), Hay (1923, 1924, 1927), Stock (1929), and other early students of American vertebrate paleontology. Summaries of these data are presented by Romer (1933), Colbert (1942), and Hibbard (1958). After the discovery in 1927 of projectile points in association with the bones of extinct bison near Folsom, New Mexico, it has become apparent that many species of this megafauna became extinct after the advent of man (ca. 12,000 years ago) in the New World. Within the last decade, as a result of the dating of late survivors of these extinct species by radiocarbon (Jelinek, 1957; Hester, 1960), it has been demonstrated that many of them became extinct within a short period between 8,000 and 6,000 years B.C. This has led to renewed interest into the cause or causes of extinction (Martin, 1958; Newell, 1963; Slaughter, n.d.; Sellards, 1952; Wormington, 1957; Jelinek, 1957), culminating in the present symposium.

Numerous causes contributing to late-Pleistocene extinction have been suggested, including climatic change, disease, poor synchronization of breeding habits with climate, and predation by man. The possibility that human agency was a serious contributing factor to late-Pleistocene extinction has been discussed in detail by Martin (1958). He reached the conclusion that the presence of man was the only clearly identified unique feature present at the end of the Pleistocene. Martin also points out as evidence that this particular period was unique in that the extinction was not followed by replacement of forms within each specific ecologic niche, an unusual occurrence in the paleontologic record. Refutations of Martin's thesis are based primarily on the scantiness of archaeological evidence for human responsibility for late-Pleistocene extinction. This in part is the result of infrequent preservation of early animal kills and in part because of low human population density in North America during the terminal Pleistocene. However, the latter inference is based on the small number of archaeological sites known for the period. Such reasoning has an element of circularity that does not do justice to Martin's thesis. It seems that one profitable approach to the study of this problem is the examination of historical records of the effect of civilized European man on animal populations in North America. Such an examination may be useful, by analogy, in the understanding of the nature of the late-Pleistocene extinction.

HISTORIC MAN'S ROLE IN EXTINCTION

The colonization of North America by Europeans, beginning in the sixteenth century, initiated a period of widespread decimation of native animal populations. This culminated late in the nineteenth century in the near extinction of several major species, notably the American bison (Hornaday, 1887), and the threatened extinction of many additional species (Hornaday, 1913; Allen, 1942; Greenway, 1958). During this period the public viewed the animal populations as inexhaustible. As a result, numerous species declined in numbers, but little pertinent information was recorded. A summary of the information available about the causes of this extinction or severe reduction is presented by species in Table 1. These data are not strictly comparable in validity, because some species are not well recorded historically. Frequently, in the latter case, the stated causes are only the opinion of the author long after extinction had actually occurred. In spite of this limitation, a pattern of man's influence on animal populations emerges from examination of the table.

Descriptions and historical accounts of the decimation of individual species are not included in the present paper; they may be obtained from numerous works, notably those of Greenway (1958) and Allen (1942). Analysis of the records indicates that similar causes affected both land mammals and birds, and this discussion will therefore not distinguish these major groups. The paleontologic record of amphibians, fishes, and reptiles is less well known for the Pleistocene (Blair, 1965; Miller, 1965; Auffenberg and Milstead, 1965); therefore these groups will not be discussed here.

In order of frequency, the major factors that have led to historic animal extinctions are hunting by primitive man, hunting by civilized man with repeating firearms, hunting for food, isolation of populations, destruction of habitats, small size of population, hunting for a commercial market, natural concentration in herds or flocks, introduction of predators, forest fire, overspecialization, eradication of animals detrimental to crops, disease, cycles of population, slow reproduction, killing by storms, excessive inbreeding, and an unbalanced sex ratio. The number of these factors contributing to the reduction of an individual species (according to the sources consulted) ranges from one to ten.

TABLE 1.

Causes Attributed to the Reduction in Numbers or Extinction of Animal Species in North America and Adjacent Islands Since European Settlement

Extinct and vanishing animals	Destruction of habitat	Predators and competitors Introduced by man	Hunted for market	Hunted for feathers	Forest fires	Hunted by civilized man	Hunted by primitive man	Isolated population	Concentrated in herds or flocks	Never abundant	Hunted for food	Flightless	Considered pests	Cyclical population	Disease	Excessive inbreeding	Unbalanced sex ratio	Killed by storm	Slow reproduction	Natural change in habitat	Natural predators	Date of extinction or of greatest reduction	Conservation attempted
North America																							
Alca impennis, great auk						×	×		×	×	×	×						×				1844	
Camptorhynchus labradorius, Labrador duck		cat	×			×			×	×	×											1875	
Tympanuchus cupido cupido, heath hen	×		×		×	×			?		×			×	×	×	×	×				1932	×
Ectopistes migratorius, passenger pigeon			×			×			×		×								×			1914	
Conuropsis carolinensis, Carolina parakeet	×					×			×	×			×									1914	
?*Branta canadensis maxima*, Canada goose							×				×											1929	

Species	Date	Notation
*Numenius borealis, Eskimo curlew	1937	×
Sciurus niger, fox squirrel		fur
Castor canadensis, beaver		fur
Microtus californicus scirpensis, desert vole		
*Microtus nesophilus, Gull Island vole	1898 ?	fur
Canis lupus, gray wolf	1920–1930	fur
Canis niger, red wolf		fur
Vulpes macrotis macrotis, kit fox	1903	fur
Vulpes velox hebes, swift fox		fur
Ursus americanus, black bear		fur
Ursus horribilis, grizzly bear	1900–1920	fur
Martes americana, marten	1910	fur
Martes pennanti, fisher	1910	fur
*Mustela macrodon, sea mink	1890–1894	fur
Mustela nigripes, black-footed ferret		fur
Gulo luscus, wolverine	1860–1920	fur
Felis onca, jaguar		
Felis concolor, mountain lion	1905	fur
Cervus canadensis, elk	1840	wolf
Odocoileus hemionus, mule deer	1910	wolf
Rangifer tarandus arcticus, reindeer		wolf
Rangifer tarandus dawsoni, Queen Charlotte Islands caribou		
Rangifer tarandus caribou, woodland caribou		wolf
Rangifer tarandus montanus, mountain caribou		wolf

TABLE 1.—continued

Extinct and vanishing animals	Destruction of habitat	Predators and competitors Introduced by man	Hunted for market	Hunted for feathers	Forest fires	Hunted by civilized man	Hunted by primitive man	Isolated population	Concentrated in herds or flocks	Never abundant	Hunted for food	Flightless	Considered pests	Cyclical population	Disease	Excessive inbreeding	Unbalanced sex ratio	Killed by storm	Slow reproduction	Natural change in habitat	Natural predators	Date of extinction or of greatest reduction	Conservation attempted
Antilocapra americana, pronghorn antelope	×				×	×	×		×		×							×			wolf		×
Bison bison bison, buffalo	×		×			×	×		×		×							×			wolf	1879–1884	×
Bison bison athabascae, woodland bison						×	×		×		×							×	×		wolf	1907	×
Ovibos moschatus, musk-ox								×	×	×	×				×				×		wolf	1905+	×
Ovis canadensis, mountain sheep					×	×	×	×			×							×	×		coyote		×
Ovis dalli, Dall's sheep											×										wolf		×
West Indies																							
?*Monophyllus frater*, Puerto Rican long-nosed bat	×									×													
Phyllonycteris aphylla, Jamaican long-tongued bat										×												1898	
?*Ardops nichollsi*, tree bat	×									×			×										

Species		Predator	Date
Phyllops haitiensis, Hispaniolan falcate winged bat	×	owl	1917
Ariteus flavescens, Jamaican fig-eating bat	×		
Stenoderma rufum, Puerto Rican fruit-eating bat	×		
?Phyllonycteris obtusa, Haitian long-tongued bat	×	owl	
?Natalus major primus, Cuban funnel-eared bat	×		
?Nesophontes edithae, Puerto Rican shrew	×	rat / owl	1492+?
?Nesophontes zamicrus, Hispaniolan shrew		mouse / owl	1492+? / 1910?
?Solenodon cubanus, Cuban shrew		mongoose / owl	1877
Solenodon paradoxus, Hispaniolan shrew	×	mongoose / owl	1897?
*Oryzomys antillarum, Jamaican rice rat		mongoose	
?Oryzomys victus, St. Vincent rice rat		mongoose	
*Megalomys audreyae, Barbuda muskrat	×	snake	1902
*Brotomys (2 species), Hispaniolan spiny rat		rat / owl	1492?
*Boromys (2 species), Cuban spiny rat		owl	19th Cent.
Capromys (2 species), Cuban hutia		mongoose	
*Geocapromys columbianas, Cuban short-tailed hutia		mongoose	
*Geocapromys ingrahami, Bahaman hutia	×	owl	18th Cent.

TABLE 1.—continued

Extinct and vanishing animals	Destruction of habitat	Predators and competitors Introduced by man	Hunted for market	Hunted for feathers	Forest fires	Hunted by civilized man	Hunted by primitive man	Isolated population	Concentrated in herds or flocks	Never abundant	Hunted for food	Flightless	Considered pests	Cyclical population	Disease	Excessive inbreeding	Unbalanced sex ratio	Killed by storm	Slow reproduction	Natural change in habitat	Natural predators	Date of extinction or of greatest reduction	Conservation attempted
Plagiodontia hylaeum, Haitian hutia		mongoose						X			X								X				
*Hexolobodon phenax, Hispaniolan hutia								X														1492+	
*Isolobodon portoricensis, Puerto Rican hutia							X	X			X										owl	ca. 1492	
*Isolobodon levir, Hispaniolan hutia							X	X			X											ca. 1492	
*Elasmodontomys obliquus, Puerto Rican giant rodent							X	X		X	X										owl	ca. 1550	
Dasyprocta albida, St. Vincent agouti	X	mongoose			X		X	X			X									X			
Hawaii																							
*Drepanis pacifica, mamo	X			X			X	X			X			X								1898	

Guadalupe Islands

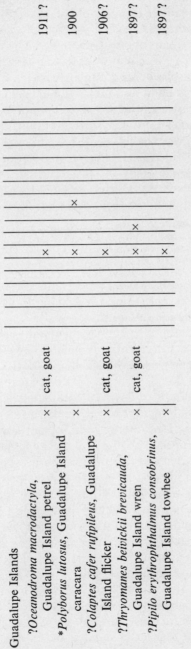

			Date
?*Oceanodroma macrodactyla*, Guadalupe Island petrel	×	cat, goat	1911?
Polyborus lutosus, Guadalupe Island caracara	×		1900
?*Colaptes cafer rufipileus*, Guadalupe Island flicker	×	cat, goat	1906?
?*Thryomanes beivickii brevicauda*, Guadalupe Island wren	×	cat, goat	1897?
?*Pipilo erythrophthalmus consobrinus*, Guadalupe Island towhee	×		1897?

* Extinct.
? Probably extinct.
Sources consulted include Allen, 1942, Greenway, 1958, and Walker et al., 1964.

An inference drawn from the records is that, no matter what initial causes reduced the numbers of a species, once they had occurred the species came under pressure from additional factors so that an adequate breeding population became increasingly difficult to maintain. This is a feature recognized many years ago by Osborn (1906, p. 859): "Following the diminution in number which may arise from a chief or original cause, various other causes conspire or are cumulative in effect. From weakening its hold upon life at one point an animal is endangered at many other points." Some of the best histories of this feature of extinction concern bird species. For example, even with the establishment of conservation practices, the heath hen was unable to survive, although for a number of years the population increased after conservation was initiated (Greenway, 1958). The whooping crane is currently in this position. The present population of forty-four could be reduced through disease, a severe winter, or many other natural causes, to a level from which it could not recover.

A second generality documented by this survey is that isolated populations, especially those in insular locations, are much more likely to have suffered a high rate of extinction.

Factors specifically attributable to the influence of man include hunting and trapping by both European colonists and the American aborigines. The destruction of habitats was accomplished through uncontrolled fires, agricultural practices, lumbering, and grazing. The introduction of predators included the population of the New World by rats, mice, and cats escaping from ships and also the deliberate introduction of predator species (i.e. the mongoose) as a part of animal-management schemes. The eradication of species considered pests was selective, emphasizing the removal of a specific animal such as the Carolina parakeet. The introduction of disease spread from domestic herds was an unintentional result of man's influence, but nonetheless detrimental. There is ample evidence that all these practices resulted in the destruction of wild animal herds and flocks during the historic period.

Factors that could have been present at the end of the Pleistocene are of interest to this study of human factors influencing animal extinction. Those appearing to be unique to the historic period include hunting and trapping for the commercial market, the introduction of predators, the introduction of diseases carried by domestic animals, and all types of habitat destruction except uncontrolled fires.

Arguments have been advanced that a primitive group would not exterminate its food supply, but no evidence to substantiate this opinion was found in the literature. On the contrary, although no individual species appears to have been exterminated by an aboriginal group in

North America, there are accounts of the complete and wasteful annihilation of individual herds, for example the narwhal and musk-ox by the Eskimo (Freuchen, 1935, p. 180–81, 212) and the buffalo by the Plains Indians (Roe, 1951, p. 334–520). This evidence, although scanty, suggests that primitive groups might occasionally kill off their local food supply.

Perhaps we may utilize these historic data by analogy to answer the question of the ability of early man to exterminate a major fauna. In nearly every case during the historic period, large herds and flocks were decimated by market hunting with commercially manufactured weapons. Although some species such as the buffalo, musk-ox, and beaver were hunted in quantity by the Indians, this hunting was the result, in part, of demands by traders for hides and meat to be shipped to markets in the eastern cities. In addition, during the period of the greatest kills, the aborigines were equipped with modern rifles and traps.

In summary, I believe that the great reduction in animal populations during the historic period was the result of the European colonists who viewed animal populations as natural resources to be exploited by an organized commercial economy. By contrast, the annual kill of animals by American Indians, estimated by Kroeber (1939, p. 143) to number one million at the time of colonization, does not seem to have endangered the herds. Our best evidence on this point is obtained from data on the buffalo. With herds estimated at thirty to forty million (Roe, 1951, p. 334–520, especially p. 518–19), the annual kill is estimated at two million, which appears to be less than the annual increase. The portion of this total killed by Indians is estimated at 405,000 per annum for the years 1872–74, including 84,000 sent to commercial markets (Roe, 1951, p. 440–41). Viewed in this perspective, it hardly seems possible that the Plains Indians would ever have exterminated the bison just to satisfy their own needs.

PREHISTORIC MAN'S ROLE IN EXTINCTION

Activities of Early Man included hunting for food (by stampede and fire drive), destruction of the habitat by fire, and differential selection of the young of the species hunted. The use of poison, documented for modern primitive peoples hunting whales (Heizer, 1943) and also for the Bushmen, is more problematical. A tabulation by species of the animals killed by Early Man, found at selected archaeological sites in the Great Plains and Rocky Mountain states (Table 2), indicates the nature of his hunting practices.

The staple food during the Llano period consisted of the mammoth; during later periods it was the bison. All other animals are poorly

TABLE 2. *Inventory of Extinct Animals Present in Selected Early Man Sites*

Selected Sites	4-Horned antelope	Bison	Camel	Horse	Mammoth	Tapir	Peccary
Llano Horizon							
Burnet Cave, N.M.*	21	16	1	62			
Clovis, N.M.		5	×	4†	12†		1†
Dent, Colo.					12		
Domebo, Okla.					2		
Lehner, Ariz.		1		1	9	1	
McLean, Tex.					1		
Miami, Tex.					5		
Naco, Ariz.		1			1		
Folsom Horizon							
Clovis, N.M.		10 + †					
Folsom, N.M.		23					
Lindenmeier, Colo.		9	×				
Linger, Colo.		5					
Lipscomb, Tex.		14					
Lubbock, Tex.		×					
MacHaffie, Mont.		×					
Midland, Tex.	1			×			
San Jon, N.M.		3†					
Zapata, Colo.		5					
Parallel Flaked Horizon							
Agate Basin, Wyo.		×					
Allen, Wyo.		15					
Clovis, N.M.		100†					
Custer Co., Neb.		1					
Finley, Wyo.		×					
Horner, Wyo.		200					
Lime Creek, Neb.		×					
Lone Wolf Creek, Tex.		1					
MacHaffie, Mont.		×					
Meserve, Neb.		3					
Milnesand, N.M.		27					
Olsen-Chubbock, Colo.		65					
Plainview, Tex.		100					
Red Smoke, Neb.		×					
San Jon, N.M.		20†					
Scottsbluff, Neb.		30					

* Also present at Burnet Cave: musk-ox, 9; *Arctodus*, 4; cervid, 9; caribou, 2.
† Personal estimate based on bones present at site today.
× Unknown number present.

represented. The size of the kills varies from site to site, and it is not always possible to determine if the animals present were killed at one time or on several different occasions. In addition, it is not always possible to determine if the animals present at a site were killed by man or some other agency. The mammoth sites vary in nature. The Dent and Lehner sites seem to record kills of several animals at one time, whereas the specimens at the Clovis site represent individual kills of single animals. In any event, the largest number found at any one site was twelve. Although young mammoth were killed, there seems to have been no deliberate age selection. Apparently the animals were taken at a waterhole.

The Folsom hunters selected small herds of bison, as many as twenty-three individuals (in the sites known), seemingly without preference as to age. In three separate kills at the Clovis site, the number of individuals was either two or three. In every case, the kills took place while the animals were knee deep in a pond, presumably drinking.

In the Parallel Flaked Horizon is the first evidence for stampedes caused by Early Man. At three sites—Horner (Wyoming), Olsen-Chubbock (Colorado), and Plainview (Texas)—large herds were driven across unfavorable terrain, such as arroyos or stream channels, and many animals were trampled to death (Wendorf and Hester, 1962). The uppermost animals were butchered, the remainder left untouched, our earliest evidence of extreme waste. The composition of these kills seems to indicate a normal ratio of young to adult individuals.

Evaluation of these data is difficult, as the preserved and excavated sites represent only a small proportion of the kills that must have occurred. Nonetheless, several statements can be made.

1. Of the species that became extinct, Early Man hunted only two to any great degree—mammoth and bison.

2. Prior to about 7000 B.C., all kills were of small size, usually less than a dozen individuals.

3. The stampede, with its accompanying waste, probably was not used until after 7000 B.C.; it thus developed later than the major period of extinction.

Another way to present these data is to plot comparative graphs illustrating the cumulative extinction of late-Pleistocene species with the North American glacial fluctuations and the incidence of Early Man sites (Fig. 1). The data are based by species on terminal radiocarbon dates, which must be examined for validity, proper association, and possible contamination. As a result, the dates accepted vary with the individual researcher; dates accepted for this paper are summarized in Table 3. Examination of the plots suggests that the appearance of man occurred well after extinction was underway. The similarity between the graphs for cumulative extinction and glacial retreat suggests to me that these two events are related.

The series of dates listed in Table 3 will, of course, change with additional analyses. Bone dates, often considered unreliable, form only a small portion of those utilized here. Perhaps of greater importance is the undue weight attached to the date used for the large fauna from La Brea. Future dating should outline more precisely the age of the various La Brea pits with their differing faunas.[1]

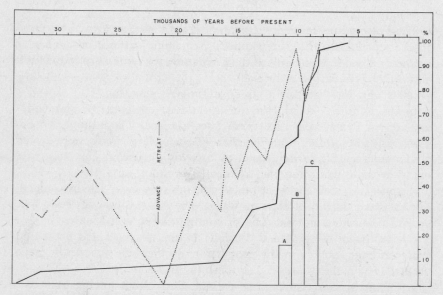

FIG. 1. Graphs comparing late-Pleistocene extinction with glacial events and populations of Early Man: Dotted line records glacial advances and retreats according to Flint (1963, Fig. 2). The solid line is the cumulative graph of North American faunal extinction; each species equals 2.5% (40 species in sample); terminal dates for each species are taken from Table 3. Bars are the percentage representation of 88 sites of Early Man from the Llano Estacado by cultural period (Wendorf and Hester, 1962). A, Llano period (16.5%); B, Folsom period (35.8%); C, Parallel Flaked period (47.7%).

An additional problem concerns the proper terminal date for mastodon. The date in Table 3 and in Figure 1 is from Russel Farm, Michigan, on tusk: 5,950 ± 300 years B.P. This would seem to some ▸

1. Dates received too late to incorporate in Figure 1 are 10,000±175 B.P. for *Platygonus* sp. from Zone II, Levi Rock Shelter, Texas, and 10,900±190 on bone for *Symbos* sp. from Cave Without a Name, Texas (Lundelius, 1967, Table I).

TABLE 3. Terminal Radiocarbon Dates for Extinct Species Utilized in Figure 1

Species	Years B.P.	Source, Reference, and Material Dated
Arctodus sp.	11,500	Burnet Cave, N.M.; age based on association with Clovis point
Bison alleni	31,000	Wilson Ford, Kans., M-997 (Flint et al., 1962, p. 186); shell
B. antiquus	8,274	Allen Site, Neb., C-108a (Hester, 1960, Table 5); charcoal
B. crassicornis	16,400	Fairbanks, Alaska, M-38 (Hester, 1960, Table 5); horn sheaths; 11,950 ± 135 (St 1633) obtained on fur and hide from a "super-bison" from Fairbanks Creek, Alaska (Péwé, 1965, p. 33), presumably *B. crassicornis*, but reference does not make assignment to species absolutely clear
B. latifrons	32,000	American Falls Reservoir, Idaho, W-358 (Hester, 1960, Table 5); charcoal
B. occidentalis	7,350	North Saskatchewan River, S-107 (Flint et al., 1962, p. 74); wood
Boreostracon sp.	23,000	Sims Bayou, Texas (Slaughter, personal communication)
Breameryx minor	11,170	Blackwater Draw, N.M., A-481 (Wendorf and Hester, n.d.); carbonized plant remains
Camelops sp.	8,240	Whitewater Draw, Ariz., A-184c (Flint et al., 1962, p. 244); charcoal
Canis dirus	8,527	Gypsum Cave, Nev., C-222 (Hester, 1960, Table 5); sloth dung
C. furlongi	13,890	Rancho La Brea, Calif., Y-354b (Hester, 1960, Table 5); log
C. milleri	13,890	Rancho La Brea, Calif., Y-354b (Hester, 1960, Table 5); log
Capromeryx sp.	11,000	Scharbauer Site, correlation A (Wendorf and Krieger, 1959)
Castoroides sp.	9,550	Ben Franklin local fauna, Tex., SM-532 Slaughter and Hoover, 1963); charcoal
Dasypus bellus	9,550	Ben Franklin local fauna, Tex., SM-532 (Slaughter and Hoover, 1963); charcoal
Equus sp.	8,240	Whitewater Draw, Ariz., A-184c (Flint et al., 1962, p. 244); charcoal
Euceratherium collinum	11,500	Burnet Cave, N.M.; age based on association with Clovis point
Felis bituminosa	13,890	Rancho La Brea, Calif., Y-354b (Hester, 1960, Table 5); log
F. daggetti	13,890	Rancho La Brea, Calif., Y-354b (Hester, 1960, Table 5); log
Holmesina septentrionalis	16,000	Ben Franklin local fauna, Tex. (Slaughter, personal communication)

TABLE 3. (continued)

Species	Years B.P.	Source, Reference, and Material Dated
Mammut americanum	5,950	Russell Farm, Mich., -M-347 (Hester, 1960, Table 5); tusk
Mammuthus columbi	8,240	Whitewater Draw, Ariz., A-184c (Flint et al., 1962, p. 244); charcoal
M. exilis	11,800	Arlington Canyon, Santa Rosa Island, Calif., UCLA-106 (Flint et al., 1962, p. 110); charcoal
M. imperator	11,003	Santa Isabel Iztapan, Mex., C-205 (Hester, 1960, Table 5); peat
M. primigenius	10,200	Kassler Quadrangle, Colo., W-401 (Hester, 1960, Table 5); molar
Megalonyx sp.	9,400	Evansville, Ind., W-418 (Hester, 1960, Table 5); wood
Mylohyus sp.	9,550	Ben Franklin local fauna, Tex., SM-532 (Slaughter and Hoover, 1963); charcoal
Nothrotherium shastense	8,527	Gypsum Cave, Nev., C-222 (Hester, 1960, Table 5); sloth dung
Oreamnos harringtoni	10,050	Rampart Cave, Ariz., L-473a (Hester, 1960, Table 5); sloth dung
Ovibos sp.	9,700	Pictograph Claim, S.D., W-223 (Hester, 1960, Table 5); bone
Panthera atrox	13,890	Rancho La Brea, Calif., Y-354b (Hester, 1960, Table 5); log
Paramylodon sp.	13,890	Rancho La Brea, Calif., Y-354b (Hester, 1960, Table 5); log
Platygonus sp.	13,890	Rancho La Brea, Calif., Y-354b (Hester, 1960, Table 5); log
Preptoceros sinclairi	11,500	Burnet Cave, N.M.; age based on association with Clovis point
Rangifer fricki	11,500	Burnet Cave, N.M.; age based on association with Clovis point
Sangamona sp.	11,500	Burnet Cave, N.M.; age based on association with Clovis point
Smilodon californicus	13,890	Rancho La Brea, Calif., Y-354b (Hester, 1960, Table 5); log
Stockoceros onusrosagris	11,500	Burnet Cave, N.M.; age based on association with Clovis point
Symbos cavifrons	11,100	Scotts Musk-ox, Mich., M-1402 (Flint et al., 1964, p. 2); bone
Tanupolama	8,527	Gypsum Cave, Nev., C-222 (Hester, 1960, Table 5); sloth dung
Tapirus sp.	9,400	Evansville, Ind., W-418 (Hester, 1960, Table 5); wood
Ursus optimus	13,890	Rancho La Brea, Calif., Y-354b (Hester, 1960, Table 5); log

researchers to be too late, since it postdates the spruce–fir vegetational climax in Michigan, presumed to be the habitat of the mastodon (Oltz and Kapp, 1963). Skeels (1962, p. 124–25) states that this spruce–fir forest was replaced about 8,000 years ago with a mixed forest of pine, oak, and hickory, which changed to a pine climax about 6,000 years ago, indicating a warm phase. The present type of deciduous forest was established about 5,000 years ago.

If Skeels is correct in the statement that the mastodon was adapted to life in a mixed coniferous and broad-leafed forest, and that a mature coniferous forest would provide poor browse (Skeels, 1962, p. 125), the extinction of the mastodon may be correlated with the pine maximum. Such a correlation suggests that the date of 5950 B.P. in Figure 1 could be a reasonable terminal date for the mastodon. A few investigators believe the mastodon became extinct earlier. Martin (1967) claims there are no well-dated mastodons younger than 10,000 B.P. At this point the dating of terminal mastodons is still a problem to be resolved.

Another type of analysis concerns the relative numbers of the extinct species that were actually killed and eaten by Early Man. Unfortunately, these data are seldom comparable. The published information is so varied that in some cases we are comparing "combined ungulates" with individual "species." In other cases, we cannot determine whether species were introduced into the sites by *human* or other agencies. Another problem consists of inadequate stratigraphic control, so that some doubt may exist whether certain species came from layers containing human remains (for example, Burnet Cave). Finally, we must determine whether the sample represents all the bones from the site or only a selection.

If we can resolve these problems within rather broad limits we are then faced with interpreting the meaning of the data. No single diet pattern appears to have been present. In some cases, large extinct mammals seem to have been the main food (for example, Ventana Cave; Colbert, 1950). In others, such as Gypsum Cave, recent mammals outnumbered the extinct species. In Gypsum Cave the most common species was mountain sheep, followed by the desert tortoise, extinct sloth, and rabbits, with birds and reptiles poorly represented (Harrington, 1933, p. 80–81). Unique data are present at the Lime Creek site, Nebraska. Here the most common species in the lowest level, Zone 1, dated just after 7600 B.C., were pronghorn antelope and beaver, with additional bones of bison, elk, deer, jackrabbit, cottontail, coyote or dog, raccoon, prairie dog, field mole, and pocket gopher (Davis, 1962, p. 69). The overlying Zones II and III contain only bones of extinct bison. According to Davis (1962, p. 74), the excavator, these facts indicate that the game animals changed as the climate changed and that the prehistoric peoples hunted whatever was available. This point of view considers that man is conditioned by the environment rather than vice versa.

The horse population at Ventana Cave was much greater in the pre-artifactual Conglomerate level than it was in the overlying Folsom Age Volcanic Debris level (Haury, 1950, Fig. 17). This fact implies that the cause or causes reducing the horse population were effective before the advent of man in the area.

Additional evidence concerning man's influence may be obtained by comparing the number of known finds of extinct animals in North America with the incidence of demonstrated kills of these species by Early Man. For example, a total of 195 finds of proboscideans is recorded for Ohio (Forsyth, 1963), yet none of these was associated with artifacts. Hay (1923, 1924, 1927), Hibbard (1958), and Hibbard et al. (1965, Fig. 1) also present data on Pleistocene faunal remains without associated artifacts.

In summary, examination of the archaeological evidence for the reduction of animal populations by the *hunting* of early man reveals that evidence to be inadequate and ambiguous. Although man in the New World did kill species that ultimately became extinct, he also frequently relied heavily on species that survived. The absence of a single dietary standard suggests that Early Man ate whatever he could. Evidence available from Europe is much more dramatic. In some European sites, horses and mammoths were killed by the thousands, yet, surprisingly enough, the European horse survived, whereas the North American horse became extinct. The North American evidence, although limited, does not indicate that man's hunting was an important factor in the late-Pleistocene extinction.

THE IMPORTANCE OF FIRE DRIVES

Man's use of fire, either controlled or uncontrolled, is considered by some authors to have been a major factor in the alteration of the environment: "The unrestricted burning of vegetation appears to be a universal culture trait among historic primitive peoples and therefore was probably employed by our remote ancestors" (Stewart, 1956, p. 129). Other authors, including Sauer (1944) and Eiseley (1946), have considered the importance of man-made fires in mammalian extinction. "By the use of fire alone, it would seem, could such animals be caught, crippled, and made ready for destruction in number" (Sauer, 1944, p. 543). "The fire drive provides an apparently adequate explanation of the extinction of the large mammals in post-Pleistocene time" (Sauer, 1944, p. 544).

Evidence of widespread fires is rare in the archaeological record. A few sites, such as those on Santa Rosa Island and near San Diego, possess bands of charcoal that could be the remains of fires set by man, although the possibility of natural fires cannot be ruled out. This means that we cannot use archaeology to evaluate the role of fire in Pleistocene extinction. One theory proposes that frequent burning encourages the spread of a grassland. "The formation of grasslands requires a process that operates against long-lived perennials. The one widespread agency certainly known to act in this direction is recurrent burning" (Sauer, 1944, p. 551). Although this statement may well be true, the corollary does not necessarily follow, that an increase in grassland resulted in widespread extinction. The record of species that became extinct hardly supports this point of view. Many of the species that became extinct, such as the horse, four-horned antelope, extinct bison, and possibly the Columbian mammoth, were grazing herbivores that should have profited by an expansion of their habitat. Historic evidence of the effect of burning indicates the animals most affected were those that relied on long-lived perennial plants. For example, the great decimation of caribou and reindeer in the 1940s in Alaska can be traced to the burning of the lichen ranges. Lichen is a plant that forms long-enduring mats of growth, which, if not consumed one year, may be eaten whenever the migratory herbivores visit the range at a later date. Burning the lichen is disastrous to these herds, as the range needs several decades to recover. By contrast, burning, according to some authorities, improves the cover on grasslands.

It appears that, although we cannot assess the effect of fire drives or uncontrolled human-set fires on extinction in quantitative terms, we may assume a qualitative difference; i.e. uncontrolled fires should have been detrimental to some species that became extinct and favorable to other species that also became extinct. If this assumption is valid, we then may infer that other factors contributing to extinction were also at work.

The following hypothesis may be examined: A human population estimated at a few thousand individuals, with hunting methods including the stampede and fire drive, was a serious factor influencing the late-Pleistocene extinction of a widespread fauna of more than forty species having an aggregate total of several million individuals. Data derived from my historical study concerning this hypothesis are:

1. The human population of North America during the historic period was perhaps a thousand times greater than the population during the terminal Pleistocene.
2. The great kills were primarily a result of market hunting.

3. Even market hunting took more than a hundred years to exterminate some species.

4. Destruction of habitat, resulting from the introduction of intensive agriculture, was extremely detrimental to animal populations.

5. The introduction of predators and disease by man were also important causes of herd reduction.

The contrast between the historic evidence and that available from the terminal Pleistocene leads me to infer that early man lacked the numbers and means to threaten the herds of the Pleistocene fauna. Only in a time of extreme ecologic stress could the presence of human predators have been of importance.

According to Martin (1958, p. 413), four factors are unique in the record of the late-Pleistocene extinction. "(1) Differential loss of large animals, (2) lack of evidence of major climatic change during the extinction period, (3) the narrow chronological range in which extinction occurred, and (4) the phenomenon of removal without replacement."

Since Martin's report in 1958, a great deal of information has been accumulated concerning the nature of late-Pleistocene vegetational communities. To cite only one example, the pollen record from the southern high plains of Texas and New Mexico clearly indicates that a major change in regimen occurred at the end of the Pleistocene. Before this change, these plains featured gallery forests of pine and spruce trees along running streams or adjacent to ponds with intervening areas of grassland and sagebrush. After the Valders advance (the last major glacial episode) dated ca. 9000 B.C., the region became a short-grass prairie without surface water (Haynes, n.d., Fig. 7). Pollen diagrams indicate that this transition occurred very rapidly (Wendorf and Hester, n.d.). Therefore, in relation to the unique factors cited by Martin, we have evidence that:

1. A major climatic change did occur.

2. This climatic change occurred rapidly and contemporaneously with the demise of the Pleistocene species (Fig. 1).

3. In a change of the type recorded for the southern high plains, a major resulting feature was less total vegetal cover and the disappearance of surface water. Both of these were environmental alterations that could have been selective factors against large mammals requiring large quantities of food and water. "The mastodon, for example, needing great quantities of herbage for its food supply, might, in cases of severe drought, succumb to the

food competition of the rabbit, or some still more insignificant
creature, which, spreading in vast numbers over the country,
devoured the sparse herbage and left its huge competitor to
starve" (Morris, 1895, p. 254).

Instead of indicting man as the major factor influencing late-Pleistocene animal extinctions, I suggest as an alternative explanation the
following hypothesis: The late-Pleistocene North American megafauna
was well adapted to ecological conditions present in the periglacial zone
and adjacent vegetational zones south of the late-Wisconsin ice sheet.
The progressive retreat of this ice, beginning 18,000 to 20,000 years ago,
resulted in widespread vegetational changes in these ecological niches—
changes that in themselves brought selective pressure to bear on their
animal populations. The halts and minor readvances of the ice-sheet
border between 18,000 and 11,000 years B.P. were of insufficient magnitude to restore the ecologic balance and served only to slow temporarily
the rate of progressive extinction of species under stress.

In conclusion, we may observe that, owing to the great number of
variables influencing the relative proportions of the members of any fauna
through time, our hypothesis is far from specific. In fact, it is little more
specific than the hypothesis advanced over a century ago by Charles
Lyell (1863, p. 374):

> It is probable that causes more general and powerful than the
> agency of Man, alterations in climate, variations in the range of
> many species of animals, vertebrate and invertebrate, and of plants,
> geographical changes in the height, depth, and extent of land and
> sea, or all of these combined, have given rise, in a vast series of
> years, to the annihilation, ... of many large mammalia.

I wish to express my appreciation to the following for their constructive
criticism of this paper: James Lee, Richard Manville, Paul S. Martin, Herbert
E. Wright, Jr., Fred Wendorf, and the Editorial Board of the National
Institutes of Health.

References

Allen, G. M., 1942, Extinct and vanishing mammals of the western hemi-
sphere: *New York, Amer. Comm. Int. Wild Life Protection, Spec. Publ. 11,*
620 p.
Auffenberg, W., and Milstead, W. W., 1965, Reptiles in the Quaternary of
North America, p. 557–68, *in* Wright, H. E., Jr., and Frey, D. G., *Editors,*
The Quaternary of the United States: Princeton Univ. Press, 922 p.

Blair, W. F., 1965, Amphibian speciation, p. 543–56, *in* Wright, H. E., Jr., and Frey, D. G., *Editors, The Quaternary of the United States:* Princeton Univ. Press, 922 p.

Colbert, E. H., 1942, The association of man with extinct mammals in the western hemisphere: *8th Amer. Sci. Cong. Anthropol. Sci. Proc.*, v. *11*, p. 17–29

———— 1950, The fossil vertebrates, p. 126–48, *in* Haury, E. W., and collaborators, *The stratigraphy and archaeology of Ventana Cave, Arizona:* Tucson, Univ. Arizona Press, 599 p.

Cope, E. D., 1899, Vertebrate remains from the Port Kennedy bone deposit: *Acad. Nat. Sci. Philadelphia Jour.*, v. *11*, p. 193–267

Davis, E. M., 1962, Archaeology of the Lime Creek site in southwestern Nebraska: *Univ. Nebraska Spec. Publ. 3*, 106 p.

Eiseley, L. C., 1946, The fire drive and the extinction of the terminal Pleistocene fauna: *Amer. Anthropol.*, v. *48*, p. 54–59

Flint, R. F., 1963, Status of the Pleistocene Wisconsin Stage in central North America: *Science*, v. *139*, p. 402–04

Flint, R. F., Deevey, E. S., and Rouse, I., *Editors*, 1962, *Radiocarbon*, v. *4; 1964*, v. *6*

Forsyth, J. L., 1963, Ice Age census: *Ohio Conservation Bull.*, v. *27*, p. 16–19, 31, 33

Freuchen, P., 1935, *Arctic adventures—My life in the frozen north:* New York, Farrar and Rinehart, 467 p.

Greenway, J. C., 1958, Extinct and vanishing birds of the world: *New York, Amer. Comm. Int. Wild Life Protection, Spec. Publ. 13*, 518 p.

Harrington, M. R., 1933, Gypsum Cave, Nevada: *Southwest Museum Papers*, no. 8, 197 p.

Haury, E. W., 1950, *The stratigraphy and archaeology of Ventana Cave, Arizona:* Tucson, Univ. Arizona Press, 599 p.

Hay, O. P., 1923, 1924, 1927, The Pleistocene of North America and its vertebrated animals . . .: *Carnegie Inst. Publ. 322, 322A, 322B (3 vols.)*

Haynes, C. V., n.d., Pleistocene and recent stratigraphy of Blackwater Draw, New Mexico and Rich Lake, Texas, *in* Wendorf, D. F., and Hester, J. J., *Assemblers, Paleoecology of the Llano Estacado*, v. *2:* Santa Fe, Fort Burgwin Res. Center

Heizer, R. F., 1943, Aconite poison whaling in Asia and America, an Aleutian transfer to the New World: *Bur. Amer. Ethnol., Anthropol. Papers*, *24*, p. 415–68

Hester, J. J., 1960, Late Pleistocene extinction and radiocarbon dating: *Amer. Antiq.*, v. *26*, p. 58–77

Hibbard, C. W., 1958, Summary of North American Pleistocene mammalian faunas: *Michigan Acad. Sci. Papers*, v. *43*, p. 3–32

Hibbard, C. W., Ray, D. E., Savage, D. E., Taylor, D. W., and Guilday, J. E., 1965, Quaternary mammals of North America, p. 509–26, *in* Wright, H. E., Jr., and Frey, D. G., *Editors, The Quaternary of the United States:* Princeton Univ. Press, 922 p.

Hornaday, W. T., 1887, The extermination of the American bison: *U.S. Nat. Museum Ann. Rept.*, p. 367–548

——— 1913, *Our vanishing wild life; Its extermination and preservation:* New York, Scribner's, 411 p.

Jelinek, A. J., 1957, Pleistocene faunas and early man: *Michigan Acad. Sci. Papers*, v. *42*, p. 225–37

Kroeber, A. L., 1939, Cultural and natural areas of native North America: *Univ. California Publ. Amer. Archeol. Ethnol.*, no. *242*, 38 p.

Lundelius, E. L., Jr., 1967, Late-Pleistocene and Holocene faunal history of central Texas (this volume)

Lyell, C., 1863, *Geological evidences of the antiquity of man*, 2nd ed., rev.: London, Murray, 528 p.

Martin, P. S., 1958, Pleistocene ecology and biogeography of North America, p. 375–420, *in* Hubbs, C. L., *Editor, Zoogeography: Amer. Assoc. Adv. Sci.*, Publ. *51*, 509 p.

Miller, R. R., 1965, Quaternary freshwater fishes of North America, p. 569–81, *in* Wright, H. E., Jr., and Frey, D. G., *Editors, The Quaternary of the United States:* Princeton Univ. Press, 922 p.

Morris, C., 1895, The extinction of species: *Acad. Nat. Sci. Philadelphia, 3rd ser., Proc.* (1895), p. 253–63

Newell, N. D., 1963, Crises in the history of life: *Sci. Amer.*, v. *208*, p. 77–92

Oltz, D. F., Jr., and Kapp, R. O., 1963, Plant remains associated with mastodon and mammoth remains in central Michigan: *Amer. Midland Naturalist*, v. *70*, p. 339–46

Osborn, H. F., 1906, The causes of extinction of mammalia: *Amer. Naturalist*, v. *40*, p. 769–95, 829–59

Peters, J. A., 1950, Extinction; Its causes and results: *Biologist*, v. *32*, p. 4–8

Péwé, T. L., 1965, Guidebook for field conference F, Central and South Central Alaska: *VII INQUA Cong.*, 141 p.

Roe, F. G., 1951, *The North American buffalo; A critical study of the species in its wild state:* Univ. Toronto Press, 957 p.

Romer, A. S., 1933, Pleistocene vertebrates and their bearing on the problem of human antiquity in North America, p. 47–83, *in* Jenness, D., *Editor, The American aborigines, their origin and antiquity:* Univ. Toronto Press, 396 p.

Sauer, C. O., 1944, A geographical sketch of early man in America: *Geog. Rev.*, v. *34*, p. 529–73

Sellards, E. H., 1952, *Early man in America:* Austin, Univ. Texas Press, 211 p.

Skeels, M. A., 1962, The mastodons and mammoths of Michigan: *Michigan Acad. Sci. Papers*, v. *47*, p. 101–33

Slaughter, B. H., n.d., An ecological interpretation of the Brown Sand Wedge local fauna, Blackwater Draw, New Mexico; And a hypothesis concerning late Pleistocene extinction, *in* Wendorf, F., and Hester, J. J., *Assemblers, Paleoecology of the Llano Estacado*, v. *2*: Santa Fe, Fort Burgwin Res. Center

Slaughter, B. H., and Hoover, B. R., 1963, Sulphur River formation and the Pleistocene mammals of the Ben Franklin local fauna: *Southern Methodist Univ., Grad. Res. Cent., Jour.*, v. *31*, p. 132–48

Stewart, O. C., 1956, Fire as the first great force employed by man, p. 115–33, *in* Thomas, W. L., *Editor, Man's role in changing the face of the earth:* Univ. Chicago Press, 1193 p.

Stock, C., 1929, A census of the Pleistocene mammals of Rancho La Brea, based on the collections of the Los Angeles Museum: *Jour. Mamm.,* v. *10,* p. 281–89

Tolmachoff, I. P., 1930, Extinction and extermination: *Smithson. Inst. Ann. Rept. for 1929,* p. 269–84

Walker, E. P., Warnick, F., Lange, K., Uible, H., Hamlet, S., Davis, M., and Wright, P., 1964, *Mammals of the world:* Baltimore, Johns Hopkins Press, 3 vols., 2269 p.

Wendorf, D. F., and Hester, J. J., 1962, Early man's utilization of the Great Plains environment: *Amer. Antiq.,* v. *28,* p. 159–71

———— n.d., *Assemblers, Paleoecology of the Llano Estacado,* v. *2,* Santa Fe, Fort Burgwin Res. Center

Wendorf, D. F., and Krieger, A. D., 1959, New light on the Midland discovery: *Amer. Antiq.,* v. *25,* p. 66–78

Wormington, H. M., 1957, *Ancient man in North America:* Denver, Denver Museum Nat. Hist., 322 p.

ARTHUR J. JELINEK

Department of Anthropology
University of Arizona
Tucson, Arizona

MAN'S ROLE IN THE EXTINCTION
OF PLEISTOCENE FAUNAS

Abstract

The extinction of the Pleistocene megafauna appears to have been due to a variety of causes. The lack of replacement of most forms suggests that unusual patterns of selection were responsible for the large-scale disappearance of these animals. In the periglacial areas of the Old World the lack of plant resources stimulated the development of a human technology oriented almost exclusively to animal resources. It appears likely that toward the end of the glaciation many elements of the megafauna in this area may have disappeared as a result of hunting practices in increasingly constricted habitats. In Africa, however, less severe climatic restrictions and abundant sources of vegetal foods for human populations may have aided the survival of the megafauna. A totally different situation may have prevailed in the New World, where the megafauna had evolved independent of human contact. Here the intrusion of man, as a new and formidable predator, into a natural community unprepared for this stress may have been basic to the extinction of many genera of the Pleistocene megafauna. A major obstacle, however, to attributing a key role to man in the extinction of the New World megafauna is the lack of direct evidence of human association with many of the extinct genera.

> We may infer from these facts, what havoc the introduction of a new beast of prey must cause in a country, before the instincts of the indigenous inhabitants have become adapted to the stranger's craft or power [Darwin, *Voyage of the Beagle*, 1845, p. 386].

The particular nature of the large-scale extinction of the Pleistocene Nearctic megafauna continues to present several puzzling aspects. The problem, in brief, is this: Why, after surviving a succession of four or more major glacial advances, do most elements of the large mammalian faunas of the Northern Hemisphere disappear during the retreat of the last glaciation? In the New World alone this extinction encompasses at least twenty-nine genera (Hibbard, 1958; Hibbard et al., 1965) including probiscideans, edentates, ungulates, and carnivores. In contrast, small mammals (in continental populations) are conspicuous by their apparent continuity, with only a few species known from the late Pleistocene that did not survive into the Holocene.

A further puzzling factor accompanying this extinction is the lack of ecological replacement of most forms (Martin, 1958, p. 402), indicating that biological competition was probably not the reason for their disappearance. Recent pollen evidence from western America (Kapp, 1965) seems to indicate that in at least some areas occupied by the extinct fauna the conditions following the retreat of an earlier glaciation (Illinoian) were probably more arid and as warm or warmer than at present. Thus conditions of temperature and aridity do not appear likely as direct causes of extinction.

The few areas of the world in which this late-Pleistocene extinction is not present include most of Africa, especially sub-Saharan Africa, and portions of southern Asia. In fact, in discussions of a possible human cause of the extinction, Africa is frequently cited as negative evidence: the long cultural record of peoples known to hunt large animals, coexistent with the abundant megafauna, seems to indicate that human effects on the fauna should have been seen there, if anywhere. The lack of such effects is taken as evidence that the presence of humans where extinction occurred is not sufficient cause to postulate their major role in the extinction. I believe that this argument for the elimination of man as a factor can be shown to be invalid on cultural and paleobotanical grounds, and that for these reasons Africa is not comparable to the areas of the Northern Hemisphere in which extinction occurred. The relatively even topography of much of Africa meant that, whereas broad areas of vegetational zones shifted with varying temperature and precipitation, the dramatic climatic pressures on the megafauna found in the Northern Hemisphere probably did not exist there, and the faunal elements followed these shifts more or less undisturbed. The relatively lush flora, especially of the pluvial periods and throughout the sequence, as compared to Europe and northern Asia, provided an important food source for African hominids in the course of their cultural development. The continuing importance of these food sources into the historic period is confirmed by ethnological accounts of food-collecting groups. In contrast, Europe during the Würm glaciation provided few resources other than meat for man north of the Mediterranean. As a result, with the appearance of *Homo sapiens* in Europe, perhaps 40,000 years ago, we see the development of a series of cultures (the Upper Paleolithic) overwhelmingly oriented to a carnivorous subsistence pattern. The technological complexity of these cultures for hunting and processing animals had no equal at that time, and in many ways it compares favorably with the native technology of the modern Eskimo, the only surviving culture with a carnivorous orientation as extreme as the Eurasian Upper Paleolithic.

The efficiency of these hunters is attested by such sites as Solutré in east-central France, where a late-Perigordian level is estimated to contain the remains of over 100,000 horses (MacCurdy, 1933, p. 173). The restricted orientation of subsistence activities is evinced by a concentration on animal representation in the art forms of these cultures to the exclusion of virtually all other naturalistic motifs (with the exception of women). The effectiveness of this cultural adjustment is attested by the relatively dense population of Europe by the time of the terminal Upper Paleolithic Magdalenian Culture (Sonneville-Bordes, 1963, p. 354), a density unparalleled in any earlier period. Certainly in Europe the role of man in the depletion of the Pleistocene fauna as it followed the retreating ice merits serious consideration.

This same adaptation to an Upper Paleolithic way of life spread across northern Asia and made it possible for *Homo sapiens* to cross the tundra of the Bering land bridge into the New World (Jelinek, 1965). On the basis of currently acceptable dated cultural evidence, this entry may have taken place as late as 15,000–12,000 B.C. (see also Haynes, 1967).

The first men to enter the New World were most certainly hunters, because no likely food sources but game existed for most of the year in the tundra areas they traversed. It seems most likely that they were of the same Upper Paleolithic cultural tradition as the very proficient and highly meat-oriented cultures of the rest of northern Eurasia. They were entering a continent where, in contrast to virtually all of the Old World north of Australia, the large herbivorous fauna *had evolved removed from human contact*. Is it not reasonable to expect, then, that the initial presence of man among the New World faunas caused no more fear among these animals than the initial presence of man among the Antarctic fauna? Is it not also probable that throughout man's initial expansion over the New World, the fauna was exceedingly vulnerable to human attack? To go one step farther, does this difference in cultural history explain the survival of species of horses, elephants, large pigs, and large camels in the Old World, groups totally lost in the New?

Charles Darwin, who was very much interested in extinction as an inevitable concomitant of evolution, was deeply impressed by the evidence of the extinction of the large fauna of the New World. In his journal of *The Voyage of the Beagle* he gives striking accounts of the tameness of

the animals which the expedition found in isolated areas such as the
Falklands and the Galápagos. He describes a fox that could be coaxed
into range of a knife held in the hand, or birds that could be killed with
a hat or walking stick. These same animals were skilled in avoiding their
natural predators, but showed only curiosity at the presence of man.
It is of interest to note that a species of goose was very easily captured in
the Falklands, where man was absent, but was as elusive in Tierra del
Fuego as the wild goose of England (Darwin, 1845, p.383–86). How long
does it take, through natural selection, for an animal species to acquire
the ability to flee or seek concealment at the sight or scent of man?

Given this unique relationship with native resources and his carnivorous cultural heritage, as well as a total lack of demographic or
serious climate barriers (with a dependence on the ubiquitous large
animals), it is not strange to find man at the southern tip of South
America well before 8000 B.C. (Rubin and Berthold, 1961, p. 96), four
to seven thousand years after his entry via the Bering Strait. All these
factors, in conjunction with known data on the spread of successfully
introduced mammalian species through an available habitat (such as
European *Equus* in the New World and *Lepus* in Australia), even if
differences in generational span and locomotive patterns, are considered,
suggest that if the date of 8760 B.C. ± 300 from Fells Cave on the
Strait of Magellan really represents the earliest arrival of man at the
tip of South America, his entry into Alaska may well have been closer
to 12,000 than 15,000 B.C.

The nature of the hunting implements of the earliest well-defined
New World cultures seems also to favor the hypothesis of a vulnerable
fauna. Most characteristic of these horizons are stone points which
show marked grinding of the sharp lateral edges adjacent to the base.
Grinding would prevent the edges of the point from cutting the lashing
that bound it to a shaft if the point was subjected to repeated lateral
stress. The most likely circumstance in which such stress could occur
would be in a point on a thrusting spear or lance whose shaft remained
in the hand of the hunter after it penetrated the animal—a technique
that would be most effective against a relatively easy quarry and of
little use against a skittish and fearful prey. The problem of the extinction
of the larger elements of this fauna, however, cannot be explained,
except perhaps in part, by this initial spread of man through the New
World.

Man's entry coincides with a period of climatic stress accompanying
the retreat of the glacial ice. By about 10,000 B.C. the cooler flora of the
glacial period was noticeably on the wane in most of North America
south of Canada, being replaced by plant communities similar to the

present potential vegetation. In several areas of arid western America we have evidence of large Pleistocene animals surviving into this period and existing on a diet of vegetation essentially similar to that in the area at present (Martin et al., 1961). This fact has been offered as evidence that the conditions of natural habitat in the West at 8,000 to 10,000 B.C. represent a favorable environmental setting for such forms as *Nothrotherium*† (Martin et al., 1961), when in fact they may represent the barely tolerable conditions under which a few survivors were able to continue to exist as remnants of an originally larger population. A similar claim can be made for mammoth and other large forms surviving as relict populations in the vicinity of the few remaining sources of water and relatively lush vegetation. The introduction of man into such an assemblage of animals not yet adapted to his presence might well have resulted in the extermination of local populations of animals.

It should also be stressed that it is difficult to anticipate the way the presence of humans might have interfered with the normal life cycle of animals whose evolution had proceeded independently of such contact. The basic necessity for water, and man's presence at water sources, for instance, might well have created considerable difficulties for western faunas. While these animals had certainly been adjusted to the activities of carnivores, man's patterns of activity were quite different. The use of fire could have had drastic effects, especially in grassland areas. The narrative of Cabeza de Vaca, the first account of the southern and southwestern Indians, relates that peoples of what is now southwest Texas were in the habit of burning favorable areas of range, not to drive the game, but to deprive the animals of forage and force them into areas where they could be hunted more successfully (Covey, 1961, p. 81). Use of fire in this manner contrasts sharply with the effects of natural random firing of the range encountered by the game prior to the arrival of man; it is of interest to speculate how early such a practice might have been followed by New World hunters, and in what areas. This kind of burning could have had a significant effect on the populations of large herbivores in these areas.

Such explanations are not so useful in accounting for extinction in eastern North America in this period. The wide range of many eastern forms, such as mastodon, mammoth, and the peccary *Mylohyus*†, as well as the relatively abundant vegetation and water in this area, seems to preclude any climatic explanation for their disappearance. It is probable that within the great breadth of the natural habitat of these forms there occurred some zone in which the effects of Slaughter's

† Extinct genus.

proposed heat sterility and cold infanticide would not have been effective (Slaughter, 1966, 1967).

Throughout the New World one major puzzle exists with regard to linking man with the extinction. This is the absence of direct evidence of human activity associated with the remains of extinct animals. In fact, we have kill sites with implements in association with partially articulated skeletons for only one of the many genera that disappeared in western North America and Mexico—*Mammuthus*†. Several extinct species of bison are known in the same context; however, here the genus was successful in survival. Several extinct genera are linked to human activity on somewhat less secure grounds. These include *Equus*, *Nothrotherium*, *Paramylodon*, *Camelops*, *Tetrameryx*, and *Breameryx*, of which teeth or isolated bones have been reported from deposits containing cultural materials. Inasmuch as this type of vertebrate material has been shown to be derived from an earlier context in at least one instance (Jelinek and Fitting, 1963, p. 534) and exists in some abundance in earlier deposits near other postulated contexts of association, I feel that a demonstration of man's utilization of these forms is still to be made.

In this connection it may be helpful to consider the problems of preservation. Through much of western North America, deposits of the approximate age of the extinction are widely scattered and are frequently confined to a restricted set of environmental origins. The nature of the geologic record in most areas precludes the possibility of preservation of sediments and fossils from all but the lowest topographic levels of any particular age. Thus most of the events of this period are lost, and those recorded are only a very small selection, in deposits containing fossils that were fortuitously covered shortly after deposition and have recently been exposed by erosion or excavation. One area favorable for the preservation of relatively large numbers of animals is the northeastern United States, where an abundance of bogs in glaciated terrain serves as a repository for the remains of numerous mastodons and lesser numbers of mammoths, musk-oxen, and other forms. Although over 160 finds of *Mammut*† have been reported from Michigan alone (Skeels, 1962), not one specimen from Michigan or any other area of the eastern United States has yet been demonstrated in clear association with human artifacts or evidence of human alteration. The contemporaneity of man and these animals seems assured, however (Griffin, 1965, p. 657), and the most likely explanation for the lack of association of artifacts would seem to be that in this area man deliberately ignored the megafauna in favor of smaller, less dangerous, and perhaps more palatable animals. The only identifiable bone fragment from a site approximating this age in the East is from southern Michigan and assigned to *Rangifer arcticus*, the barren-ground caribou (Cleland, 1965). While the extinction of the larger forms in the east is puzzling, the extinction of *Mylohyus*†, a woodland peccary of smaller size than the surviving Cervidae, is even more so. Its ubiquitous distribution and probably omnivorous diet would also seem to have favored survival. It is possible that its habitat and diet coincided too closely with that of man.

In summary, while recent evidence indicates that climate was undoubtedly an important element in the extinction of the Nearctic Pleistocene megafauna, such factors as the survival of these genera under generally comparable conditions in earlier periods suggests that climate alone cannot be responsible. There is still no conclusive evidence that seasonal temperature fluctuations and annual temperature averages following the last glaciation were unlike those of earlier interglacials, and it is apparent that not all accessible habitats were made untenable by the factors that Slaughter suggests. The different times of extinction of the same species in different areas (Jelinek, 1957, p. 232) can be cited against any possible invoked catastrophe. Barring as yet undiscovered evidence of general pathologies, except in the case of the Old World Ursus arctos spelaeus *(Kurtén, 1958),* Homo sapiens *remains as a new element in the environment, with a formidable potential for disruption, whether directly as an extremely efficient and rapidly expanding predator group, against whom no evolved defense systems were available, or indirectly as the source of profound changes in ecology already under the process of adjustment as a result of considerable climatic stress.*

References

Benninghoff, W. S., and Hibbard, C. W., 1961, Fossil pollen associated with a late-glacial woodland musk ox in Michigan: *Michigan Acad. Sci. Pap.*, v. *46*, p. 155–59

Cleland, C. E., 1965, Barren ground caribou (*Rangifer arcticus*) from an early man site in Southeastern Michigan: *Amer. Antiq.*, v. *30*, p. 350–51

Covey, C., *Editor*, 1961, *Cabeza de Vaca's adventures in the unkown interior of America:* New York, Collier, 152 p.

Darwin, Charles, 1845, *The Voyage of the Beagle:* London, Everyman ed., 1906, 496 p.

Griffin, J. B., 1965, Late Quaternary prehistory in the northeastern woodland p. 655–67, *in* Wright, H. E., Jr., and Frey, D. G., *Editors, The Quaternary of the United States*: Princeton Univ. Press, 922 p.

Haynes, C. V., Jr., 1967, Carbon-14 dates and Early Man in the New World (this volume)

Hibbard, C. W., 1958, Summary of North American Pleistocene mammalian local faunas: *Michigan Acad. Sci. Pap.*, v. *43*, p. 3–32

Hibbard, C. W., Ray, C. E., Savage, D. E., Taylor, D. W., and Guilday, J. E., 1965, Quaternary mammals of North America, p. 509–25, *in* Wright, H. E., Jr., and Frey, D. G., *Editors, The Quaternary of the United States:* Princeton Univ. Press, 922 p.

Jelinek, A. J., 1957, Pleistocene faunas and early man: *Michigan Acad. Sci. Pap.*, v. *42*, p. 225–37

——— 1965, The Upper Paleolithic revolution and the peopling of the New World: *Michigan Archaeologist*, v. *11*, p. 85–88

Jelinek, A. J., and Fitting, J. E., 1963, Some studies of natural radioactivity in archaeological and paleontological materials: *Michigan Acad. Sci. Pap.*, v. *48*, p. 531–40

Kapp, R. O., 1965, Illinoian and Sangamon vegetation in southwestern Kansas and adjacent Oklahoma: *Univ. Michigan Mus. Paleont. Contrib.*, v. *19*, p. 167–225

Kurtén, B., 1958, Life and death of the Pleistocene cave bear: *Acta Zool. Fennica*, v. *95*, p. 1–59

MacCurdy, G. G., 1933, *Human origins*, v. *1:* New York, Appleton–Century, 440 p.

Martin, P. S., 1958, Pleistocene ecology and biogeography of North America, p. 375–420, *in* Hubbs, C. L., *Editor, Zoogeography:* Amer. Assoc. Adv. Sci. Publ. *51*, 510 p.

Martin, P. S., Sabels, B. E., and Shutler, D., Jr., 1961, Rampart cave coprolite and ecology of the Shasta ground sloth: *Amer. Jour. Sci.*, v. *259*, p. 102–27

Rubin, M., and Berthold, S., 1961, U.S. Geological Survey radiocarbon dates VI: *Radiocarbon*, v. *3*, p. 86–98

Semken, H. A., Miller, B. B., and Stevens, J. B., 1964, Late Wisconsin woodland musk oxen in association with pollen and invertebrates from Michigan: *Jour. Paleont.*, v. *38*, p. 823–35

Skeels, M. A., 1962, The mastodons and mammoths of Michigan: *Michigan Acad. Sci. Papers*, v. *47*, p. 101–33

Slaughter, B. H., 1966, An ecological interpretation of the Brown Sand Wedge local fauna, Blackwater Draw, New Mexico, and a hypothesis concerning late Pleistocene extinction, *in* Wendorf, F., and Hester, J. J., Assemblers, *Paleoecology of the Llano Estacado*, vol. *2:* Santa Fe, Fort Burgwin Res. Center

——— 1967, Animal ranges as a clue to late-Pleistocene extinction (this volume)

Sonneville-Bordes, D., 1963, Upper Paleolithic cultures in western Europe: *Science*, v. *142*, p. 347-55

REGIONAL ASPECTS AND CASE HISTORIES

ESTELLA B. LEOPOLD

United States Geological Survey
Denver, Colorado

LATE-CENOZOIC PATTERNS
OF PLANT EXTINCTION[1]

Abstract

Rates of plant evolution as well as of extinction are slow compared to those of higher animals. During the late Cenozoic, most species extinctions took place in the late Tertiary, and rather few are recorded for the Quaternary epoch.

Evidence presented here concerning rates of plant extinction during the late Cenozoic is based on seed floras of Europe. An analysis of data from Poland and from northwestern Europe demonstrates that extinction rates in vascular plants are very much influenced by changes of climate and are not merely a simple function of time. Rates of species extinction tend to be faster during warm, non-glacial times than during cold, glacial periods. A large number of extinctions can be linked with a cooling climate in Pliocene time, but it is not certain whether the Pleistocene extinctions occurred during warming or cooling climates.

Geographic differences in rates of plant extinctions are seen in Europe; in Western Europe the impoverishment of the British and coastal European flora took place very rapidly during the Pliocene and provided a flora much like the present by the first Quaternary interglacial; extinction rates during this episode were rapid compared to those occurring contemporaneously in Poland. Modernization of the Polish flora involving a floristic sequence similar to that of western Europe took place more gradually, occupying late-Miocene, Pliocene, and early-Pleistocene time; the Polish flora was modern in aspect by the last interglacial (Eemian). Extinction rates during this interval oscillated, but on the whole were slower than those that occurred in the Pliocene of Western Europe.

Floristic modernization earlier than that of Europe occurred in the Aral Sea area of the Russian steppes, where evidence suggests that a modern semi-arid vegetation and flora probably existed as early as Pliocene time. The semiarid nature of the climate implied by the Pliocene floras there contrasts with the humid warm-temperature Pliocene climates of Europe. The east-west gradient in climate

1. Publication authorized by the Director, U.S. Geological Survey.

and vegetation suggests that a distinctly continental climate characterized parts of interior Asia at the same time that an oceanic climate existed in Europe.

Shrinkage of geographic ranges representing regional extinctions of many temperate plant genera was characteristic in the Neogene and probably reflects the progressive restriction of temperate climates throughout the world. Generic losses of late-Oligocene and Neogene floras on a percentage basis were especially rapid and early in areas that are now arid and have continental climates with large seasonal extremes of temperature. As many of the genera lost from these areas are temperate forms, the losses suggest a development of increased continentality and decreased temperateness. Regional extirpations of plant genera in the sense used here occurred mainly in the Tertiary in the United States, but in Europe losses of several genera are recorded during the Pleistocene.

Supporting evidence for a development of continentality during the early postglacial is found from plant records of Europe and the United States. The climatic changes of the early Holocene may have destroyed peculiar late-glacial plant assemblages of the midwestern United States and Europe; counterparts of late-glacial vegetation may occur along the eastern seaboard near Labrador, and perhaps nowhere do they persist in Europe. If the late-glacial plant assemblages from Minnesota are in place, they indicate that a more oceanic climate with warm but short summers occurred in the Great Lakes area. During the early postglacial that followed, an increased continentality of climate may have occurred, and at that time the strange assemblages of late-glacial plants may have sorted themselves out and moved toward their present ranges.

Some important postglacial vegetation changes occurred earlier at arid mid-continent sites than in eastern United States. Migrations of certain genera were more extended in mid-continent than in coastal areas.

Worldwide cooling affected the general pattern of the developing Neogene (Mio-Pliocene) floras of middle latitudes; the climatically induced changes in the composition of floras, however, were apparently more extreme in some geographic regions than in others. The purposes of this paper are (1) to compare major compositional changes in some late-Cenozoic floras of different regions in the Northern Hemisphere, and (2) to describe from the plant and geological record possible effects of regional geography on climatic expression in selected areas.

The secular cooling of the late Cenozoic brought about an overall impoverishment of mid-latitude floras, as well as a replacement of extinct species by living species and of warm-temperate by cool-temperate or boreal forms. These modifications are the main earmarks of floristic modernization. The rates and kinds of change may have been mainly functions of climate, but they were probably also controlled by rates of evolution and migration.

If climatic change was a primary controlling factor, then the degree of floristic change should have followed closely the regional patterns and vacillations of climate. During a single climatic change, isotherms may be expected to have moved a short distance along flat coastal plains and a greater distance along mid-continent plains, merely because coastal isotherms in flat terrain now tend to be closer together than mid continental ones. Thus we might expect that plant migrations under the same climatic change would in general have been more extensive in midcontinental areas than in coastal flatlands.

The effects of local geography must have been superimposed on the patterns of floristic modernization caused by worldwide cooling, and local geography must have altered the expression of climate in the past much as it does now. Inferred influences of local geography, e.g. elevation and rain shadows, on the expression of late-Cenozoic climate and on vegetation have been spelled out by many able authors, including Axelrod (1966, 1959, 1950), Szafer (1961), Reid (1920), and others. However, regional summaries that compare in graphic form the compositional changes of floras in relation to geography and age have not been previously attempted to my knowledge.

In an effort to portray a broad geographic picture, I have had to gloss over local conditions, such as soil type and the existence of relict environments for individual floras. The broad picture, however, will still have meaning because of its portrayal of climate and major geographic features as the dominant controlling factors in floristic modernization in the late Cenozoic.

DEFINITIONS AND BACKGROUND

Blanket, partial, and regional extinction. Although there are many kinds of extinction, I should like to distinguish among three types that are of particular importance in considering the Cenozoic fossil record: the first two, blanket and partial, may be considered to be kinds of true extinction, and the third, regional extinction, is only local extermination. As Guilday (1967) pointed out, true extinction may include blanket extinction, or the total loss of a phyletic line, such as a genus, family, or higher category, leaving no near relatives; and partial extinction, or the loss of a group incidental to the evolution of a phyletic line.

In opposition to true extinction, regional extinction refers to the extermination of a group in one area though it continues to thrive elsewhere. Regional eliminations of taxa were termed "extirpations" by Martin (1958), who used the word to refer to various examples of shrinkage in geographic range, and by Reid (1920a), who referred to Pliocene extermination in Britain of plants that never returned again. I shall restrict the use of extirpation to mean regional extinction, or regional shrinkage of range that under the test of late-Cenozoic time appears to be essentially irreversible.

Most contributors to this volume are concerned mainly with blanket extinction, which is best illustrated by the fate of many large mammals toward the end of the Pleistocene.[2] For the paleobotanist who deals with late-Cenozoic plants, the problem of blanket extinction scarcely exists, as I shall attempt to show. Instead, examples of species extinctions in living genera and broad-scale shrinkage of range are abundant in the late-Cenozoic plant record.

Some notes on rates of evolution. In the plant kingdom, at least, the more highly evolved groups display more rapid evolutionary rates than the more primitive plants. One way to judge relative rates of evolution is to determine the geologic age of most living species within different groups. In this regard, the floras reviewed here indicate that fossil records of several living fern species and most living species of conifers go back to the Miocene, and that most living angiosperm species of which we have fossil records date from the Pliocene.

In the animal kingdom, the geologic age of most living species of mammals is only about a million years (mid-Pleistocene), but most living species of mollusks can be traced back about 25 million years (early Miocene) (Briggs, 1966). Hence angiosperms, which appear to be the most rapidly evolving plants, have evolutionary rates only a tenth those of mammals.

For examples of generic extinction, zoologists need look back only 10,000 years in the fossil beds of the Americas to find them in abundance, but botanists have to penetrate the record of at least three million years to find them. The Oligocene dicot *Fagopsis* (MacGinitie, 1953) is to my knowledge the youngest extinct woody genus of angiosperm, but some extinct herbaceous genera are much younger; for example, *Hemitrapa* of the Hydrocaryaceae (Miki, 1959) and *Stipidium* of the Gramineae (Elias, 1935) are of Pliocene age. The youngest extinct fern and conifer genera are much older.

Some living plant genera may well be as young as Pliocene or Pleistocene. This is a good guess because, according to fossil pollen and leaf records, the most highly evolved plant family, the Compositae, apparently originated in the Oligocene; because there are now about 20,000 species in this family, evolution of new genera probably occurred during the Ice Age.

Although several new vertebrate genera appeared for the first time in the Quaternary, fossil evidence has not yet demonstrated a Quaternary

2. Pleistocene as here used refers to that part of the Quaternary that precedes the postglacial (after Wilson et al., 1959).

origin for any plant genus. Because the most rapid evolution is in herbaceous angiosperm groups that do not preserve well, the record is doubtless very incomplete.

Schmalhausen (1946) and Stebbins (1950) have concluded that the slower evolution rates of plants and nonmammalian animals may be attributed to their lack of mobility and to their biology of reproduction. Stebbins (1950) pointed out that the most slowly evolving plants are the algae, which rely for survival solely on the large number of zygotes and gametes they produce. The bryophytes and homosporous pteridophytes and many kinds of wind-pollinated plants also fit into this group. The middle position is occupied by many plants ranging from algae to flowering plants, which produce large, heavily coated, highly resistant resting spores or seeds. The highest position is occupied by those rapidly evolving angiosperms that are cross-pollinated by insects or animals or that have fruits adapted for dispersal by animals. Stebbins believed that his analysis explained to a large extent the evolutionary diversity and plasticity of such families as Compositae, Orchidaceae, Gramineae, and Leguminoseae.

Some limitations of the plant record. In many cases, the plant record gives us an idea mainly of the dominant or larger forms in the fossil flora; herbs, for example, are seldom well preserved. Even though sometimes found in abundance, leaves are often the least diagnostic for generic or specific identification. Although more abundant than leaves, fossil pollen grains have a morphology that seldom allows identification below the generic level. More precise identification can usually be made from seeds and fruits, and in many cases from leaves. The fossil record includes only a minute sample of the vegetation that existed on earth, and it is far from a random sample. Also, our records mainly come from the north-temperate zone, so we must remember that we may be recording only times when plants from low latitudes immigrated to higher ones.

Even though the paleobotanist has available a reasonably rich fossil record preeminently suited to climatic interpretations, he has not been able to describe changes in diagnostic plant features comparable to those in dental morphology that have enabled the vertebrate paleontologist to infer evolutionary rates with great precision.

Exceptions in the botanical literature are outstanding. For example, Florin's (1944) work in establishing the evolutionary succession between Cordaitales, Taxales, and Coniferales describes gradual changes in nearly all organs. Based on seed morphology, nine successive species between Eocene and Pleistocene of the water plant *Stratoides* have been described by Chandler (1923).

Built-in shortcomings are not only intrinsic to the fossils, but they lie also in the fact that taxonomic treatments vary among workers. For example, most American paleobotanists have followed the inclination of Chaney in placing all Tertiary plant species in fossil taxa as though they were extinct, even if they seemed identical with living species, and in presuming that all Quaternary forms belong to living species; if the fossils did not fit into living species, they were cited as comparing with (cf.) or having an affinity with (aff.) the nearest living form. On the other hand, many European workers, for example Marjorie E. J. Chandler, Dr. and Mrs. Clement Reid, and Dr. and Mrs. W. Szafer, have placed late-Tertiary fossils in living species if the relationship was sure; if they were uncertain, they related the fossils to the nearest living form by cf. or aff. And they considered the plant extinct and gave it a new species name only if the evidence was compelling. The European late-Tertiary megafossil floras are largely based on seeds and fruits that give a better basis than leaves for identification with living species; the American late-Tertiary floras are based mainly on leaf evidence. The result is that most late-Cenozoic records of extinct versus living plant species come from the work of European botanists.

Continental versus oceanic climates. Continental climate, as here used, is that climate now typical of most midcontinental areas in which the seasonal variations of temperature and humidity are extreme, annual rainfall tends to be low, and orographic rain shadows are sometimes well developed. In contrast, oceanic or coastal climates at the same latitudes are more moderate in these respects, and orographic rain shadows, if present, tend to be less severe. In having a greater range of seasonal temperatures, midcontinental areas tend to be less temperate than coastal areas at the same latitude.

Some coastal localities with oceanic climates, as well as inland sites with continental or semicontinental climates at roughly the same latitudes are listed in Table 1. The annual range of temperature and humidity is two to three times greater at the midcontinental sites than along the coasts in Eurasia and western United States, whereas annual rainfall is much reduced inland.

The continental nature of the climate in the interior basins of the western United States is probably accentuated by the presence of north–south mountain chains along the west coast as well as in the Great Basin itself. The most effective barriers are the high Sierra Nevada and the Cascade Mountains, which lower the rainfall in downward areas (eastward) by factors ranging from 2 to 20, depending on the relief, and at times they block air circulation in the basins, heightening seasonal extremes in temperature. Mild rain shadows are developed in Eurasia

TABLE 1. *Localities with Oceanic, Continental, or Semicontinental Climates, and Characteristic Climate Features*

Data from Gerasimov et al., 1964.

	Degrees N. Lat.	Locality	Average monthly temperature °C			Average annual precipitation (mm)	Average monthly relative humidity	
			Warmest mo.	Coldest mo.	Range		Annual range (%)	Annual deviation (%)
Coastal localities with oceanic climate	45	Bordeau, France	17	5	12	820	78–90	12
	50	Plymouth, England	16	5	11	1,095	78–83	5
	40	Eureka, Calif., U.S.A.	14	9	5	1,013	82–90	8
	10	Moyamba, Sierra Leone, Africa	24	16	8	2,693	17–23	6
Mid-continent localities with continental climate	50	Kazalinsk, U.S.S.R.	22	−11	33	118	45–80	35
	50	Krakow, Poland	18	−2	20	686	65–85	20
	40	Winnemucca, Nev., U.S.A.	22	−2	24	209	20–43	23
Mid-continent locality with semicontinental climate	10	Fort Archambault, Chad, Africa	32	25	7	1,175	30–85	55

east and southeast of the Norwegian highlands and the Urals, but in general a maritime and mild transitional semioceanic climate prevails far inland, because central Europe itself lacks a north–south mountain chain along its west coast.

The fact that not all mid-continent regions today have strongly developed continental climates is evident from climatic data of central and southern Africa. South of about 10° N. Lat., the continent has surprisingly small seasonal variations in temperature, and it has rather consistent or abundant annual rainfall. The climate of Fort Archambault, Chad, cited in Table 1, is a case in point. The lack of well-developed rain shadows in central and southern Africa is undoubtedly attributable to the absence of mountain chains along its west and southwest coast. As mid-continental climates go, the climate of southern Africa is distinctly uncontinental.

BROAD COMPOSITIONAL CHANGES IN SOME LATE-CENOZOIC FLORAS

Species extinctions in selected Eurasian and United States floras. This section presents data on numbers of species extinctions and proportions of extinct species in selected floras of Europe (west of the Ural Mountains) and the United States. The floristic data are listed according to geologic age in Tables 2–4. The meaning of these comparative data depends largely on the reliability of the correlations as well as on the size of the floras. In all cases the correlations given here, which are the best estimates of the authors cited, are the basis for comparison of floras. In Europe, isotope dates are so far available only from late-Quaternary deposits; in addition, evidence from fossil invertebrates and mammals and field relations of local stratigraphic units have aided in correlations of older Quaternary and late Tertiary sections. The American Tertiary floras cited in Table 3 are dated only by plant evidence and by field relations of the matrix sediments with nearby deposits.

In the dating of the Polish floras, Szafer (1961, 1954) indicates that his correlations with late-Tertiary and early-Quaternary floras of northern Europe are tentative. They appear to depend in part on his assumption that datable marine transgressions in Poland correlate with climatic oscillations recorded by his floras, and in part on the makeup of the floras. Szafer's correlations with north European events were accepted by Godwin (1956) and Grichuk (1960), and the age of the Russian Plain floras relating to both are summarized by Grichuk (1960).

In Szafer's (1961, 1954) data from the Mizerna deposits, I have accepted his geologic age assignments, including that of Günz-Mindel interglacial for the Mizerna III flora but have rejected his correlation of

Mizerna III with the Tiglian (pre-Günz; Pannekoek, 1956) flora of the Netherlands.

It should be noted here that the geologic age of the Donau glaciation in the Alps has been termed middle Pliocene by Szafer (1961) and late Pliocene by Pannekoek (1956). It is not known whether the Castle Eden flora of middle Pliocene age coincided with the time of the Donau. Both are considered middle Pliocene here.

Most of the floras listed include from 40 to 100 species (Tables 2, 3, and 4). No percentages were used unless the sample included twelve or more species.

Numbers of species extinctions. A tally of the last recorded fossil occurrences of extinct species in Pliocene and younger floras of southern Poland and coastal floras of northwestern Europe is given in Table 2. These data, which essentially record times and numbers of plant extinctions, are arranged according to plant groups. Extinctions of vascular species as percentages of the total identified species in each flora are plotted against geologic age in Figure 1; for all practical purposes this graph records percentages of plant extinctions in the regional floras.

Several interesting aspects of plant extinction are evident from these data.

1. The bulk of recorded late-Cenozoic extinctions are of late-Tertiary, not Quaternary, age. However, extinctions of some species occurred in

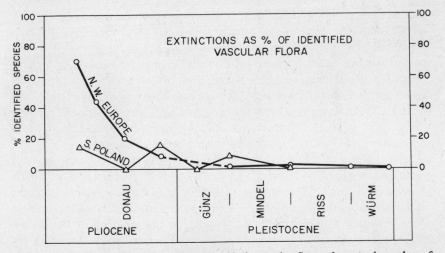

FIG. 1. Extinctions as percent of the identified vascular flora; the actual number of last occurrences of extinct species are calculated as a percentage of the total identified vascular species in each flora (data from Table 2). The data include Pliocene and Pleistocene floras of southern Poland (Szafer, 1954) and northwestern Europe (Reid and Reid, 1915).

TABLE 2. Numbers of Recorded Extinctions of Plant Species in Pliocene and Pleistocene Floras of Southern Poland and North-western Europe

	Pliocene			Pleistocene						
	Early	Middle	Late		Cromerian			Hoxnian	Eemian	Late Glacial
Southern Poland (Szafer, 1953)	Kroscienko	Huba	Mizerna Late	Günz			?			
No. of recorded extinctions		I, I–II	II	II–III	III	III–IV	IV			
Mosses	3									
Ferns and Fern Allies										
Gymnosperms	2		1			1				
Dicots	12	1	17		5	1				
Monocots	2		2		2					
Total	19	1	20	0	7	2	0	0	0	0
Total excluding mosses	16	1	20	0	7	2	0	0	0	0
No. of identified vascular species	114	45	117	51	87		18	46	55	97
Extinctions as % of identified vascular flora	15	0	17	0	10		0	0	0	0

Northwestern Europe (Reid and Reid, 1915)	Pont de Gail	Reuver	Castle Eden	Tiglian	Cromer	Hoxne	Eem	Late Glacial
No. of recorded extinctions								
Mosses		2				1		
Ferns and fern allies		1		1				
Gymnosperms			9		1	2		
Dicots	9	34		6				1?
Monocots	2	7		1				
Total	11	44	9	8	1	3	0	1?
Total excluding mosses	11	42	9	8	1	3	0	1?
No. of identified vascular species	16	92	46	85	122	110	49	47
Extinctions as % of identified vascular flora	69	46	20	9	1	2	0	2

the first Quaternary interglacial (Cromerian), and three are recorded in the second (Hoxnian); the latter are based on fruiting parts of *Crataegus reidii* and *C. clactonensis* at Clacton (Reid and Chandler, 1923) and of *Azolla filiculoides* at Neede (Pannekoek, 1956). Several English records of allegedly extinct plants as recent as late glacial in age have been discounted by Godwin (1956), but Clement Reid (1915) and Godwin (1956) both thought that *Linum praecursor* Reid in the late glacial is an extinct variety of common flax. A supposedly extinct species, *Picea omorikoides* Web., was found by Zagwijn (1961) in beds of Würm age at Amersfoort, Netherlands, but it is somewhat questionable because the pollen evidence is not backed up with megafossils.

2. The highest mortality in both geographic regions occurred among species of angiosperms, especially among the dicots.

3. Even though the first local glaciations in Poland and England were contemporaneous (Mindel), the times of greatest extinction were different in the two areas. Along the European coast the major losses occurred in the early Pliocene, while in Poland there were equal losses in early- and late-Pliocene time.

4. Most plant extinctions appear to have occurred during relatively warm, rather than cold periods; but whether they happened under a cooling or warming climate cannot be ascertained in most cases. The early Pliocene, which certainly was a period of general cooling in Europe, saw far more plant extinctions than any period since. Small increases in numbers of extinctions are recorded during nonglacial periods (Donau-Günz and Günz-Mindel) in Poland. In some cases extinctions probably occurred under a cooling climate; for example, *Azolla tegeliensis* at the end of the Tiglian and *A. filiculoides* toward the end of the Needian (Pannekoek, 1956).

Early-Pliocene extinctions were far greater at European coastal sites than in such areas as southern Poland; however, it should be borne in mind that the pre-Pliocene flora of the coast may have been more diverse to begin with than was that of Poland. Although as a whole the flora of Europe is now depauperate in species, coastal floras of many areas tend to be richer in species and more diverse than inland floras of the same latitudes. It has long been thought by many (e.g. Reid, 1935) that the severe impoverishment of the European flora was directly related to the climatic cooling and advent of the Ice Age, and to the fact that European lowlands suffered glaciations that were like a pincers movement from two glacial centers. In this connection, it is interesting to note that most of the species extinctions in the European flora were complete by middle Pliocene time (Fig. 1).

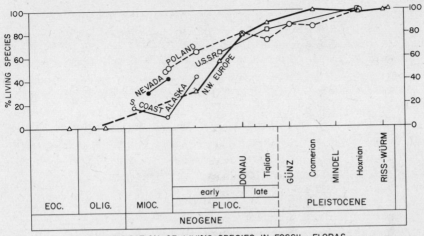

PARTICIPATION OF LIVING SPECIES IN FOSSIL FLORAS

FIG. 2. Proportions of living species in Cenozoic floras of Europe and western United States; percentages of living vascular species are calculated in relation to the total identified species in each flora. The differences between these percentages and 100% are represented by extinct species. The horizontal scale indicates relative geologic age (data from Tables 2 and 3).

Proportion of living species in regional floras. The declining role of extinct species and the increasing proportions of living species are modernizing trends in late-Cenozoic floras. Though quantitative data portraying this change do not directly represent rates of plant extinction, they are nevertheless one measure of compositional change in successively younger floras.

The graph in Figure 2 (based on data in Tables 3 and 4) is intended to evaluate the changing proportions of living and extinct species in late-Cenozoic floras of different regions. Included are data from the Russian Plain (Grichuk, 1960), from western United States (Wolfe, 1964, and Wolfe et al., 1966), as well as from southern Poland and northwestern Europe. The percentages of living species among the total vascular species identified in the fossil floras (vertical axis) are plotted against geologic age (horizontal axis); the differences between the percentages plotted and 100 per cent represent extinct species. In general the graph in Figure 2 demonstrates that pre-Pliocene floras are largely composed of extinct species, whereas middle Pliocene and younger floras are chiefly comprised of living forms. Specific features of the data are as follows:

1. During any given time in the Miocene or early Pliocene, inland floras were more modern in aspect (i.e. contained more living species) than the coastal ones. However, by the late Pliocene and in the early

TABLE 3. *Relative Importance of Extinct and Living Vascular Plant Species in Floras of Northwestern Europe*

Per cent of total identified vascular species.

Age	Deposit	Author	Total identified vascular species Number	Living species Number	Living species %	Extinct species Number	Extinct species %	Living Native species %
Late-Glacial	Lea Valley, England	Reid, 1915	47	46	98	1?	2?	
Eem and Brørup	Amersfoort, Netherlands	Zagwijn, 1961	10	9		1?		
Eemian	Ipswich, England	West, 1957	49	49	100	0	0	92
Needian–Hoxnian	Clacton, England	Reid & Chandler, 1923	110	108	98	2	2	96
Cromerian	Cromer, England	Godwin, 1956	122	121	99	1	1	95
Late Pliocene	Teglen, Netherlands	Reid, 1920a; Zagwijn, 1963a	86	79	92	7	8	60
Middle Pliocene	Castle Eden, England	Reid, 1920a	46	34	80	9	20	35
Early Pliocene	Reuver, Netherlands	Reid, 1920a, b	92	47	58	39	42	12
	Pont de Gail, France	Reid, 1920a	16	5	31	11	69	6
Miocene?								
Oligocene	Hamstead beds, Isle of Wight	Chandler, 1963	14			13	100	
	Bembridge beds, Isle of Wight	Chandler, 1963	54			53	100	
	Osborne beds, Isle of Wight	Chandler, 1963	4			2		
Late Eocene	Upper Headon beds, Isle of Wight	Chandler, 1963	27			27	100	

Quaternary the situation was reversed: living species played a greater role in the coastal floras than they did in Poland or western Russia.

2. On a species level, the origin of the modern flora in the northern mid-latitudes is Neogene, or even Oligocene. The data (Fig. 2; Tables 3 and 4) demonstrate that many living species of vascular plants have fossil records going back to the Neogene. Hence we cannot assume a Quaternary age for living species, although their distribution patterns have undoubtedly been influenced by Quaternary events.

3. The data on decreasing percentages of extinct species in Polish and northwestern European floras (Table 4) show parallel trends with the data (from Table 2) on declining percentages of extinctions in these floras; hence the former appear to be a good index of the latter.

The American data in Figure 2, which include only two Tertiary floras from Nevada and three from the south coast of Alaska, compare with the European data in an interesting way. First, Alaskan floras suggest a Miocene oscillation not recorded in the European data. Although this curve may have climatic implications, not enough detailed floras are available from the Alaskan sequence to interpret the curve at this time. Second, the Miocene datum points from Nevada fall near those from Poland, while the Alaskan Mio-Pliocene data, even though scattered, are closer to the European curve. In spite of the great differences in latitude between the American localities, the following relationship is still of interest: at least in the late Miocene, the Nevada (inland) flora was more modern in aspect, i.e. contained more extant species, than the coastal Alaskan flora.

In Pleistocene American floras from which only scant megafossil evidence is available so far, the following compose the entire fossil record of extinct plant species in the Quaternary of the United States: an extinct variety, *Juniperus californica* var. *breaeasis* Frost (1927) of late-Pleistocene age recorded at Rancho la Brea, Calif., is discounted by Mason (1944) but not by Axelrod (1966); an extinct species, *Larix* (?) *pleistocenicum* (Beals and Milhorn, 1961), was described from the second interglacial (Yarmouth) of Indiana.

The data from Tables 3 and 4 are not broken down into plant groups, but such an analysis might be revealing. A percentage comparison of extinct species by plant groups in localities of widely different latitudes might also be interesting. Lumped totals of various Japanese Pliocene seed floras studied by Miki (1952, 1955, 1956a, 1956b, 1958, 1960, 1961) show that a third of the conifer species and half of the angiosperm species are extinct, whereas in Miki's Pleistocene floras only 3% of the conifers and 7% of the angiosperms are extinct. Miki's data, which are all from lowland floras, point toward a slower rate of extinction among the

TABLE 4. *Relative Importance of Extinct Vascular Plant Species in Floras of Europe and Western United States* Percent of total identified vascular species.

	Russian plain Grichuk, 1960		S. Poland Szafer, 1961 Grichuk, 1960		N.W. Europe Reid, 1920a, b Reid & Reid 1915		S. Alaska Wolfe et al., 1966	Nevada Wolfe, 1964
		(%)		(%)		(%)	(%)	(%)
Pleistocene Riss-Würm interglacial	Mikulinian (4 loc., 120 spp.) Odintsovian (4 loc., 64 spp.)	0 ?	Masowian II (55 spp.)	0	Eem	0		
Mindel-Riss interglacial	Likhvinian (3 loc., 117 spp.)	3	Masowian I (46 spp.)	0	Neede– Hoxne	2		
Günz-Mindel interglacial			Mizerna III, III–IV (87 spp.)	13	Cromer	1		
Günz			Mizerna II–III (51 spp.)	12				

Epoch	Age					
Pliocene	Late	Akchagylian (4 loc., 107 spp.) 15	Mizerna II (117 spp.) 25	Tegelen 8		
	Middle		Mizerna, I, I–II (45 spp.) 20	Castle Eden 20		
	Early	Kimmerian (3 loc., 115 spp.) 37	Kroscienko Huba (114 spp.) 37	Reuverian 42 — Pont de Gail 69	Clamgulchian (12 spp.) 58	
Miocene	Late		Starych Gliwic (103 spp.) 55		Homerian (27 spp.) 89	Stewart Springs (31 spp.) 58
	Middle		Rypin (14 spp.) 50			Finger-rock (22 spp.) 68
	Early				Seldovian (44 spp.) 81	

conifers than among the angiosperms. Pliocene floras of the Japanese low-lands (Miki, various works) and of coastal Alaska (Wolfe et al., 1966) show similar percentages of extinct angiosperm species and total extinct species in the floras (close to 50% in each case). The coastal Alaskan localities lie about 15° in latitude farther north than the Japanese localities, and 10° farther north than all other fossil floras referred to in Figure 2. Although in composition the Pliocene floras of Japan and Alaska are very different, the similarity of the percentages of participating extinct species is of interest. The data might suggest that in Pliocene time the difference of 10° or 15° of latitude along the coast did not make so much difference in the progress of modernization as distance inland; i.e. through early and middle Pliocene time the inland floras of Europe maintained a more modern aspect (contained a smaller percentage of extinct species) than coastal floras at the same latitude.

The time scale along the horizontal axis of Figure 2 is only a relative scale according to geologic age. If we wish to know whether the increasing participation of living species is a function of time, we must plot the same data on a cartesian graph, using absolute age. Because absolute ages for the European floras are not available, I have utilized the general age spans accepted by the U.S. Geological Survey (Wilson et al., 1959) except for the length of the Pleistocene, which is now thought to be about 2.5 million years (Malde, 1965). I have roughly divided the spans of these epochs into thirds for plotting "early," "middle," and "late" age categories. By this presumptuous technique, I arrived at the graph in Figure 3, in which I have plotted data from Table 4 for Polish and coastal European floras against inferred absolute age.

A glance at Figure 3 will show that the data do not plot as straight lines at all. Just as one may have suspected from Figure 2, the time of greatest loss through extinctions in the coastal European floras (early Pliocene) was the time of most rapid loss. On the other hand the losses by extinction occurred at a slower pace in Poland than along the coast and were spread out during the late Miocene and early Pliocene. Both areas saw a slackening of losses in the mid-Pliocene, and after this in Poland there were stepwise changes, probably relating to climatic oscillations through the early Quaternary, and probably corresponding to higher extinction rates during nonglacial intervals (Fig. 1). In neither area were these trends a mere function of time; in both areas, extinction rates may have been accelerated by climatic cooling.

Changing geographic elements in late-Cenozoic floras of Europe. Analyses of late-Tertiary and Quaternary floras of Europe, with a grouping of taxa according to the present distributions of their nearest living relatives, have been compiled for southern Poland by Szafer (1961)

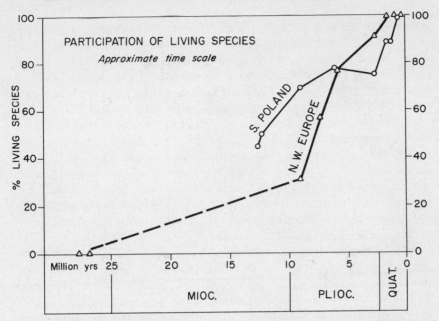

FIG. 3. Proportions of living species in late-Cenozoic floras of southern Poland and northwestern Europe, showing approximate time scale (see text). The differences between these percentages and 100% are represented by extinct species (data from Tables 2, 3).

and for Britain and northwestern Europe by Reid and Reid (1915). These analyses illuminate the changing biogeographic patterns.

In two charts (Figs. 4, 5) the relative importance of various geographic elements in southern Polish floras is plotted with geologic age. These data are based mainly on species but include some forms that were identified only to genus. The graphs show shaded vertical bars to mark the times of European glaciation, which are labeled as such in Figure 4.

The graph in Figure 4 shows the declining role of plants that are now foreign to the native flora of southern Poland. Subtropical taxa, such as *Mastixia*, which were abundant in the Polish Oligocene, were gradually eliminated before the Quaternary. East Asian elements, such as *Eucommia*, which are mainly warm-temperate today, became important in southern Poland during the late Oligocene and Miocene but waned and died out locally by the first local glaciation (the Krakowian or Mindel). North American species, such as *Tsuga* aff. *canadensis* and *Abies fraseri*, rose to prominence during the Miocene, and hung on until the second Polish glaciation (Srodkow, or Riss). Mediterranean species, such as *Pinus salinarum*, became abundant in the Miocene but then gradually

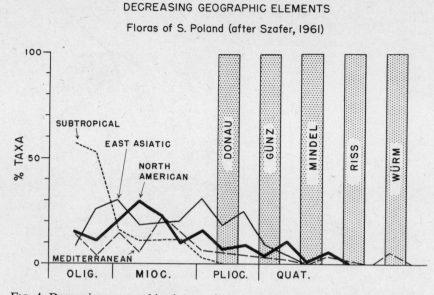

FIG. 4. Decreasing geographic elements in late-Cenozoic floras of southern Poland (after Szafer, 1961); the data include generic as well as specific identifications. The present distributions of the nearest living relatives of the fossils are the basis for classifying the geographic elements.

faded; in Poland this group was last seen in the Eemian (last, or Riss-Würm) Interglacial. The slight relative increases of East Asian and North American forms during older interglacials mark reimmigrations of a few species. The overall aspect of these curves is like the damping curve of a bouncing ball; after each glaciation, relatively fewer of these geographic elements returned to the Krakow area from their glacial refuges.

Figure 5 indicates the elements of the flora that were increasing during the late Cenozoic. These groups are mainly temperate European forms that now compose the local flora—arctic–alpine species excepted, many of which grow today in the Carpathian Mountains near Krakow. The data (Figs. 4, 5) have several fascinating implications.

1. Increases in participation of European temperate elements during interglacials alternate in the late Pleistocene with enormously increased participation of arctic–alpine elements during glacials (compare upper with lower parts of Fig. 5). Such changes, of course, have great value to the paleoclimatologist.

2. Arctic and alpine species first appear in Poland (and in Germany; Zagwijn, 1963b) in the Krakowian (Mindel). Records from elsewhere

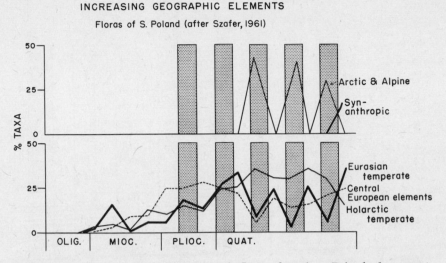

FIG. 5. Increasing elements in late-Cenozoic floras of southern Poland; the present distributions of the nearest living relatives of the fossils are the basis for classifying these elements (after Szafer, 1961).

suggest that modern alpine and arctic species did not evolve until late Pliocene (Wolfe and Leopold, 1967).

3. The adventive weed flora, which Szafer (1961) terms "synanthropic," is not evident until the postglacial, during which time it climbs from 0 to 17%. Szafer's data suggest the possible origin of the weed species in the Neolithic or else a Neolithic introduction of weeds to the Krakow area.

Certain contrasts can be drawn between the Polish succession just described and that of coastal Europe during the same period. From the data provided by Reid and Reid (1915) for Britain and northwestern Europe with that from Poland, the following contrasts emerge. (1) Whereas the subtropical Malayan element departed from Poland just before the Pleistocene, it disappeared from the coastal European sites before the Tiglian (late Pliocene). (2) The North American element, which Szafer reports as strong in Poland through the Mindel-Riss, appears to have dwindled along the European coast to a mere one or two species by the late Pliocene. (3) The strictly Chinese element, which Szafer claims returned to Poland during the Günz-Mindel (Cromerian) interglacial, was still well represented in the coastal floras during the late Pliocene; except for *Eucommia*, Chinese taxa did not re-enter coastal Europe during the Quaternary. (4) Mediterranean elements were apparently represented in most interglacials in both areas.

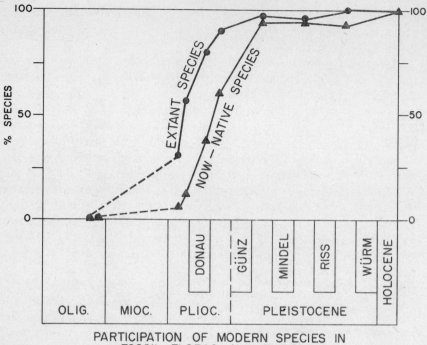

PARTICIPATION OF MODERN SPECIES IN
FOSSIL FLORAS, N. W. EUROPE

FIG. 6. Proportions of total living species ("extant species") in comparison with percentages of the species that now live locally ("now-native species"); late-Cenozoic floras of northwestern Europe (data from Table 3).

The graph in Figure 6 shows the increasing role of living species in the coastal floras of Europe (extant species, Fig. 6) and indicates what proportion of these species still live in northwestern Europe (now-native species, Fig. 6; from the data of Reid, 1920a). The differences between the two curves plotted in Figure 6 represent percentages of species now exotic to Europe, and include living species now native to China and North America. It is evident from Reid's (1920a) data that the incoming flora of early-Pliocene time was chiefly comprised of Chinese–North American species, but these were largely diluted by modern European elements by mid-Pliocene, and were ousted from Europe by the Pleistocene.

In summary, a vast alteration of the coastal European flora occurred during the Pliocene, with the result that by Cromerian time and before the first English glaciation the flora was much like that of today. Losses of a few genera occurred during the Quaternary, as discussed in the next section. In contrast to the British sequence, the Polish flora underwent a slow, drawn-out alteration probably during the entire Neogene and through at least Mindel-Riss time, providing a modern flora by the Eemian, or last interglacial.

Extirpation of plant genera. The decreasing role of Tertiary relicts and the increasing role of modern native elements—trends with time that Szafer's floras demonstrate primarily on a species basis—are evident on a generic basis in various late-Cenozoic floras. Mrs. E. M. Reid (1920a, 1935) first suggested that the steadily increasing role of the "native" plants in progressively younger floras was a reliable factor of age; she discussed the origin of the British flora in terms of the increasing proportion of now-native species as well as genera. Barghoorn (1951) and Wolfe and Barghoorn (1960) utilized Mrs. Reid's idea and plotted the proportion of now native genera in Tertiary floras against geologic age. The resulting "age curves" are in their words "a graphic expression of a phenomenon long recognized in the study of late Cenozoic floras, viz., the older the flora, the greater the deviation from the plant assemblages which feature the modern vegetation of the area."

WESTERN UNITED STATES DATA. Axelrod (1957) criticized Barghoorn's (1951) generic age curves for dating floras. Axelrod stated that though there may be real use for the method on a local geographic basis, data lumped from large geographic areas did not take into account local factors, such as elevation, soil type, and the existence of relict environments, which may cause discrepancy among data from different regions. Generic age curves discussed below utilize somewhat smaller geographic provinces than were used by Barghoorn (1951) or Wolfe and Barghoorn (1960) and demonstrate the degree of difference of age curves for various topographic areas.

Wolfe and Barghoorn's (1960) data covered a large region, utilizing United States floras from west of the Rocky Mountain front range; actually more than half their localities were from the Pacific drainage west of the Cascades and Sierra Nevada, and about three fourths were from Washington, Oregon, and California. The actual data had a scatter of plus or minus 8% around their smoothed age curve. I have selected from their material a number of floras that are located on or west of the Cascades and Sierra Nevada and have plotted a modernization curve for floras of Washington and Oregon (Fig. 7).

Generic modernization curves for a number of other western United States floras are shown in Figure 7 along with the selected data from Wolfe and Barghoorn (1960); time on a log scale is plotted against the percentages of the identified genera that still grow in the region. For the present purposes I use "native" and "exotic" in a very broad geographic sense; for example *Tsuga,* which now grows in northern Idaho, is now

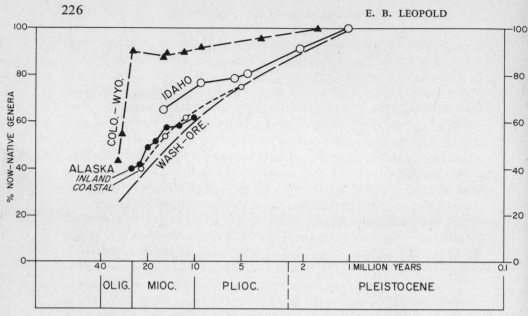

% OF NOW-NATIVE GENERA IN FOSSIL FLORAS

FIG. 7. Generic age curves; percentages of now-native genera in relation to total identified genera in late Cenozoic floras of western United States; data for Washington–Oregon curve are selected floras from Wolfe and Barghoorn (1960). Horizontal axis is a log time scale.

"native" to the Rocky Mountain province, but *Juglans*, which has species in California and Texas (Fig. 8b), is now "exotic" to the Rocky Mountain region. The method of graphing the data is similar to that used by Wolfe and Barghoorn except that this plot utilizes an absolute time scale, includes new data, and separates the data by geographic area.

It should be mentioned here that, in a series of six floras listed by Wolfe and Barghoorn (1960), where both megafossil and microfossil evidence was available for determining percentages of native genera in the floras, the average difference between the independent calculations based on pollen and those based on leaves was only 2%. Hence their findings indicate that pollen data yield close to the same result as megafossil evidence for computing late-Cenozoic age curves.

Although Barghoorn (1951) and Wolfe and Barghoorn (1960) have suggested that floras from different regions modernized at about the same rates, i.e. follow the same generic age curve, the new data included in Figure 7 indicate that this was not the case. Oligocene and Miocene rates of generic change proceeded more rapidly at midcontinental sites than at coastal ones. The composition of individual Rocky Mountain floras was probably affected by local volcanic and regional orogenic activity, for the resulting curves do not correspond to those for West Coast floras. The data are explained below.

The source material used in Figure 7 includes floras from Idaho, Colorado, Wyoming, and Alaska. Data for the Idaho floras are from my own pollen work (Leopold, *in* Weber, 1965, p. 456) and from leaf evidence of R. W. Brown (unpublished) and Axelrod (1964). Several of the Idaho floras plotted in Figure 7 are dated by the K/Ar isotope method (Evernden and Curtis, 1965; Malde, 1965), and the rock units with which we are concerned are mostly in superposition. The Colorado and Wyoming floras are from vertebrate-dated localities; these are based chiefly on pollen (Leopold, unpublished data), but leaf evidence from two localities was included (R. W. Brown, unpublished data; MacGinitie, 1953). Alaskan data are from Cook Inlet (Wolfe et al., 1966) and from the Alaskan Range (Wolfe and Leopold, unpublished MS).

The chart in Figure 7 demonstrates the increasing role of now native genera compared to extirpated genera found as fossils in the Tertiary but no longer native to the region. The generic age curves reveal that, at any given time during the Mio-Pliocene, the Rocky Mountain floras contained a higher percentage of now native genera (i.e. are more modern in aspect) than the West Coast or Alaskan floras. Idaho floras are intermediate. Conversely, for any given level of "modernization," the Rocky Mountain floras are older than the West Coast floras. While the Washington–Oregon data indicate a relatively gradual loss of genera between the Oligocene and earliest Pleistocene, Colorado–Wyoming floras show a sharp loss of genera during the late Oligocene and gradual losses during the Neogene.

The more "modern" aspect of the inland floras is indicated by some of the Miocene floras of Alaska. Early and middle Miocene floras from the foothills of the Alaskan Range about 200 miles inland are consistently more modern in aspect than those from along the coast at Cook Inlet; the relationship is not evident in the late-Miocene floras, however.

The significance of these curves (Fig. 7), which indicate local rates of generic extirpation, depends on the nature of the genera lost. As in Europe, the Neogene climates of western United States brought about a general loss of floristic diversity, an early loss of subtropical forms, and a later severe restriction in number of temperate forms.

In general, the Colorado-Wyoming floras lost their subtropical forms by late Oligocene, and they lost all warm-temperate broadleaved tree genera during the Neogene. Between the Oligocene, when the dominant conifers were *Sequoia* and its near relatives, and the Miocene, the role of pinaceous genera suddenly increased. Oligocene species and genera connote a higher summer rainfall and a climate much more temperate than the present one (MacGinitie, 1953). Miocene forms suggest that the regional climate was still temperate but less so than before, and increasingly arid.

In the Washington–Oregon floras, subtropical taxa were eliminated during the middle Miocene, and the diversity of hardwood tree genera diminished during the Neogene. In California, and probably to a certain extent in eastern Washington and Oregon, summer-wet conifers of the Miocene became more restricted in their range in the Pliocene, and summer-dry Pinaceae became more widespread (Axelrod, 1959).

The floristic changes in mid-continent and West Coast regions indicate a decrease in temperateness and a relative decrease in summer rainfall during the late Oligocene and Neogene; these changes appear to be much more pronounced in Wyoming and Colorado than in Washington and Oregon and suggest the development of a distinctly continental climate in the mid-continent during the late Oligocene.

An example of a late-Tertiary relict that once ranged over the western U.S. is walnut, which has five living U.S. species. Four of these are black walnuts; their modern ranges are shown in Figure 8b. The fifth species is a white walnut, *Juglans cinerea*, which occurs in eastern U.S., and its range (not shown in Fig. 8b) is similar to that of *J. nigra* but extends a hundred miles or so farther north.

Plotted above Figure 8b is a map taken from Bailey (1964), showing the extent of temperate climates in the U.S. (Fig. 8a). According to Bailey's (1960, 1964) definition, temperateness is dependent upon the severity of seasonal changes in temperature. In order to express temperateness quantitatively, he ranked total seasonal deviations of monthly temperatures above or below 14°C and assigned them a value M. Bailey's M indicates by a numerical index scaled from 0 (greatest deviation) to 100 (least deviation) the degree to which temperature distributions depart from earth's average temperature (14°C). By this index, climates that are typically warmer or colder than 14°C or those that have pronounced seasonal changes are judged less temperate and have low values of M; climates in which the thermal behavior approaches 14°C are more temperate and have higher values of M.

Bailey's maps (Figs. 8a and 9) indicate that the highly temperate climates of the U.S. now occur only in three isolated areas: along the west coast; in parts of Arizona, New Mexico, and Texas; and in southeastern U.S.

Except for the northward extension of range for two eastern species, the restriction of U.S. *Juglans* species to Bailey's temperate zones is remarkable; the maps show that the western species are primarily

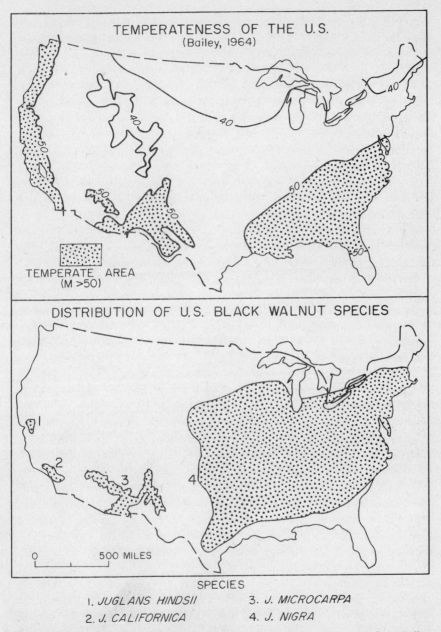

SPECIES

1. *JUGLANS HINDSII*   3. *J. MICROCARPA*
2. *J. CALIFORNICA*   4. *J. NIGRA*

FIG. 8. Distribution of temperateness in the United States (8a, upper map; from Bailey, 1964, with permission of Geographical Review), compared with the distribution of United States black walnut species (8b, lower map; after Munns, 1938; Jepson, 1925; and Benson and Darrow, 1954).

temperate and that the group is temperate to subtemperate in its U.S. distribution. Note that the northward extension of *Juglans* into subtemperate climates occurs in humid, not dry regions.

In the light of the fossil record, these data suggest that the extirpation of *Juglans* from the greater part of the western U.S. resulted from a climatic event marked by a decrease in temperateness and by the development of a more continental climate. Fossil pollen indicates that this event occurred earlier in Colorado and Wyoming (mid-Miocene) than in Idaho or in the Pacific Northwest (Pliocene). Species of *Juglans* persisted through the Ice Age, of course, in the temperate Great Valley of California, and in temperate parts of the Southwest. Climatic barriers now separate the ranges of four black walnut species of the U.S. (Fig. 8).

Other relatively temperate tree genera that had a similar history in the U.S. Neogene include at least *Cercis, Platanus, Tilia, Liquidambar, Ilex,* and *Nyssa*; of these only *Cercis* and *Platanus* stayed on in California after the Pliocene, but all still have relict species in temperate highlands of northeastern Mexico.

In summary, the data concerning generic extirpation in western United States floras demonstrate that several kinds of floristic changes happened earlier or were more extreme at mid-continent localities than along the coasts at the same latitudes; for example, (1) loss of subtropical elements; completed during the mid-Miocene along the west coast but during the mid-Oligocene in Colorado and Wyoming; (2) loss of temperate forms; a partial loss of temperate broadleaved tree genera occurred in the West Coast area during the Neogene, but was fairly complete in Colorado and Wyoming during the same interval; (3) rates of generic change (extirpation); generic losses along the coast during the entire Neogene; relatively rapid losses during the late Oligocene of Colorado and Wyoming, followed by gradual slow change during the Neogene. Recorded generic extirpations in both areas were all of Tertiary age.

Because of the wide apparent difference between curves showing generic change in floras of different areas (Fig. 7), it is clear that such curves for dating floras pertain respectively only to floras within limited geographic areas.

EUROPEAN DATA. In spite of the relatively complete Pliocene modernization on a species basis of the coastal European flora (Figs. 2, 3, 6), a limited number of generic extirpations occurred there during the Quaternary. In northwestern Europe, losses of the following temperate and boreal genera are recorded (strictly American genera are marked with an asterisk).

Liquidambar, Magnolia, Phellodendron, Nyssa at end of Tiglian (late
   Pliocene)
Eucommia, Carya, Tsuga at end of Cromerian (first interglacial)
Azolla at end of Needian-Hoxnian (second interglacial)
Brasenia, Dulichium* at end of Eemian (last interglacial)
Trapa by late Würm or early postglacial

In Poland, where there were losses of geographic elements recorded
through at least the Riss glaciation, gradual losses of genera also continued throughout most of the Quaternary.

Liquidambar, Corylopsis, Nyssa at end of Cromerian (first inter-
   glacial)
Azolla, Carya, Juglans, and probably *Pterocarya* at end of Needian-
   Hoxnian (second interglacial)
Dulichium at end of Eemian (last interglacial)
Trapa by late Würm or early late-glacial.

Although there may have been a lingering of some "exotic" elements
in Poland and a few more Pleistocene species extinctions in Poland than
in Britain, the general Quaternary pattern of modernization follows
roughly the same overall pattern in both areas. In my view, the main
differences between the late-Cenozoic records of the two areas are seen in
the late Tertiary; during that time the patterns of change, and especially
rates of change, were very different. The European coastal flora was
principally modernized during the Pliocene, and the changes were
relatively rapid; on the other hand, analogous changes in the Polish
flora were more gradual and were spread out through late Miocene,
Pliocene, and early-Quaternary time.

INFERRED NEOGENE CLIMATE AND GEOGRAPHIC FACTORS

During the late Tertiary, broad differences in both rates and times of
floristic change can be distinguished between midcontinental and coastal
floras of the same latitudes. Regional geographic factors that helped
shape these patterns in western United States and Europe are discussed
below.

In western United States, regional differences in floras during the late
Tertiary may be due to at least two geographic factors: effects of orogeny
on local climate, and possible physical effects of volcanism.

Orogeny would have brought about a cooling of the uplifted area by
greater elevation; orogeny of nearby mountain ranges would have
increased the severity of local rain shadows. Both would have caused a
more continental (less temperate and probably more summer-dry) climate
in the basins.

Some complex interactions of secular climatic change and late-Cenozoic uplift have been inferred from floras of various West Coast, Great Basin, and Idaho localities (Axelrod, 1964, 1950, and other works; Chaney and Axelrod, 1959), but little is yet known about these interrelationships in Colorado and Wyoming.

Geologists believe that the present lineation of mountain systems in the Sierra Nevada, Cascades, Great Basin, and Rocky Mountains was developed by mid-Miocene time. Regional uplift had taken place earlier in these areas but was especially pronounced in late-Tertiary and early-Quaternary time. Barriers formed or elevated during the late Cenozoic may have had far-reaching effects on the climate and floras of the mid-continent region (Chaney and Axelrod, 1959) because in this area rain shadows are cumulative eastward, downwind from the prevailing north-westerly winds. For example, rainfall in the lowlands of eastern Oregon depends largely upon the height of the Cascades, whereas rainfall in eastern Colorado is influenced by all the intervening mountain ranges.

Local volcanism may have impoverished nearby vegetation so that preserved plant assemblages may not represent the regional flora. An example may be the Creede flora of Colorado, recently dated by the K/Ar method as latest Oligocene (T. A. Steven, written communication, 1966). This flora, which is represented by the most aberrant point on Figure 5, accumulated in the heart of an active volcanic field during a time of recurrent ash-flow eruptions (Steven and Ratté, 1965). The plants best represented at this locality, *Populus*, *Acer*, and genera of the pine family, have seeds that are especially well adapted for wind dispersal and therefore should have had a selective advantage in reseeding devastated areas. Fossil plants having heavier seeds, as *Carya*, *Tilia*, and *Juglans*, are extremely rare in the Creede flora, although these are common in other Rocky Mountain floras of late-Oligocene age.

The final elimination of temperate hardwood tree genera from the Rocky Mountains was doubtless controlled by climate, but their scarcity or absence in individual floras may have been the result of local volcanic activity.

In central Europe and western Asia, where there was hardly any late-Cenozoic volcanic activity, and where, at least west of the Urals and in the Aral Sea area, there are no orographic barriers accenting climatic patterns, regional differences in rates of generic loss should mainly reflect growing continentality of climate under the secular cooling. At the beginning of the Pliocene, there was already an east–west climatic gradient across Europe and Asia. While the early-Pliocene floras of Europe were still subtropical to warm-temperate, the flora of the upper Irtysh Basin and the Aral Sea area southeast of the Ural Mountains was arid and subtemperate, to judge from Zaklinskaya's (1962) summary of Neogene succession there. (The Aral Sea area is not in the rain shadow of the Urals, but its inland position puts it at the end of a 2,500-mile-long storm path along which total precipitation becomes progressively lower from northwest to southeast (Gerasimov et al., 1964). Pliocene and Miocene floras of the Aral district lacked temperate tree genera and indicate a dry steppe landscape with abundant grasses; the floras connote a climate of continental and arid aspect. A steppe climate in the Irtysh lowlands east of the Caspian Sea formed an effective climatic barrier between the Pliocene flora of Europe and its counterpart in China; the barrier has apparently existed ever since.

Because continental Europe is temperate today in at least a wide belt along the coasts west of Germany (see Fig. 9) several of the Tertiary plants that were eliminated might survive there now if planted; for example *Azolla* now thrives as an introduced plant in Britain (Seward, 1959).

Reid and Reid (1915) felt that the severe impoverishment of the European flora was strongly related to the topography as well as to the climatic history of the region. They pointed out that, throughout the length of Europe and Asia, retreat to the south was cut off by "one unbroken

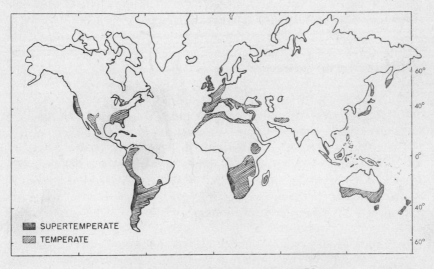

TEMPERATENESS OF THE CONTINENTS
(after Bailey, 1964)

FIG. 9. Distribution of temperateness over the world (excluding Antarctica); based on seasonal extremes of temperature (after Bailey, 1964, with permission of Geographical Review).

barrier of seas, deserts and mountains through which no rivers opened to the south" (p. 18). They also observed that mountainous areas may serve as good refugia, for in such terrain plants may change altitude to find "a change of climate necessary for their survival." The plants may survive in mountainous terrain "so long as the climate never becomes colder than it was when this flora entered the valley. If this happens, the warmer elements succumb" (p. 20). Apparently the late-Cenozoic cooling passed this threshold for many European plants during the Pliocene.

The east–west barriers that blocked passage to the south, combined with the development by Pliocene time of a continental climate in interior Asia, cut off escape routes to the south or east, isolating the temperate European flora. Under these circumstances the Pliocene cooling was then very effective in causing a severe impoverishment of European floras.

Overall sequences of modernization in Polish and northwestern European floras are strikingly similar; this would be expected because there are no intervening mountain ranges, and the two areas shared the same flora during most of the late Cenozoic. The present climatic differences between the two areas may not have developed until the Quaternary, yet two differences in their Neogene histories stand out: the representation of modern, local species was consistently higher during the early Pliocene in Poland than in northwestern Europe (Figs. 2, 3); and the modernization of the Polish flora was a long, slow process encompassing the late Miocene, Pliocene, and earliest Pleistocene. However, the same changes in the coastal floras of Europe were more rapid and were completed during the Pliocene (Fig. 3).

CONTINENTALITY OF LATE-QUATERNARY CLIMATE

The evidence reviewed and presented in the foregoing section indicates that there was a major increase in continentality of climate in western United States and western Asia during the late Oligocene and Neogene. In western United States this development was aided by orogenic uplift, which accentuated rain shadows and heightened seasonal extremes in the lowlands of the Rocky Mountain region. In Europe and Asia, mid-continent areas of low relief without orographic barriers became more continental just the same.

In order to determine whether further increases in continentality occurred after the Tertiary and to evaluate what effects these might have had on the vegetation, I should now like to review certain climatic data from the Quaternary, and especially from the postglacial.

Comparison of Pleistocene with postglacial climate. In an effort to characterize the degree of climatic continentality during the Quaternary, one must seek evidence from organisms whose present ranges are limited by summer temperature and from others that are limited by winter temperature. For this kind of information, the available Pleistocene plant records are not the best source for, as Dahl (1964) points out, most of the American botanical inferences concerning Quaternary climate are based on tree genera, and the distributions of these are mainly a function of growing-season temperatures. Hence, inferences from these would probably give only half the climatic picture. Somewhat the same is the case for the European Quaternary plant records, except for the floras of late-Quaternary age from which a wide variety of plant fossils has been identified—many of them on the species level.

Although, as a whole, Pleistocene plant records do not yet yield information concerning winter temperature, the Pleistocene evidence provided by fossil vertebrates and invertebrates is more helpful in this respect. Different taxa of nonmarine mollusks and vertebrates, which reflect summer and winter temperatures respectively, have furnished a nearly complete record of Quaternary climates in southwestern Kansas and northwestern Oklahoma. In a summary of these records, Taylor (1965) has described a chronology of late-Cenozoic climate for that area. He concludes that the climates of all the major interglacials were typified by mild climates, with relatively frost-free winters, and with summers that were much cooler than now. According to Taylor, glacial climates of the Kansan, Illinoian, and early Wisconsin (second, third, and fourth glacials) had cool summers, and at least the Illinoian had winters about as now. He points out that the climate of the late Wisconsin and early postglacial were essentially unique in the Quaternary history of this region, being characterized by aridity and strong seasonal contrasts. Taylor considers that the Pliocene and Quaternary climates before the late Wisconsin were much less continental than now, and those of the late Wisconsin and early Holocene were much more continental than at present.

Hibbard et al. (1965, p. 515), in summarizing vertebrate records for North America, more or less agree with Taylor's analysis, pointing out that "present-day climates with their seasonal extremes of temperature and aridity are geologically atypical, even of the Pleistocene."

A surge of increasing continentality in the late Wisconsin and early postglacial is shown by mammals (Slaughter, 1967; Hibbard et al., 1965) as well as by mollusks (Taylor, 1965). Because the interval during which this development occurred is one that is well documented by fossil plants, I should like to discuss evidence from the plant record that might reflect a development of increased continentality in the late Quaternary.

Expressions of continentality in postglacial vegetation changes. At least three kinds of plant evidence appear to reflect the existence of a continental climate in the early postglacial: comparative distances of migration, differential timing of vegetation changes in different regions, and re-arrangement of plant "associations." It is important to note that, although plant evidences of continental climate are very obvious in the postglacial, the degree of these postglacial changes cannot be readily compared to those of middle or early Quaternary age, mainly for lack of older records.

DISTANCES OF DISPLACEMENT OF INDIVIDUAL GENERA. A parameter of floristic change that reflects differences in climatic patterns between mid-continental and coastal regions of about the same latitude is the distance that genera had to migrate during a single climatic change. During the postglacial these distances were quite different for individual genera, depending on the region. In the following comparisons, I tried to find some examples of one genus that could be traced in its movements during the postglacial warming, but some cases are cited in which the movement of treeline in separate areas involves different genera.

Europe vs. the Russian Plain. With reference to oak (*Quercus*), I compared the position of its northern border at the end of the late-glacial with its most northerly position during the Thermal Maximum. In Europe, the distance between southern Germany, where oak occurred at the end of the late glacial, and central Sweden, oak's northern outpost in the Thermal Maximum (Faegri and Iversen, 1964), is about 1,500 km. However, on the Russian Plain, where the northern limit of oak was along the Caspian Sea during the late-glacial, oak moved northward and almost reached the Barents Sea by the Thermal Maximum (Neustadt, 1959), a distance of 2,400 km in 5,000 years! The comparative distances of these migrations are shown by the length of the arrows in Figure 10.

Coastal areas of Alaska vs. eastern Siberia. During the late-glacial, the eastern limit of treeline in continental eastern Siberia lay in the Aldan Basin, but by Thermal Maximum time the treeline (larch and pine) had migrated eastward about 2,000 km toward the Bering seacoast (Chukotsk Peninsula) (Petrov, 1963; Vangengeym, 1961). During the same period treeline (spruce) moved only a few miles along the coasts of Alaska (Arctic Slope: Livingstone, 1955; Seward Peninsula: Hopkins et al., 1960; Colinvaux, 1963) (Fig. 10).

Northwest coast vs. midcontinental United States. Postglacial migrations of conifer genera, including spruce in the Puget Sound area and nearby mountains, were probably limited to less than 70 km, and from the work of Hansen (1947) I infer that they probably involved less than 1,000-m elevation. In the flat Midwest the southern limit of the genus *Picea* moved from at least southern Indiana in the late-glacial to about central Michigan in the Thermal Maximum—a distance of 500 km (Fig. 10).

SOME EARLY HOLOCENE MIGRATIONS

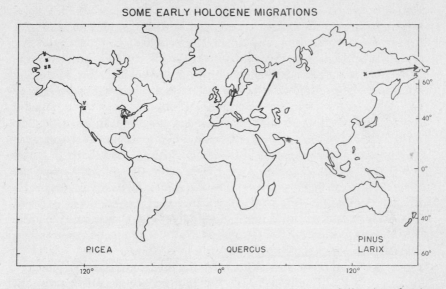

FIG. 10. Some early-Holocene migrations; arrows show extent and direction of early postglacial migrations; x's mark fossil localities referred to in text. Migrations indicated here involve change of plant range boundary between late glacial and Thermal Maximum (from various sources).

During the early postglacial, displacements of individual tree genera were much greater in midcontinental than in coastal areas in North America and Eurasia. The low-relief mid-continent areas saw extended horizontal migrations of *Picea* in the United States and of *Quercus* in central Soviet Union—both migrations involving much greater geographic displacement than in either mountainous or flat west-coastal areas.

SOME DRAMATIC VEGETATION CHANGES. Although the Allerød oscillation and the early-Holocene warming were probably worldwide secular climatic trends, the most dramatic vegetational changes during this interval were manifested at different times and in different ways in various regions. In the overall warming that ensued after the Older Dryas, the Allerød oscillation was a small wiggle that brought about relatively minor changes (compared to later ones) of woodland composition in central, eastern, and northwestern United States and in Europe. In all these areas, which are mesic and forested today, the early postglacial vegetational changes of 9,000 years ago were much more dramatic than those that took place during the Allerød and Younger Dryas, 12,000 to 10,000 years ago. However, in the now arid southwestern United States, including the high plains of Texas (Hafsten, 1961), the most dramatic vegetational change took place about 12,000 years ago, when the Allerød oscillation and the Two Creeks interstadial were just beginning. The return to moister conditions, about 11,000 to 10,000 years ago, was indicated in the Southwest by comparatively weak vegetational changes.

An abrupt postglacial change, consistently shown in the postglacial pollen sequence of midwestern and eastern United States, is the replacement of dominant spruce pollen in zone A by dominant pine pollen in zone B. This feature of postglacial pollen diagrams is generally considered to represent a switch from predominantly spruce forest or parkland to pine or mixed pine forest (without spruce). According to Ogden (1967), the event took less than 1,000 years and probably occurred around 9,000 or 9,500 years ago (Davis, 1965). It probably marks the movement of spruce forest northward into southern Canada and probably upslope in New England. To the palynologist, this event signifies the beginning of postglacial vegetation.

In late- and postglacial pollen diagrams of the American Southwest, a decrease in tree pollen indicates the change from pollen assemblages like those of the full-glacial to those like the postglacial. Martin and Mehringer (1965) placed the boundary between the late- and postglacial at this horizon, which is between 13,000 and 12,000 years in age. Evidence that this change marks the beginning of an uphill migration of treeline is documented by numerous pollen diagrams and by wood from fossil rat middens (Wells, 1966). The total vertical displacement of treeline was between 800 and 1,200 m in the Mohave and Chihuahua deserts (Wells, 1966). Between 12,000 and 10,000 years ago, the Pleistocene lakes of the Great Basin dried up, and surface water, which had been plentiful in the late Wisconsin (Snyder et al., 1964), became scarce. In the arid west the climatic change brought increasing aridity and summer heat— both of which are marks of increasing continentality.

In summary, the most dramatic changes in vegetation during the late- and postglacial occurred a full 2,000 to 3,000 years earlier in the arid west than in eastern United States. In both areas, these changes mark the beginning of postglacial vegetation, they are well documented over wide areas, and they are deemed to have been rapid changes under a warming climate. In the Southwest the climatic change brought increased continentality.

DEFUNCT PLANT "ASSOCIATIONS" OF THE LATE-GLACIAL. An expression of continentality development in the postglacial, as described by Slaughter (1967), suggests that extant animal species that do not live together today occurred together in the early postglacial. Slaughter described the fossil occurrence of species that are intolerant of cold winters in the same deposits with species that are intolerant of hot summers, and he suggests that the late- and early-postglacial climate of Texas was less continental than now.

Some degree of support for his thesis is found from late-glacial and younger plant records of the Northern Hemisphere. Fossils of plants whose ranges are regionally separated today are found together in late-glacial deposits of the Middle West, in eastern United States, and in Europe. Some of these assemblages are comprised of megafossils, and some of pollen. In neither case can we assume that the associated fossil plants necessarily grew in the same plant communities or occupied the same niche or habitat. However, the records suggest that their ranges overlapped geographically during the late-glacial. Essentially, this evidence is in agreement with that described by Slaughter (1967) for the late-glacial rodents of Texas.

In the Great Lakes region, a spruce-dominated assemblage character-ized the first vegetation during and after the wasting of Wisconsin ice. Typical pollen assemblages of late-glacial age include spruce (dominant), ash (often *Fraxinus nigra* type; common), oak, elm, and *Ostrya-Carpinus* (usually rare) (Cushing, 1965). Recent findings by Baker (1965) at a late-glacial site in northern Minnesota include megafossil remains of the following arctic-alpine, maritime-boreal, and temperate plants:

Arctic-alpine

Dryas integrifolia
Salix herbacea
Vaccinium uliginosum var. *alpinum* (= *V. gaultherioides*)
Rhododendron lapponicum
Potentilla nivea
Carex capillaris

Southern boreal but boreal and maritime in S.E. Canada

Juncus balticus

Temperate and boreal

Typha latifolia
Ranunculus trichophyllus

Fossil remains of *Juncus balticus* and of several arctic–alpine forms were common in the late-glacial zones, according to Baker. Through the late-glacial zones in which the megafossils were found was pollen of spruce (20–40%), ash (2–18%), oak (2–4%), elm (0–2%), and *Ostrya-Carpinus* (1–2%). According to Baker (oral communication, 1967) the ash pollen in this late-glacial deposit was chiefly of the *Fraxinus nigra* type. Though late-glacial pollen of thermophilous trees has been interpreted by some authors as reworked from older deposits, the consistency of its frequency and composition over wide areas suggests that it may be in place.

This fossil assemblage of thermophilous and boreal trees and of arctic–alpine species has no apparent counterpart in the modern vegetation of North America, for the ranges of these plants do not all overlap anywhere today. More of them overlap along the southern coast of Labrador and on Anticosti Island than anywhere else on the continent. The southern limit of occurrence for six species of arctic–alpine plants along with the present range of *Fraxinus nigra* are shown in Figure 11; two other ash species have ranges that extend farther to the northwest than *F. nigra*, but none of these overlaps with the ranges of the six arctic species except along the eastern seaboard near Anticosti Island.

Possible climatic implications of the fossil assemblage might be as follows: the presence of the arctic-alpine forms *Rhododendron lapponicum*, *Carex capillaris*, and *Dryas integrifolia* suggests that the climate had short summers with less than four months above 10°C. On the other hand, the presence of plants from the southern part of the boreal forest (*Fraxinus*, which is usually common in the pollen count, but also *Quercus*, *Ostrya-Carpinus*, and *Ulmus* are consistently present) suggests

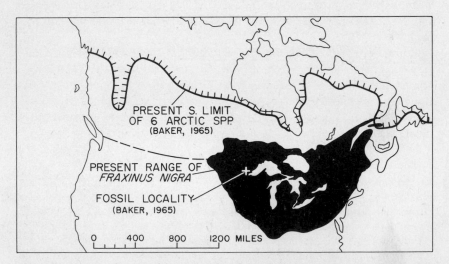

FIG. 11. Ranges of selected late-glacial species at present time, and fossil locality where they have been found together. Southern limit of 6 species of arctic plants includes the following: *Dryas integrifolia, Potentilla nivea, Carex capillaris, Vaccinium uliginosum* var. *alpinum, Rhododendron lapponicum,* and *Salix herbacea* (after Baker, 1965 and personal communication, 1967).

that the growing season was moderately warm; in the present ranges of *Fraxinus, Ulmus, Ostrya-Carpinus*, the July temperature is above 16°C, and the warmest month is cooler than 22°C. The only place in eastern North America where one can find a short growing season combined with warm summer temperatures is along the East Coast in the environs of Labrador and Anticosti Island, and it is only in this area where some (but not all) of these plants actually grow together today (Fig. 11). Hence, a relict late-glacial habitat may occur in a small area along the eastern seaboard in this region (climatic data from Brooks at el., 1936).

Pollen of warm-growing-season plants and megafossils of short-growing-season (arctic) plants are also found in late-glacial deposits of New England (Argus and Davis, 1962; Davis, 1958). The possibility that the thermophilous types are represented only by reworked pollen is countered by the consistent presence of these forms in late-glacial deposits.

In Europe, Iversen (1954) has described similar strange assemblages of arctic, heliophytic, and thermophilous plants in the late-glacial. Reid (1915) has found megafossils of *Carpinus betulus, Apium, Potamogeton crispus, Valeriana dioica*, which are strictly temperate plants requiring a warm growing season, along with megafossils of *Salix reticulata, Armeria arctica*, and *Thalictrum alpinum*, which are strictly arctic, short growing-season plants. These plants do not occur together today. The fact that their ranges overlapped in the late-glacial suggests that the climate and habitat of the European late-glacial essentially has no modern counterpart.

Plants that have widely separated ranges in the mid-continent today may have had overlapping ranges in Minnesota during the late-glacial. The fact that many of the late-glacial plants now grow together only along the eastern seaboard suggests that the late-glacial climate of Minnesota was less continental than now. Evidence of decreased continentality during the late-glacial is also provided by records from Europe.

CONCLUSIONS

An analysis of European seed floras of late Cenozoic age leads to the following conclusions: (1) The highest mortality of species during the late Cenozoic of Europe was among the angiosperms, especially among the dicots. (2) Changing proportions of extinct species in successive floras of the same area are a good index of changes in extinction rates in the regional flora. Percentages of actual extinctions in the regional floras compare favorably and show parallel trends with data on proportions of extinct species occurring in the same floras. (3) Extinction rates were altered by climate in given areas, and were not a simple function of time.

In the United States, evidence on a generic basis indicates that developing continentality of climate brought about earlier and more extensive floristic changes at inland than at west coastal localities. The high degree of continentality in present climates of western interior United States is related to the existence of north–south mountain ranges that create rain shadows and accentuate seasonal extremes of temperatures in the basins. Mountain and regional uplift during the Oligocene and late Tertiary accentuated the continental nature of the late Tertiary climate in the western interior. A surge of continentality during the late Wisconsin and early postglacial is indicated by vertebrate, invertebrate, and plant records of the central United States as well as by plant records of central northern Europe.

The importance of geographic factors in influencing kinds and rates of floristic change is clear from a wide variety of Tertiary and Quaternary plant records. Differences in scale of floristic change occurring contemporaneously in various areas of about the same latitude undoubtedly relate to geographic factors that caused climate to be expressed differently in the respective regions. Because of the fundamental relation between floristic change and factors that affect local climate, modernization curves of the types presented here are mainly meaningful on a local basis (e.g. the smaller the geographic base, the better); the curves dramatically illustrate regional differences or similarities in broad patterns of floristic change.

References

Argus, G. W., and Davis, M. B., 1962, Macrofossils from a late-glacial deposit at Cambridge, Massachusetts: *Amer. Midland Naturalist*, v. *67*, p. 106–17

Axelrod, D. I., 1950, *Studies in Late Tertiary paleobotany:* Carnegie Inst. Wash. Publ. 590, 323 p.

———— 1957, Age-curve analysis of angiosperm floras: *Jour. Paleont.*, v. *31*, p. 273–280

———— 1959, Late Cenozoic evolution of the Sierran bigtree forest: *Evolution*, v. *13*, p. 9–23

———— 1964, The Miocene Trapper Creek flora of southern Idaho: *Univ. California, Publ. Geol. Sci.*, v. *51*, 148 p.

———— 1966, The Pleistocene Soboba flora of Southern California: *Univ. California, Publ. Geol. Sci.*, v. *60*, 79 p.

Bailey, H. P., 1960, A method of determining the warmth and temperateness of climate: *Geografiska Annaler*, v. *42*, p. 1–16

———— 1964, Toward a unified concept of the temperate climate: *Geog. Rev.*, v. *54*, p. 516–45

Baker, R. G., 1965, Late-glacial pollen and plant macrofossils from Spider Creek, southern St. Louis County, Minnesota: *Geol. Soc. America Bull.*, v. *76*, p. 601–10

Barghoorn, E. S., 1951, Age and environment; A survey of North American Tertiary floras in relation to paleoecology: *Jour. Paleont.*, v. *25*, p. 736–44

Beals, H. O., and Milhorn, W. N., 1961, An extinct conifer *Larix*(?) *pleistocenicum* n. sp. from Yarmouth interglacial deposits of Indiana: *Jour. Paleont.*, v. *35*, p. 667–72

Benson, Lyman, and Darrow, R. A., 1954, *Trees and shrubs of the southwestern deserts:* Tucson, Univ. Arizona Press, 437 p.

Briggs, J. C., 1966, Zoogeography and evolution: *Evolution*, v. *20*, p. 282–89.

Brooks, C. F., Connor, A. J., and others, 1936, *Climatic maps of North America:* Cambridge, Harvard Univ. Press, 27 p.

Chandler, M. E. J., 1923, The geological history of the genus *Stratiotes*; An account of the evolutionary changes which have occurred within the genus during Tertiary and Quaternary times: *Geol. Soc. London Quart. Jour.*, v. *79*, p. 117–38

——— 1963, Revision of the Oligocene floras of the Isle of Wight: *Brit. Mus. Nat. Hist. Bull.*, v. *6*, p. 321–84

Chaney, R. W., *Editor*, 1944, *Pliocene floras of California and Oregon:* Carnegie Inst. Washington, publ. 553, 407 p.

Chaney, R. W., and Axelrod, D. I., 1959, *Miocene floras of the Columbia Plateau:* Carnegie Inst. Washington, publ. 617, 237 p.

Colinvaux, P. A., 1963, A pollen record from Arctic Alaska reaching glacial and Bering land bridge times: *Nature*, v. *198*, p. 609–10

Cushing, E. J., 1965, Problems in the Quaternary phytogeography of the Great Lakes region, p. 403–16, *in* Wright and Frey, 1965

Dahl, Eilif, 1964, Present-day distribution of plants and past climate, p. 52–61, *in* Hester, J. J., and Schoenwetter, J., *Assemblers, Reconstruction of past environments:* Santa Fe, Fort Burgwin Res. Center, no. *3*, 89 p.

Davis, Margaret B., 1958, Three pollen diagrams from central Massachusetts: *Amer. Jour. Sci.*, v. *256*, p. 540–70

——— 1965, Phytogeography and palynology of northeastern United States, p. 377–402, *in* Wright and Frey, 1965

Elias, M. K., 1935, Tertiary grasses and other prairie vegetation from the high plains of North America: *Amer. Jour. Sci.*, v. *29*, p. 24–33

Evernden, J. F., and Curtis, G. H., 1965, The potassium-argon dating of late-Cenozoic rocks in East Africa and Italy: *Curr. Anthropol.*, vol. *6*, p. 343–385

Faegri, Knut, and Iversen, Johs, 1964, *Textbook of pollen analysis:* Copenhagen, Munksgaard, 237 p.

Florin, Rudolf, 1944, Evolution in Cordaites and Conifers: *Acta Horti Bergiani*, v. *15*, p. 285–388

Frost, F. H., 1927, The Pleistocene flora of Rancho La Brea: *Univ. Calif. Publ. Bot.*, v. *14*, p. 73–98

Gerasimov, I. P., Baranov, A. H., and others, 1964, *Physical geographic Atlas of the world* (in Russian): Moscow, Akad. Nauk SSSR i glavnoe upravlenie geodezii i kartographii, 298 p.

Godwin, H., 1956, *The history of the British flora:* Cambridge Univ. Press, 383 p.

Grichuk, V. P., 1960, Stratigraphic division of the Pleistocene on the basis of paleobotanical materials: 21st Internat. Geol. Cong. (Copenhagen, 1960), Rept., Pt. 4, p. 27–35

Hafsten, Ulf, 1960, Pollen-analytic investigations in South Norway, p. 434–62, *in Geology of Norway*, Norges Geologiske Undersøkelse, no. 208

———— 1961, Pleistocene development of vegetation and climate in the southern High Plains as evidenced by pollen analysis, p. 59–91, *in* Wendorf, Fred, *Assembler, Paleoecology of the Llano Estacado:* Santa Fe, Fort Burgwin Res. Center, no. 1, 144 p.

Hansen, H. P., 1947, Postglacial forest succession, climate, and chronology in the Pacific Northwest: *Amer. Phil. Soc. Trans.*, no. *37*, 130 p.

Hibbard, C. W., Ray, C. E., Savage, D. E., Taylor, D. W., and Guilday, J. E., 1965, Quaternary mammals of North America, p. 509–25, *in* Wright and Frey, 1965

Hopkins, D. M., MacNeil, F. S., and Leopold, E. B., 1960, The coastal plain at Nome, Alaska; A type section for the Quaternary chronology of the Bering Sea region: 21st Internat. Geol. Cong. (Copenhagen, 1960), Rept., Pt. 4, p. 46–57

Iversen, J., 1954, The late-glacial flora of Denmark and its relation to climate and soil: *Danmarks Geol. Unders.*, ser. II, no. *80*, p. 87–119

Jepson, W. L., 1925, *A manual of the flowering plants of California:* Berkeley, Univ. California Press, 1237 p.

Livingstone, D. A., 1955, Some pollen profiles from Arctic Alaska: *Ecology*, v. *36*, p. 587–600

MacGinitie, H. D., 1953, *Fossil plants of the Florissant beds, Colorado:* Carnegie Instn. Publ. 599, 198 p.

Malde, H. E., 1965, Snake River plain, p. 255–63, *in* Wright and Frey, 1965

Martin, P. S., 1958, Pleistocene ecology and biogeography of North America, p. 375–420, *in* Hubbs, C. L., *Editor, Zoogeography:* Amer. Assoc. Adv. Sci., Publ. 51, 509 p.

Martin, P. S., and Mehringer, P. J., Jr., 1965, Pleistocene pollen analysis and biogeography of the southwest, p. 433–51, *in* Wright and Frey, 1965

Mason, H. L., 1944, *A Pleistocene flora from the McKittrick asphalt deposits of California:* Calif. Acad. Sci. Proc., 4th ser. v. *25*, p. 221–34

Miki, Shigeru, 1952, *Trapa* of Japan with special reference to its remains: *Jour. Inst. Polytechnics*, Osaka City Univ., Ser. D, v. *3*, p. 1–29

———— 1955, Nut remains of Juglandaceae in Japan: *Jour. Inst. Polytech.*, *Osaka City Univ.*, Ser. D, v. *6*, p. 131–44

———— 1956a, Seed remains of Vitaceae in Japan: *Jour. Instit. Polytech.*, Osaka City Univ., Ser. D, v. *7*, p. 247–71

———— 1956b, Endocarpal remains of Alangiaceae, Cornceae and Nyssaceae in Japan: *Jour. Inst. Polytech., Osaka City Univ.*, Ser. D, v. *7*, p. 141–46

———— 1958, Gymnosperms in Japan with special reference to their remains: *Jour. Inst. Polytech., Osaka City Univ.*, Ser. D, v. *9*, p. 125–52

———— 1959, Evolution of *Trapa* from ancestral *Lythrum* through *Hemitrapa*: *Japan Acad. Proc.*, v. *35*, p. 289–94

———— 1960, Nymphaeaceae remains in Japan with new fossil genus *Eoeurale: Jour. Inst. Polytech., Osaka City Univ.,* Ser. D, v. *11,* p. 63–78

———— 1961, Aquatic floral remains in Japan; *Jour. Biology, Osaka City Univ.,* v. *12,* p. 91–121

Munns, E. N., 1938, The distribution of important forest trees of the United States: U.S. Dept. Agric. Misc. Publ. no. *287,* 170 p.

Neustadt, M. I., 1959, Geschichte der Vegetation der USSR im Holozän: *Grana Palynologica,* v. *2,* p. 69–76

Ogden, J. G., III, 1967, Radiocarbon and pollen evidence for a sudden change in climate in the Great Lakes region approximately 10,000 years ago, *in* Cushing, E. J., and Wright, H. E., Jr., *Editors, Quaternary Paleoecology:* New Haven, Yale Univ. Press (in press)

Pannekoek, A. J., *Editor,* 1956, *Geological history of the Netherlands:* Geologische Stichting, Geologische Dienst, 'S-Gravenhage Staatsdrukkerij en Uitgeveribedrijf, Netherlands, 147 p.

Petrov, O. M., 1963, *The stratigraphy of the Quaternary deposits of the southern parts of the Chukotsk Peninsula:* Moscow, Acad. Sci. USSR, Bull. Commission for the study of the Quaternary, no. 28. Trans. by M. C. Blake, U.S. Geol. Survey, Menlo Park, California

Reid, Clement, 1915, The plants of the Late Glacial deposits of the Lea Valley: *Geol. Soc. London Quart. Jour.,* p. 155–63

Reid, C., and Reid, E. M., 1915, *The Pliocene Floras of the Dutch–Prussian Border:* The Hague, Mededeelingen van de Rijisopsporing van Delfstoffen, no. 6, 178 p.

Reid, E. M., 1920a, A comparative review of Pliocene floras, based on the study of fossil seeds: *Geol. Soc. London Quart. Jour.,* v. *76,* p. 145–61

———— 1920b, On two preglacial floras from Castle Eden: *Geol. Soc. London Quart. Jour.,* v. *76,* p. 104–44

———— 1935, Discussion on the origin and relationship of the British flora; British floras antecedent to the Great Ice Age: *Roy. Soc. Proc.,* B, v. *118,* p. 197–202

Reid, E. M., and Chandler, M. E. J., 1923, The fossil flora of Clacton-on-Sea: *Geol. Soc. London Quart. Jour.,* v. *79,* no 4, p. 619–23

Schmalhausen, I. I., 1946, *Factors of evolution: the theory of stabilizing selection,* trans. by I. Dordick: Philadelphia, Blakiston, 327 p.

Seward, A. C., 1959, *Plant life through the ages:* New York, Hafner, 602 p.

Slaughter, B. H., 1967, Animal ranges as a clue to late-Pleistocene extinction (this volume)

Snyder, C. T., Hardman, George, and Zdenek, F. F., 1964, Pleistocene lakes in the Great Basin: U.S. Geol. Survey Misc. Geol. Publ., Map I-416

Stebbins, G. L., 1950, *Variation and evolution in plants:* New York, Columbia Univ. Press, 643 p.

Steven, T. A., and Ratté, J. F., 1965, Geology and structural control of ore deposition in the Creede district, San Juan Mountains, Colorado: U.S. Geol. Survey Prof. Paper 487, 90 p.

Szafer, Wladyslaw, 1953, *Pleistocene Stratigraphy of Poland from the Floristical*

Point of View: Polskie Towarzystwo Geologiczne, Krakow. Roznik tom 22, p. 1-60

———— 1954, *Pliocene flora from the Czorsztyn area, and its relation to the Pleistocene* (in Polish): Krakow, *Inst. Geol. Prace,* v. *11,* 238 p.

———— 1961, *Miocene flora from Old Gliwice in Silesia* (in Polish): Krakow, *Inst. Geol. Prace,* v. *32,* 205 p.

Taylor, D. W., 1965, The study of Pleistocene nonmarine mollusks in North America, p. 597–612, *in* Wright and Frey, 1965

Vangengeym, E. A., 1961, Paleontological basis of the stratigraphy of the Anthropogene deposits of northeast Siberia: *Acad. Sci. USSR, Vipusk 48,* Geol. Inst. Proc. (trans.)

Weber, W. A., 1965, Plant geography in the southern Rocky Mountains, p. 453–68, *in* Wright and Frey, 1965

Wells, P. V., 1966, Late Pleistocene vegetation and degree of fluvial climatic change in the Chihuahuan Desert: *Science,* v. *153,* p. 970–75

West, R. G., 1957, Interglacial deposits of Bobbitshole, Ipswich: *Roy. Soc. Phil. Trans.,* B, no. 676, v. *241,* p. 1–31

Wilson, Druid, Keroher, G. C., and Hansen, B. E., 1959, Index to the geologic names of North America: *U.S. Geol. Surv. Bull. 1056-B,* 622 p.

Wolfe, J. A., 1964, Miocene floras from Fingerrock Wash, southwestern Nevada: U.S. Geol. Surv. Prof. Paper 454-N, 36 p.

Wolfe, J. A., and Barghoorn, E. S., 1960, Generic change in Tertiary floras in relation to age: *Amer. Jour. Sci.,* v. *258-A,* p. 388–99

Wolfe, J. A., Hopkins, D. M., and Leopold, E. B., 1966, Tertiary stratigraphy and paleobotany of the Cook Inlet region, Alaska: U.S. Geol. Surv. Prof. Paper 398-A, 29 p.

Wolfe, J. A., and Leopold, E. B., 1967, Tertiary and Early Quarternary vegetation of northeast Asia and northwest North America, *in* Hopkins, D. M., *Editor, The Bering Land Bridge:* Stanford Univ. Press (in press)

Wright, H. E., Jr., and Frey, D. G., *Editors,* 1965, *The Quaternary of the United States:* Princeton Univ. Press, 922 p.

Zagwijn, W. H., 1961, Vegetation, climate and radiocarbon datings in the Late Pleistocene of the Netherlands. Pt. I. Eemian and early Weichselian: *Geol. Foundation Netherlands, Mem.,* n.s. *14,* p. 15–45

———— 1963a, Pollen-analytic investigations in the Tiglian of the Netherlands: *Geol. Stichting, Meded.,* n.s. *16,* p. 49–71

———— 1963b, Pleistocene stratigraphy in the Netherlands, based on changes in vegetation and climate: *Ver. k. Nederlands Geol. Mijn. Genootschap, Verhandl. Geol. ser.,* v. *21–2,* p. 173–96

Zaklinskaya, Ye. D., 1962, Outline of the geology and paleogeography of the Pavlodar section of the Irtysh Valley, the northern Aral region and the Turgai Lowland: *Internat. Geol. Rev.,* v. *4,* p. 310–35

PETER J. MEHRINGER, JR.

Geochronology Department
University of Arizona
Tucson, Arizona

THE ENVIRONMENT OF EXTINCTION OF THE LATE-PLEISTOCENE MEGAFAUNA IN THE ARID SOUTHWESTERN UNITED STATES [1]

Abstract

The same types of habitat that are widespread today in the western United States were occupied by the late-Pleistocene megafauna. The arid regions of today were less widespread even after the major wave of extinction, which ended by 10,000 years ago. About 12,000 years ago some areas of the Southwest probably became marginal habitat, but there were no major barriers to migration into more favorable regions.

If climatic change is to be considered the principal cause of extinction, the extreme glacial climates of Wisconsin age should have exerted a detrimental effect on the extinct fauna. Excluding nonclimatic factors, the period of rapid deglaciation should have resulted in the expansion and not the demise of the megafaunal populations. At the present time there is a greater area and probably a wider variety of habitats available to herbivores than existed in North America during the major Wisconsin ice advances. Large herbivore biomass should have increased, not declined, as the ice retreated.

Because different species of the extinct late-Pleistocene megafauna occupied habitats ranging from warm semiarid to periglacial, it seems unlikely that a single climatic cause alone is responsible for extinction.

Climatic change to warmer and drier conditions has been suggested as the major or only cause of extinction of the Wisconsin megafauna in North America. The only thing that can be said with assurance, however,

---

1. Contribution 125, Program in Geochronology, University of Arizona.

is that extinction in North America coincided with deglaciation. The
environment in which the megafauna existed must be determined as
well as the change in this environment during the time of extinction, if
a cause-and-effect relationship is to be demonstrated.

I shall examine the chronology of Wisconsin vegetational change
before and during the apparent period of extinction, and in a general
way attempt to reconstruct the environment in which the megafauna
existed. The evidence includes the available fossil pollen and plant
records and the present distribution of extant animals associated with
the extinct megafauna. The discussion will be limited to the now arid
and semiarid areas of the Mohave and Sonoran deserts of Nevada,
California, and western Arizona and the desert grassland and Chihuahuan
Desert of southeastern Arizona (Fig. 1). If extinction was the result of

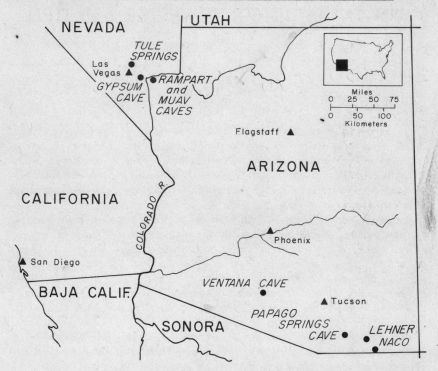

FIG. 1. Localities mentioned in text.

desiccation associated with deglaciation, such a cause should be obvious
in these currently desert areas. The main theme of this paper argues
against warming and drying climate as an important factor in late-
Pleistocene extinction in North America.

THE TIME OF EXTINCTION

Most authors have followed Martin (1958) and Hester (1960) in placing the time of extinction of most of the megafauna at about 8,000 radiocarbon years ago. By rejecting what he considers less reliable radiocarbon dates, Martin (1967, Fig. 1) has revised his estimate for the extinction of most of the megafauna to about 11,000 B.P. Hester (1967, Table 3, Fig. 1) still accepts dates of about 8,000 to 8,500 B.P. for several species. He uses a solid carbon date (C-222, 8,527 ± 250, average of three dates; Arnold and Libby, 1951) on sloth dung from Gypsum Cave, Nevada (Harrington, 1933), as the terminal date for *Nothrotherium shastense*, *Canis dirus*, and *Tanupolama*. If the date is correct, it implies that *Nothrotherium shastense* survived longest in one of today's most arid regions of North America. The date does not necessarily apply to other fossil remains from Gypsum Cave, however. In any case, it is a solid carbon date and may not be reliable. A second date from Gypsum Cave, 10,455 ± 340 (Arnold and Libby, 1951), is also a solid carbon date. A third, however, is not, and this is probably the most reliable of the three: 11,690 ± 250 (Hubbs et al., 1963). Dates on sloth dung from nearby Rampart Cave are 9,900 ± 400 (surface material), 11,900 ± 500, and > 38,000 B.P. (Olson and Broecker, 1961). The sloth dung date from Aden Crater, New Mexico, is 11,080 ± 200 B.P. (Simons and Alexander, 1964).

Hester cites a date of 8,240 ± 960 (A-184C, Damon and Long, 1962) from the Double Adobe Site (Whitewater Draw), southeastern Arizona, as a terminal date for *Mammuthus columbi*, *Camelops* sp., and *Equus* sp. The radiocarbon sample was collected by P. S. Martin to date a pollen profile from Double Adobe (Martin, 1963a, Fig. 22) and is not directly associated with the extinct fauna at the site.

On the basis of currently acceptable radiocarbon dates, I consider that the major wave of extinction was completed before 10,000 radiocarbon years ago. I do not imply that the last member of each extinct species was dead by that time, but only that extinction had progressed to the point where, with a few exceptions, the cause or causes predate 10,000 B.P. *Bison antiquus* and *B. occidentalis* survived on the plains until about 7,500 B.P. (Haynes, 1966). A terminal date for the extinction of the mastodon should perhaps be left open (see Martin, 1967).

THE CHRONOLOGY OF VEGETATIONAL CHANGE

Pollen from coprolite, playa lakes, alluvium, and spring mound deposits, as well as some plant macrofossils (Martin and Mehringer, 1965; Mehringer, 1965, p. 172–74; 1967), indicate that during the maximum of Wisconsin pluvial conditions the vegetation zones in the Southwest were probably lowered by about 1,000 m. At this time the present desert areas were partly occupied by parkland, woodland, sagebrush, or grassland (Martin and Mehringer, 1965, Figs. 3, 4). Undoubtedly many changes in the vegetational patterns occurred during the Wisconsin, but I find no convincing evidence that desert conditions like those of the present existed before 7,500–7,000 B.P.

The fossil pollen and plant record from Las Vegas Valley, Nevada (Fig. 1), indicates that about 13,000–14,000 B.P., the area now occupied by creosote bush (*Larrea divaricata*) and bursage (*Franseria dumosa*) supported sagebrush (*Artemisia*) and juniper probably much like that found today in northern Nevada. At approximately 12,000 B.P. there was a rapid change to sagebrush and shadscale (*Atriplex*). From about 12,000 to 7,500 B.P. there was a trend toward warmer and drier conditions. This trend was interrupted at least twice (about 10,500–10,000 and 8,500–8,000 B.P.) by a return to moisture and cooler conditions, but these appear to be of lesser magnitude than those prior to 12,000 B.P. By 7,500–7,000 B.P., conditions were probably much like those of the present (Mehringer, 1967).

The fossil plants recovered from sloth (*Nothrotherium shastense*) dung deposits at Gypsum Cave, Nevada, indicate a vegetational type that now grows at higher elevation (see Fig. 5; Laudermilk and Munz, 1934, p. 37). However, the plants recovered from the sloth dung at Rampart Cave, Arizona, only 35 km southeast of Gypsum Cave, led Laudermilk and Munz (1938, p. 278) to conclude that "the flora and presumably the climate were essentially the same as those of the region today." The apparent differences in the floras led Wilson (1942, p. 176) to suggest the possibility that the Rampart Cave fauna was younger than the Gypsum Cave fauna.

Further studies of the Rampart Cave coprolites (Martin et al., 1961) and the addition of C^{14} dating suggest an explanation for the apparent differences in the two floras. Martin et al. (1961, p. 104) reported twigs and a seed of juniper from the 91-cm level of the sloth dung deposit at Rampart Cave. The fossil juniper remains were recovered from what was obviously a wood rat (*Neotoma*) midden, located 45 cm below an horizon dated 11,900 ± 500 B.P. (Olson and Broecker, 1961, p. 165). On the basis of the fossil pollen content of Rampart Cave sloth dung, Martin et al. (1961, p. 115) also suggested a change from a cool-moist to a warm-dry interval, beginning between the 30- and 46-cm levels and continuing to the top of the deposit.

Dr. P. A. Munz (personal communication) collected all the Rampart and Muav Cave sloth dung samples from the *surface of the deposits.*

There is one surface-material radiocarbon date of 9,900 ± 400 B.P. (Olson and Broecker, 1961, p. 165). It is very probable that sometime after the deposition of the juniper remains, before 12,000 radiocarbon years ago, and before the surface material was deposited about 10,000 years ago, there was a considerable change in the local flora and climate, and that the apparent differences in the original reports of the Rampart and Gypsum Cave floras are the result of the different ages of the material examined.

At the Lehner site in southeastern Arizona the pollen record dates from 11,200 B.P. (Mehringer and Haynes, 1965, Figs. 7, 8). Today the vegetation of San Pedro Valley near the Lehner site is desert grassland or Chihuahuan Desert (Lowe, 1964, Figs. 4, 5, 23; Mehringer and Haynes, 1965, Figs. 2, 3). My interpretation of the pollen evidence from the Lehner site and elsewhere in southeastern Arizona (Martin, 1963a) is that the Chihuahuan Desert elements were probably not an important feature of San Pedro Valley until about 7,500–7,000 years ago.

For southern Arizona there is no dated evidence for vegetational change between about 20,000 and 11,200 B.P., but the pollen record from Lake Cochise (Martin, 1963b; Martin and Mosimann, 1965) implies that in Wisconsin time the vegetational zones were about 1,000 m lower than at present, a difference as great as that in the Mohave Desert of southern Nevada. Although the changes recorded after 12,000 and before 7,500 B.P. may be coeval, they are apparently not so great. In the Mohave Desert the vegetational zones may have been 600 m lower than the present, while in southeastern Arizona a change of more than 300 m seems improbable.

If extinction was the result of desiccation the major change about 12,000 B.P. should be critical as an event related to extinction. However, conditions like those of the present in the area discussed did not occur until 7,500–7,000 B.P., about 3,000 years after extinction of most of the megafauna. After 12,000 B.P., the now extinct megafauna of the present Mohave Desert lived in an environment that was probably like that at present existing at higher elevations or in the Great Basin Desert to the north (Figs. 2-5). Pollen evidence also shows that the megafauna of southeastern Arizona occupied a grassland (Lowe, 1964, Figs. 23, 30) similar to that existing there today (Figs. 6, 7).

EXTANT SPECIES ASSOCIATED WITH THE EXTINCT MEGAFAUNA

For a review of the Wisconsin faunas of the area discussed, two localities will be emphasized, Tule Springs, Nevada, and Ventana Cave, Arizona (Fig. 1, Tables 1, 2). These sites were selected because they have been radiocarbon-dated and are from areas where there is some independent evidence for the nature of the Wisconsin environment. Other localities with extinct megafauna and associated extant species lack radiocarbon dates clearly associated with the fauna. For example, *Marmota flaviventris* from Rampart Cave (Wilson, 1942) and Papago Springs Cave (Skinner, 1942) represent range extensions southward and to lower elevations some time during the Wisconsin. However, with a few notable exceptions (Mehringer, 1966, Fig. 4), the majority of extant species associated with the extinct megafauna, from the deserts of California, Nevada, and Arizona, do not indicate major environmental changes.

FIG. 2. From the Tule Springs site, Nevada, showing the sparse vegetation dominated mainly by creosote bush (*Larrea divaricata*) and bursage (*Franseria dumosa*) (also see Mehringer 1965, Figs. 2, 9).

The faunal list from the Tule Springs site (Fig. 2) is given in Table 1. Unit B_2 has abundant large horse and bison, whereas unit E_1 is dominated by small horse and Antilocaprids. *Brachylagus idahoensis* in unit E_1 is now found in central and northern Nevada (Hall, 1946, Fig. 451), and its presence at Tule Springs agrees with the pollen evidence for the southward extension of Great Basin Desert at that time. *Microtus californicus* is at present found in east-central California on the Nevada border (Ingles, 1965, map 76). The Tule Springs record indicates moist habitats along streams or around springs but not necessarily a major change in climate. The extant species that are associated with the extinct species give no indication of conditions very different from those found today in the Great Basin north of central Nevada.

Fig. 3. How the Tule Springs site, Nevada, may have looked 12,000 to 10,000 years ago, with scattered sagebrush (*Artemisia*) and saltbush (*Atriplex*), and springs with ash trees, cattails, and other mesic species (also see Mehringer, 1965, Fig. 11; painting by Jay Matternes, reproduced with permission of the National Geographic Society).

FIG. 4. The sparse desert vegetation in the vicinity of Gypsum Cave, Nevada (arrow at cave).

FIG. 5. Vegetational type which may have grown in the Gypsum Cave area when *Nothrotherium* occupied the cave. In southern Nevada the joshua trees (*Yucca brevifolia*) shown in the photo usually occur above an elevation of 1,200 m, 600 higher than Gypsum Cave, and their remains are common as macrofossils in Gypsum Cave sloth dung (also see Laudermilk and Munz, 1934, Pl. 11).

FIG. 6. The vegetation at the Lehner Mammoth site, southeastern Arizona. The conspicuous plants are *Ephedra trifurca* (in foreground), mesquite (*Prosopis juliflora*), and white-thorn (*Acacia vernicosa*) (also see Mehringer and Haynes, 1965, Figs. 1, 3).

FIG. 7. The grassland of southeastern Arizona, with scattered *Quercus* and *Prosopis*. Such vegetation probably occupied the area of the Lehner Mammoth site when Early Man and the extinct megafauna lived there.

TABLE 1. Faunal List from Tule Springs, Nevada

From Mawby (1967). Nine species of birds are not included in this list; five of these are extant and four were not identified to species. The two birds identified as extinct species are listed. Fossils that were not identified to species but are probably extant forms are indicated by a question mark.

	Unit B₂ > 40,000 B.P.	Unit E₁ ca. 14,000- 11,500 B.P.	Extant
Fulica americana minor	×		
Teratornis merriami		×	
Nothrotherium shastense	×	?	
Megalonyx sp.	×	×	
Lepus cf. californicus	×	×	×
Sylvilagus sp.	×	×	?
Brachylagus idahoensis		×	×
Dipodomys sp.		×	?
Microtus sp.	×		?
M. californicus		×	×
Ondatra zibethica		×	×
Canis latrans		×	×
Felis (Puma) sp.		×	?
F. sp. or Lynx sp.	×		?
Panthera (Jaguaris) atrox	×		
Mammuthus columbi	×	×	
Equus sp. (large)	×	×	
E. sp. (small)	×	×	
Camelops hesternus	×	×	
Odocoileus sp.		×	?
Tetrameryx sp.		×	
Bison sp.	×		

Marmota flaviventris, recovered from a wood rat (*Neotoma*) midden near Frenchman Flat, southern Nevada, and dated by associated wood as 12,700 ± 200 B.P., is now found no closer than the mountains of central Nevada, approximately 200 km north (Wells and Jorgensen, 1964). The majority of the extant species associated with the megafauna in southern Nevada, however, do not indicate major range changes. There are two obvious reasons for this. First, some of the extant species have a wide environmental range today. Second, there is inadequate stratigraphic control on what is included in a single fauna. For example, there are at least two significant fluctuations in the pollen record from unit E₁ (*ca.* 14,000–11,500 B.P.) at Tule Springs (Mehringer, 1967) that might have been accompanied by faunal shifts. But because all the vertebrate fossils from unit E₁ have been analyzed as a single fauna it is impossible to know exactly which extant and extinct species were associated during shorter time intervals. Without finer stratigraphic control it is difficult to draw environmental generalizations from these faunal assemblages.

TABLE 2. *Faunal List from Ventana Cave, Arizona*

Modified from Colbert, *in* Haury (1950), Table 8; Lance* (1959). Those fossils that were not identified to species but are probably extant forms and those that were not positively identified to genus are indicated by a question mark.

	Conglomerate	Volcanic debris	Extant
Lepus californicus		×	×
Cynomys ludovicianus		×	×
Callospermophilus sp.	×		?
* *Neotoma sp.*		×	?
Canis dirus	×		
C. latrans	×		×
Vulpes macrotis	×		×
Taxidea taxus		×	×
Felis atrox		×	
Nothrotherium shastense		×	
Tapirus sp.	×	×	
Equus occidentalis	×	×	
Platygonus sp. or *Pecari sp.*		×	?
Odocoileus sp.		×	?
Tetrameryx sp. or *Antilocapra sp.*	×	×	?
Bison sp.	×	×	

A list of the fauna from Ventana Cave (Fig. 8) is given in Table 2. Horse is more abundant in the conglomerate level (Colbert, *in* Haury, 1950, Figs. 15, 16, Table 9). About half the mammals present in Ventana Cave are probably extant. Of those identified by Colbert only the prairie dog (*Cynomys ludovicianus*) and golden-mantled ground squirrel (*Callospermophilus lateralis*) could indicate cooler or moister climates in the Sonoran Desert at the time the extinct fauna occupied the area. The latter species would especially signify climatic change.

Colbert (*in* Haury, 1950, p. 131) points out that there was some difficulty in making a specific identification of the fossil ground squirrel: "In view of these differences, there might be some reason for supposing that the ground squirrel from Ventana Cave represents a species distinct from any of the recent ground squirrels of this general type. And such may very probably be the case. But in view of the paucity of fossil material it is hereby proposed to regard the Ventana Cave material as identifiable under the name of *Callospermophilus lateralis*."

Because the presence of the golden-mantled ground squirrel at Ventana Cave during the Wisconsin would be the best evidence of climate change presently available for the Sonoran Desert, James D. Lane (now at McNeese State College, Louisiana) and I decided to re-examine the fossil ground squirrel material, kindly loaned to us by R. H. Thompson (Director, Arizona State Museum). We did not feel that a definite identification could be made on the material but it was also examined by Robert Wilson (South Dakota School of Mines) and by Craig Black, and John Guilday (Carnegie Museum, Pittsburgh). They all agreed with Colbert's identification of *Callospermophilus*. *C. lateralis* is the only species reported from Arizona and is usually found above an elevation of 2,150 m, or 1,400 m higher than Ventana Cave. The nearest locality is about 260 km north (Cockrum, 1960, Fig. 39, p. 91).

Fig. 8. Ventana Cave in the Sonoran Desert of southern Arizona (courtesy of the Arizona State Museum).

The other member of the Ventana Cave fauna of particular paleo-ecological interest, the prairie dog, is not known from closer than about 200 km east of Ventana Cave in the desert grassland of southeastern Arizona (Cockrum, 1960, Fig. 35). Thus the Ventana Cave fossil might indicate slightly more moist conditions during the deposition of the volcanic debris in which it is found. But it is also reported from the red sand and the overlying midden layers, which do not contain extinct fauna (Haury, 1950, Fig. 17). According to Bryan (in Haury, 1950, p. 126) the red sand was deposited during a "dry and presumably warm period," and it is separated from the volcanic debris by "a marked disconformity representing strong erosion"; it is probably of Altithermal age.

A single pollen count from the volcanic debris (Martin and Mehringer, 1965, Fig. 6) is what one would expect to find in the Sonoran Desert (Hevly et al., 1965) except for the Cyperaceae, which reflects spring seepage from the cave. Although the pollen sample came from the same stratigraphic unit it is possible that it postdates the extinct fauna.

The volcanic debris level of Ventana Cave has been dated at 11,300 ± 1,200 B.P. (Damon and Long, 1962, p. 246). A Clovis-like point from this level (Haynes, 1964, p. 1410; Haury, 1950, Pl. 20, Fig. 21) is in accord with an age of about 11,500 to 11,000 B.P. (Haynes, 1964, Fig. 2). The conglomerate unit that underlies the volcanic debris is possibly of Wisconsin age, and the erosion interval separating the conglomerate and the volcanic debris may have occurred between 12,000 and 11,500 B.P. A break in the stratigraphic record occurs at about this time at the Lehner site (Mehringer and Haynes, 1965) and at Tule Springs (Haynes, 1967a) and is found in the alluvial record throughout the plains and southwestern United States (Haynes, 1966).

There are no extant mammals from the volcanic debris that offer indisputable evidence of either cooler or moister conditions. However, the pollen record from the Lehner site and Tule Springs implies conditions slightly moister and cooler during the early deposition of the volcanic debris, and they may have been similar to the present desert grassland of southeastern Arizona. The underlying conglomerate level, which contains the fossil golden-mantled ground squirrel, may represent the even cooler and moister conditions of Wisconsin time. Spring discharge from the cave was greater during the deposition of the conglomerate, implying a less arid climate (Bryan, in Haury, 1950, p. 118).

There are no small mammals in association with the mammoth, bison, tapir, horse, and Clovis points at the Lehner site (Lance, 1959). At the Naco site, 17 km southeast, mammoth and bison were also found associated with Clovis points (Lance, 1959; Haury et al., 1953). The pollen record from the Lehner site has already been discussed. A single pollen spectrum obtained from dirt adhering to a mammoth bone from the Naco site is similar to the pollen spectra associated with the extinct fauna at the Lehner site (Martin and Mehringer, 1965, Fig. 6).

COULD THE EXTINCT WISCONSIN MEGAFAUNA LIVE IN THE
SOUTHWESTERN DESERTS TODAY?

Probably the best evidence in support of the fact that the present desert areas would not necessarily be uninhabitable by at least some of the extinct megafauna is the fossil pollen and plant contents of the dung of *Nothrotherium* (Table 3). The contents of the dung from Gypsum and Rampart caves indicate that the mainstays of the sloth diet still grow today in the Mohave and southern Great Basin deserts. At least part of the extinct fauna associated with *Nothrotherium* could also occupy the same habitat.

The present distribution of feral horses and burros includes the least grassy habitats of the southwestern deserts; in western North America wild horses range from British Columbia to Baja California (McKnight, 1964, Figs. 1, 2). The wild horse of today is able to occupy a wide variety of habitats, and by analogy the extinct horse and probably others of the extinct megafauna must have had a wide range of ecological tolerance and might have been little affected by late-Pleistocene climatic change. In only four years, in the late 1940s and early 1950s, over 100,000 wild horses were removed from Nevada ranges (McKnight, 1964, p. 9). Thus, *Equus* is not only able to survive but actually flourishes in the same area in which the genus became extinct some 10,000 years ago, and under conditions more arid than existed at the time of its extinction.

WISCONSIN CLIMATE AND EXTINCTION

If climatic change was the cause of extinction of the megafauna, possibly some cause(s) other than the onset of warmer and drier conditions may have been responsible. Other alternatives include extinction as a result of the cold climates associated with the major ice advances, and the onset of greater seasonality of climate at the time of extinction.

Slaughter (1967) suggests that the presence of both southern and northern (or higher-elevation) forms in many faunal assemblages indicates a climate with cooler summers and warmer winters. Along with others (Hibbard et al., 1965, p. 515; Taylor, 1965, p. 603), Slaughter proposes that present seasonal climatic extremes were atypical of most of the Pleistocene. He suggests such a solution to explain the environment of the fauna of the Brown Sand Wedge at Blackwater Draw, New Mexico, where, for example, both the extinct armadillo (*Dasypus bellus*), considered a warm indicator, and the masked shrew (*Sorex cinereus*), considered a cold indicator, are represented. The implications of a late-glacial climate with less seasonal extremes than today would have to be given serious consideration in examining the cause(s) of extinction.

TABLE 3. Pollen and Plant Macrofossils Recovered from the Dung of
Nothrotherium shastense

From Martin et al. (1961).

Location	Aden Crater, New Mexico (Eames, 1930)	Gypsum Cave, Nevada (Laudermilk and Munz, 1934)	Rampart Cave, Arizona	
			(Laudermilk and Munz, 1938)	(Martin et al., 1961)
Polypodiaceae	d			d
Adiantum capillus-veneris			a	
Pinus				d
Juniperus		?		a,c
Ephedra nevadensis		a	a	c (genus)
Typha		d		
Gramineae		a		c
Aristida			a	
Phragmites communis			a	
Cyperaceae				d
Yucca brevifolia		a,b (genus)		
Y. baccata		a		
Y. mohavensis		a	a (Muav)	
Nolina			a	
Agave utahensis		a		
Populus			a	
Juglans				d
Betula				d
Quercus				d
Chenopodiaceae	a			d
Atriplex	a	a	a	
Nyctaginaceae				b
Cruciferae	a			
Ribes				b
Prunus			a	
Cassia			a	
Larrea tridentata		a,b	a	b
Malvaceae		b		b
Sphaeralcea	a	a	a	
Sida	?			
Petalonyx		a		
Opuntia			a	
Onagraceae		b		b
Oenothera		a		
Fraxinus			a	
Polemoniaceae				b
Physalis			a	
Caprifoliaceae				b
Compositae	a	a,b		b,c
Gutierrezia	a			
Chrysothamnus		a		
Artemisia				c

a, Plant parts identified in dung.   c, Anemogamous pollen, abundant.
b, Zoögamous pollen.   d, Anemogamous pollen or spores, not abundant.

It first must be shown, however, by extremely detailed stratigraphic and radiocarbon controls, that all the species of assemblages used as indicators of such a climate actually occupied the area contemporaneously. Such controls are especially necessary for the late Wisconsin, when rapid glacial advances and retreats and associated floral changes are well documented.

Guilday (1967) summarizes the arguments for desiccation accompanying the retreat of Wisconsin ice as the most important factor in extinction as follows: (1) decrease in suitable habitat, especially grassland; (2) increase in competition with decrease in suitable habitat; (3) animals unable to migrate to more suitable habitats because of barriers.

Guilday's arguments might be applied more convincingly to the periods of Wisconsin glaciation when there was less area, probably less diversity of habitat, and possibly more barriers to migration. The extinction that became apparent by 10,000 years ago may not have been caused by conditions of the time but by those during the main part of Wisconsin glaciation. The time of extinction might lag several thousand years behind its cause(s). Such an explanation would require some evidence that the main Wisconsin continental glaciation was accompanied by more severely cold climates than were earlier glacial episodes. This is a distinct possibility (see discussion *in* Deevey, 1965, p. 649–50). Fossil molluscan faunas indicate to Taylor (1965, p. 602) that the Wisconsin climate was more severe than at any time earlier in the Pleistocene. Hibbard (1960, p. 24) accounts for the Wisconsin extinction of the giant tortoise (*Geochelone*) by concluding that this was the first time during the Pleistocene that winter temperatures dropped below the minimum tolerance of this tortoise.

The total area of large mammal habitat now available in North America is greater than during the maximum of Wisconsin glaciation (Haynes, 1967b, Fig. 1). Although coniferous forest and woodland were more widespread than at present, and the area of the present desert regions was reduced, there is no evidence to suggest that the total area of grassland was greater than it is today. On the contrary, in many areas of North America the grasslands were reduced in area by invading ice or coniferous trees (Horr, 1955; Hafsten, 1961; Wells, 1965; Watts and Wright, 1966).

During the Two Creeks ice recession, which corresponds in time to the beginning of warmer and drier conditions in southern Nevada and southeastern Arizona about 12,000 years ago, there was a greater area available to the North American megafauna than at any time during the preceding 10,000 years. From 12,000 to 10,000 B.P., when most of the fauna became extinct, conditions in the present desert West were cooler and moister than now, and the present desert areas of the Southwest could have been at least seasonal habitat for most of the megafauna and possibly the preferred habitat for ground sloths, Antilocaprids, and horses. Despite their low herbivore-carrying capacity, the present southwestern deserts are clearly far more productive than was the surface of the Wisconsin ice sheet. If there were major barriers to large mammal migration to more suitable habitats in the western United States, they do not exist today. If they existed in the past they would certainly have been more formidable during the maximum of glacial conditions when the high mountains were covered with ice.

If one wishes to suggest that extreme Wisconsin glacial climates were the major cause of extinction, one is faced with the same contradictions inherent in the argument for the warming and drying hypothesis—that the habitats of some of the megafauna included the periglacial or boreal forest regions near the ice front. There were undoubtedly cold-adapted forms, as *Symbos* and *Mammuthus primigenius*, which flourished in the colder climates just as *Nothrotherium* flourished under semiarid conditions. These cold habitats must still have existed along the retreating ice border after extinction of the megafauna. If one event is to be given more weight than the other, I feel that the extreme glacial climates would exert a more pronounced detrimental effect on the megafauna. As suggested by Sauer (1944, p. 541), we should expect conditions during the retreat of Wisconsin ice to be beneficial. In any case, there were certainly some animals better adapted than others to cold climates. In this respect it seems unreasonable to support any single climatic change as the major cause of extinction. If the modern African megafauna can be used as a comparison (Leakey, 1965, p. 74), the extinct North American megafauna probably had a wide range of tolerances.

At the same time that warming and drying climates accompanied deglaciation and the North American continent was recovering from a 10,000 year period of glaciation, a new and deadly predator may have entered the New World. His tool kit included not only Clovis points but, more significantly, the knowledge, skill, and ability to support himself by big-game hunting (Haynes, 1967b). Hester (1967) has argued that "only in time of great ecological stress could the presence of human predators have been of importance." But Martin (1967) argues strongly for man being the sole cause of extinction of the Wisconsin megafauna. Whatever view is taken of the importance of man as a factor in extinction, or of the relative merits of extreme glacial versus warming and drying climates as factors in extinction, it cannot be denied that man became an important predator at a critical period in the history of the North America megafauna.

References

Arnold, J. R., and Libby, W. F., 1951, Radiocarbon dates: *Science*, v. *113*, p. 111–20

Cockrum, E. L., 1960, *The recent mammals of Arizona:* Tucson, Univ. Arizona Press, 276 p.

Damon, P. E., and Long, Austin, 1962, Arizona radiocarbon dates III: *Radiocarbon*, v. *4*, p. 239–49

Deevey, E. S., Jr., 1965, Pleistocene nonmarine environments, p. 643–52, *in* Wright, H. E., Jr., and Frey, D. G., *Editors, The Quaternary of the United States:* Princeton Univ. Press, 922 p.

Eames, A. J., 1930, Report on ground sloth coprolite from Dona Ana County, New Mexico: *Amer. J. Sci.*, v. *120*, p. 353–56

Guilday, J. E., 1967, Differential extinction during late-Pleistocene and Recent times (this volume)

Hafsten, Ulf, 1961, Pleistocene development of vegetation and climate in the southern High Plains as evidence by pollen analysis, p. 59–91, *in* Wendorf, F., *Compiler, Paleoecology of the Llano Estacado:* Santa Fe, Mus. New Mexico, 144 p.

Hall, E. R., 1946, *Mammals of Nevada:* Berkeley, Univ. California Press, 710 p.

Harrington, M. R., 1933, *Gypsum Cave, Nevada:* Southwest Mus. Papers no. *8*, 197 p.

Haury, E. W., 1950, *The stratigraphy and archaeology of Ventana Cave, Arizona:* Tucson, Univ. Arizona Press, 599 p.

Haury, E. W., Antevs, Ernst, and Lance, J. F., 1953, Artifacts with mammoth remains, Naco, Arizona: *Amer. Antiq.*, v. *19*, p. 1–24

Haynes, C. V., Jr., 1964, Fluted projectile points; Their age and dispersion: *Science*, v. *145*, p. 1408–13

—— 1966, Geochronology of late Quaternary alluvium: Univ. Arizona Geochronology Labs; Interim Res. Rept. 10, 35 p.

—— 1967a, Quaternary geology of the Tule Springs area, Clark County, Nevada, *in* Wormington, Marie, *Editor, Pleistocene studies in southern Nevada:* Carson City, Nevada State Mus. (in press)

—— 1967b, Carbon-14 dates and Early Man in the New World (this volume)

Hester, J. J., 1960, Late Pleistocene extinction and radiocarbon dating: *Amer. Antiq.*, v. *26*, p. 58–77

—— 1967, The agency of man in animal extinction (this volume)

Hevly, R. H., Mehringer, P. J., Jr., and Yocum, H. G., 1965, Modern pollen rain in the Sonoran Desert: *Arizona Acad. Sci. Jour.*, v. *3*, p. 123–35

Hibbard, C. W., 1960, An interpretation of Pliocene and Pleistocene climates in North America: Michigan Acad. Sci. Ann. Rept., v. *62*, p. 1–30

Hibbard, C. W., Ray, D. E., Savage, D. E., Taylor, D. W., and Guilday, J. E., 1965, Quaternary mammals of North America, p. 433–51, *in* Wright, H. E.,

Jr., and Frey, D. G., *Editors, The Quaternary of the United States:* Princeton Univ. Press, 922 p.

Horr, W. H., 1955, A pollen profile study of the Muscotah Marsh: *Univ. Kansas Sci. Bull.,* v. *37,* p. 143–49

Hubbs, C. L., Bien, G. S., and Suess, H. E., 1963, La Jolla natural radiocarbon measurements III: *Radiocarbon,* v. *5,* p. 254–72

Ingles, L. G., 1965, *Mammals of the Pacific States:* Stanford Univ. Press, 506 p.

Lance, J. F., 1959, Faunal remains from the Lehner Mammoth site: *Amer. Antiq.,* v. *25,* p. 35–42

Laudermilk, J. D., and Munz, P. A., 1934, Plants in the dung of *Nothrotherium* from Gypsum Cave, Nevada: Carnegie Inst. Washington Publ. 453, p. 29–37

———— 1938, Plants in the dung of *Nothrotherium* from Rampart and Muav caves, Arizona: Carnegie Inst. Washington Publ. 487, p. 271–81

Leakey, L. S. B., 1965, Olduvai Gorge 1951–61, Vol. 1, *A preliminary report on the geology and fauna:* New York, Cambridge Univ. Press, 118 p.

Lowe, C. H., 1964, Arizona landscapes and habitats, p. 1–132, *in* Lowe, C. H., *Editor, The vertebrates of Arizona:* Tucson, Univ. Arizona Press, 259 p.

Martin, P. S., 1958, Pleistocene ecology and biogeography of North America, p. 375–420, *in* Hubbs, C. L., *Editor, Zoogeography:* Amer. Assoc. Adv. Sci., Publ. 51, 509 p.

———— 1963a, *The last 10,000 years; A fossil pollen record of the American Southwest:* Tucson, Univ. Arizona Press, 87 p.

———— 1963b, Geochronology of Pluvial Lake Cochise, southern Arizona, II, Pollen analysis of a 42-meter core: *Ecology,* v. *44,* p. 436–44

———— 1967, Prehistoric overkill (this volume)

Martin, P. S., and Mehringer, P. J., Jr., 1965, Pleistocene pollen analysis and biogeography of the Southwest, p. 433–51, *in* Wright, H. E., Jr., and Frey, D. G., *Editors, The Quaternary of the United States:* Princeton Univ. Press, 922 p.

Martin, P. S., and Mosimann, J. E., 1965, Geochronology of pluvial Lake Cochise, southern Arizona, III. Pollen statistics and Pleistocene meta-stability: *Amer. Jour. Sci.,* v. *263,* p. 103–11

Martin, P. S., Sabels, B. E., and Shutler, Dick, Jr., 1961, Rampart Cave coprolite and ecology of the shasta ground sloth: *Amer. Jour. Sci.,* v. *259,* p. 102–27

Mawby, J. E., 1967, Fossil vertebrates of the Tule Springs site area, *in* Wormington, Marie, *Editor, Pleistocene studies in southern Nevada:* Carson City, Nevada State Mus. (in press)

McKnight, Tom, 1964, Feral livestock in Anglo-America: *Univ. California Publ. Geog.,* v. *16,* p. 1–78

Mehringer, P. J., Jr., 1965, Late Pleistocene vegetation in the Mohave Desert of southern Nevada: *Arizona Acad. Sci. Jour.,* v. *3,* p. 172–88

———— 1966, Some notes on the late Quaternary biogeography of the Mohave Desert: Univ. Arizona Geochronology Labs., Interim Res. Rept. 11, 17 p.

———— 1967, Pollen analysis of the Tule Springs area, Nevada, *in* Wormington,

Marie, *Editor, Pleistocene studies in southern Nevada:* Carson City, Nevada State Mus. (in press)

Mehringer, P. J., Jr., and Haynes, C. V., Jr., 1965, The pollen evidence for the environment of Early Man and extinct mammals at the Lehner mammoth site, southeastern Arizona: *Amer. Antiq.,* v. *31,* p. 17–23

Olson, E. A., and Broecker, W. S., 1961, Lamont natural radiocarbon measurements VII: *Radiocarbon,* v. *3,* p. 141–75

Sauer, C. O., 1944, A geographic sketch of Early Man in America: *Geog. Rev.,* v. *34,* p. 529–73

Simons, E. L., and Alexander, H. L., 1964, Age of the shasta ground sloth from Aden Crater, New Mexico: *Amer. Antiq.,* v. *29,* p. 390–91

Skinner, M. F., 1942, The fauna of Papago Springs Cave, Arizona: *Bull. Amer. Mus. Nat. Hist.,* v. *80,* p. 143–220

Slaughter, B. H., 1967, Animal ranges as a clue to late-Pleistocene extinction (this volume)

Taylor, D. W., 1965, The study of Pleistocene nonmarine mollusks in North America, p. 297–611, *in* Wright, H. E., Jr., and Frey, D. G., *Editors, The Quaternary of the United States:* Princeton Univ. Press, 922 p.

Watts, W. A., and Wright, H. E., Jr., 1966, Late-Wisconsin pollen and seed analysis from the Nebraska sandhills: *Ecology,* v. *47,* p. 202–10

Wells, P. V., and Jorgensen, C. D., 1964, Pleistocene wood rat middens and climatic change in the Mohave Desert—A record of juniper woodlands: *Science,* v. *143,* p. 1771–74

Wells, P. V., 1965, Scarp woodlands, transported grassland soils, and concept of grassland climate in the Great Plains region: *Science,* v. *148,* p. 246–49

Wilson, R. W., 1942, Preliminary study of the fauna of Rampart Cave, Arizona: Carnegie Inst. Washington Publ. 530, p. 169–85

C. VANCE HAYNES, JR.

Geochronology Department
University of Arizona
Tucson, Arizona

CARBON-14 DATES AND
EARLY MAN IN THE NEW WORLD[1]

Abstract

Critical evaluation of carbon-14 dates of Early Man in thousand-year intervals reveals an increasing diversity of projectile point types, starting with Clovis points 11,500 years ago and reaching a wide variety of types by 8,000 B.P. The earliest types appear to be fluted points that are distributed from coast to coast almost from the beginning of their appearance between 11,000 and 12,000 years ago. After 11,000 B.P. both local and regional diversity increased, but the progression of typological change is best observed at stratified sites that are more numerous in the western United States. Major typological changes appear to occur regionally within the intervals of five hundred years; they could be determined more precisely with more careful stratigraphic control and with more careful sample selection and pretreatment to reduce error. Radiocarbon dating of alluvial and eollan stratigraphic sequences provides the most precise geochronological control of cultural change and the most reliable means of correlation. Apparently some projectile point types can be used as "index fossils," and on this basis certain cave dates should be questioned.

Too many dates have been interpreted as applying to man on very tenuous, if not erroneous, grounds. Upon critical review of literature, I can find no carbon-14 date older than 12,000 B.P. that can be positively related to Early Man in the New World.

1. The radiocarbon dating and field investigations were supported by the National Science Foundation (grant GP-2330, Paul E. Damon, Principal Investigator; and GP-5548, C. Vance Haynes, Principal Investigator) and the National Geographic Society.

Comments on the manuscript by P. E. Damon, R. Shutler, Jr. and T. L. Smiley are appreciated. (This paper is a revision of one presented at International Conference on C[14] and H[3] Dating, Pullman, Washington, 1965.) Contribution 118: Program in Geochronology, University of Arizona.

Before the development of radiocarbon age determinations in the late 1940s, the dating of Early Man in the New World was done largely through the efforts of geologists like Kirk Bryan, his students, and Ernst Antevs. Their age estimates were based upon the assumption that streams filled the valleys with alluvium during glacial or moist stades and cut channels or terraces during interstades. It was also reasoned that pluvial periods (indicated by the rise of lakes in closed basins) corresponded in at least a general way to glacial stades (Bryan, 1950). Most of the Early Man deposits investigated by these geologists were believed to correspond to the last major wet phase of the Wisconsin glaciation (W3), which they variously dated at between 10,000 and 25,000 years old by a complex estimation of varve records in New England and Canada and the time for the recession of Niagara Falls. Today, chiefly because of radiocarbon dating, we know that Early Man flourished in the New World during the Valders Stade and the subsequent recessional period, the time from 11,500 to 8,000 years ago. In retrospect, we see that the geological correlations of Antevs and Bryan were essentially correct and that their age estimates were remarkably close, considering the approximations that they were forced to make.

In addition to providing an "absolute" time scale for geologic climatic events, radiocarbon dating gives a means of determining cultural change with time and of comparing cultural changes during a given period in different areas.

EARLY MAN SITES IN SPACE AND TIME

By using the index cards of C^{14} dates (Radiocarbon Dates Association, Inc.), *Radiocarbon*, site reports, and firsthand knowledge of a number of Early Man sites, I have plotted as solid dots C^{14}-dated Early Man sites (Table 1) in intervals of 1,000 years, beginning at 8,000 B.P., on maps of the hemisphere (Fig. 1, A–F, and Fig. 2). Sites with questionable association of artifacts and C^{14} samples or of questionable archaeological association are symbolized by a circle instead of a dot. On the maps of North America I have also shown in a very generalized way the possible relation of continental glaciers to land and sea at some time during each 1,000-year period except the 7,000-year period 13,000 to 20,000 B.P., for which the maximum extent of continental glaciation is shown (Fig. 1A). It should be understood that the pattern of deglaciation shown is an interpretation of the data from the literature, but other interpretations range from no contact of Cordilleran and Laurentide ice during the late Wisconsin (R. S. MacNeish, personal communication) to contact of these glaciers up to 10,000 B.P. or later (L. A. Bayrock, personal communication). A histogram of the C^{14}-dated sites is shown in Figure 3,

TABLE 1. Carbon-14 Dates of Early Man, Listed by Site

No.	Site	C^{14} dates in years B.P.	Lab. No.	Association
1	Santa Rosa Island, Calif.	10,400 ± 2,000	L-568A	Arlington Springs man
		10,000 ± 200	L-650	?
		11,800 ± 800	UCLA-106	Charcoal and mammoth bones
		11,900 ± 200	UCLA-661	Hearth (?)
		12,500 ± 250	L-290T	Hearth (?)
		12,620 ± 200	UCLA-141	Hearth (?)
		16,700 ± 1,500	M-599	Partly burned mammoth bone and artifacts (?)
		29,700 ± 3,000	L-290R	"Charred mammoth bones"
		>25,000	M-1132	"Some cultural material"
2	Texas Street site, Calif.	>35,000	L-299D	Hearth (?) and artifacts (?)
3	Scripps service yard, Calif.	21,500 ± 700	W-142	Hearth (?)
4	Lewisville site, Tex.	>37,000	O-235	Hearths (?) and artifacts
		>37,000	O-248	Same
		>38,000	UCLA-110	Same
		>40,000	B-487	
5	Scharbauer site, Midland, Tex.	13,400 ± 1,200	L-304C	Cut and scratched horse bone
6	Rancho Peludo, Venezuela	13,920 ± 200	Y-1108IV	Hypothetical Paleo-Indian horizon
7	Muaco site, Venezuela	14,300 ± 500	M-1068	Burned (?) extinct animal bone associated (?) with scraper
8	Tule Springs, Nev.	12,450 ± 230	UCLA-509	Bone needle (?) and other possible bone tools
9	Hermit's Cave, N.M.	12,270 ± 450	W-499	Log in cave
		12,900 ± 350	W-495	Hearth
		11,850 ± 350	W-498	Burned log
10	Lubbock Reservoir, Tex.	12,650 ± 250	I-246	Knife (?)
11	Blackwater No. 1, Clovis, N.M.	8,470 ± 350	A-512	Hearth and Archaic point at same level as Scottsbluff
		9,890 ± 290	A-489	Below Scottsbluff point, above Agate Basin
		10,170 ± 250	A-488	Above Agate Basin and Folsom
		10,250 ± 320	A-380-379	Folsom level
		10,490 ± 200	A-492	Above Agate Basin and Folsom
		10,490 ± 900	A-386	Folsom level
		11,040 ± 500	A-490	Clovis level
		11,170 ± 360	A-481	Clovis level
		11,630 ± 400	A-491	Clovis level
12	Dent, Colo.	11,200 ± 500	I-622	Clovis level
13	Ventana Cave, Ariz.	11,290 ± 500	A-203	Clovis (?) level
14	Lehner site, Ariz.	10,410 ± 190	A-33bis.	Next unit above Clovis
		10,900 ± 450	A-40b	Clovis
		10,940 ± 100	A-375	Clovis hearth
		11,170 ± 140	K-554	Clovis hearth
		11,240 ± 190	A-42	Clovis hearth
		11,290 ± 500	M-811	Clovis hearth
		11,600 ± 400	A-478b	Unit below Clovis level, no artifacts
15	Domebo site, Okla.	11,045 ± 647	SM-695	Clovis
		11,220 ± 500	SI-172	Mammoth bone in Clovis level

TABLE 1. Carbon-14 *Dates of Early Man, Listed by Site*—continued

No.	Site	C^{14} dates in years B.P.	Lab. No.	Association
16	Union Pacific mammoth site, Wyo.	11,280 ± 350	I-449	Bifacial artifacts and mammoth
17	Modoc Rockshelter, Ill.	7,800 ± 900	C-904	Side-notched and lanceolate projectile points in lowest occupational level
		8,546 ± 380	C-903	
		9,101 ± 440	C-908	
		10,651 ± 650	C-907	Same
		10,947 ± 900	C-904	Same
		11,200 ± 800	C-905	Same
18	Debert site, Nova Scotia	Dates unreported		Fluted points
19	Fishbone Cave, Nev.	7,830 ± 350	L-289KK	Netting
		11,400 ± 250	L-245	Occupational level (?)
20	Leonard Rockshelter, Nev.	8,660 ± 300	C-281	Guano level containing wooden artifacts
		11,199 ± 570	C-599	Guano level with Humbold culture flakes, knives, *olivella* beads
21	Gypsum Cave, Nev.	8,527 ± 250	C-222	Occupational (?) level
		10,455 ± 340	C-221	Occupational (?) level
		11,690 ± 250	LJ-452	Dung below Gypsum Cave point
22	Fells Cave, Chile	10,200 ± 400	Sa-49	Does not necessarily date culture
		10,720 ± 300	W-915	Fire pit of oldest occupational level
23	Lindenmeier site, Colo.	10,850 ± 550	I-141	Folsom level
24	Hell Gap site, Wyo.	5,740 ± 230	A-498	Above Early Man levels
		8,600 ± 600	I-245	Scottsbluff level
		8,600 ± 380	A-501	Frederick level
		10,000 ± 200	A-499	Midland level
		10,150 ± 300	A-500	Alberta–Hell Gap level
		10,200 ± 500	A-502	Unidentified occupation
		10,600 ± 500	A-504	Midland level
		10,840 ± 200	A-503	Unidentified occupation
		10,850 ± 550	I-167	Agate Basin (?) level
		13,060 ± 600	A-431	Below occupation level
25	Danger Cave, Utah	8,970 ± 150	Tx-86	Around fireplace (feature 108)
		9,050 ± 180	Tx-88	Above fireplace (feature 108)
		9,704 ± 210	Tx-89	Associated with Tx-88
		9,789 ± 630	C-611	Firepit in level II
		10,150 ± 170	Tx-87	Same stratigraphic position as Tx-85
		10,270 ± 650	M-202	Fireplace
		10,600 ± 200	Tx-85	Fireplace (feature 108)
		11,151 ± 570	C-610	Pre-occupation
		11,453 ± 600	C-609	Pre-occupation
26	Medicine Creek, Nebr.	8,274 ± 500	C-108a	Soil B
		8,862 ± 230	C-824	Hearth in middle of terrace 2A fill
		10,493 ± 1,500	C-470	Lower occupation zone in soil B
27	Bonfire Shelter, Tex.	10,230 ± 160	Tx-153	Upper of three carbonaceous layers in Bone Bed 2, containing Plainview and Folsom
28	Coontail Spin, Tex.	10,300 ± 400	Tx-80	Earliest occupational level
29	Kincade Shelter, Tex.	7,900 ± 800	Tx-63	Zone 5
		9,110 ± 155	Tx-18	Zone 5
		10,025 ± 185	Tx-17	Zone 5
		10,065 ± 185	Tx-19	Zone 5
		10,365 ± 110	Tx-20	Zone 5

TABLE 1. Carbon-14 Dates of Early Man, Listed by Site—continued

No.	Site	C^{14} dates in years B.P.	Lab. No.	Association
30	Levi Shelter, Tex.	6,750 ± 150	O-1105	Lower 6 in. of zone IV
		7,350 ± 150	O-1128	Top 6 in. of zone IV
		9,300 ± 160	O-1129	Middle 6 in. of zone IV, with Angostura–Scottsbluff
		10,000 ± 175	O-1106	Horse; Plainview points and Clovis point in zone II
31	Cerro los Chivateros, Peru	10,430 ± 160	UCLA-683	Chivateros I phase
32	Brewster site, Wyo.	9,350 ± 450	O-1252	Upper Agate Basin
		9,990 ± 225	M-1131	Middle Agate Basin
		10,375 ± 700	I-472	Folsom
33	Warm Mineral Spgs., Fla.	10,,000 ± 200	LJ-120	Charred log with human bones
34	Falcon Hill, Nev.	8380 ± 120	UCLA-672	Early occupation and basketry
		9,540 ± 120	UCLA-675	Earliest basketry
35	Lagoa Santa, Brazil	9,028 ± 120	P-519	Occupation levels 2 and 3
		9,720 ± 128	P-521	Occupation levels 6 and 7
36	Stanfield-Worley Rockshelter, Ala.	8,920 ± 400	M-1153	Dalton zone (D)
		9,040 ± 400	M-1348	Dalton zone, level 1
		9,340 ± 400	M-1347	Dalton zone, level 4
		9,440 ± 400	M-1346	Dalton zone, level 10
		9,640 ± 450	M-1152	Meserve–Dalton and Big Sandy I points; Dalton level
37	Sister's Hill, Wyo.	9,600 ± 230	A-372	Overlies Hell Gap points
		9,650 ± 250	I-221	Hell Gap
38	Frightful Cave, Mexico	8,023 ± 350	M-188	Human feces dated
		8,080 ± 450	M-187	Sandals dated
		8,870 ± 350	M-191	Bottom level
		9,300 ± 400	M-192b	?
		9,540 ± 550	M-192a	?
39	Five-Mile Rapids, Ore.	7,675 ± 100	Y-341	Early II level, with bolas, atlatls, burins, etc.
		7,875 ± 100	Y-342	Transition level between I and II
		9,785 ± 220	Y-340	Early I level "percussion flaking?"
40	Plainview, Tex.	9,800 ± 500	L-303	Bone bed
41	Lime Creek, Nebr.	9,524 ± 450	C-471	Below zone I
42	Lubbock Reservoir, Tex.	9,883 ± 350	C-558	Folsom level
43	San Bartolo, Atepehuacan, Mexico	9,670 ± 400	M-776	Mammoth, stone tools
44	Graham Cave, Mo.	7,900 ± 500	M-132	Level 4
		8,830 ± 500	M-131	Level 6
		9,700 ± 500	M-130	Base of level 6
45	Agua Hedionda Lagoon, Calif.	7,420 ± 350	LJ-961	Lowest LaJolla level
		7,450 ± 370	LJ-966	Lowest LaJolla level
		9,020 ± 500	LJ-967	San Dieguito level
46	Fraser River, Yale, British Columbia	8,150 ± 300	S-47	Occupation level
		9,000 ± 150	S-113	Projectile points
47	Double Adobe, Ariz.	8,240 ± 960	A-184C	Pollen zone VI
		8,270 ± 250	A-188C	Pollen zone VI
		9,350 ± 160	A-67bis.	Sulphur Spgs. stage
48	Fort Cave Rock, Ore.	9,053 ± 350	C-428	Sandals

TABLE 1. Carbon-14 Dates of Early Man, Listed by Site—continued

No.	Site	C^{14} dates in years B.P.	Lab. No.	Association
49	Ray Long site, S.D.	7,715 ± 740	C-454	Association doubtful
		9,380 ± 500	M-370	Angostura points
50	Diablo Cave, Mexico	8,200 ± 450	M-498	Post-Lerma
		8,540 ± 450	M-500	Post-Lerma
		9,270 ± 500	M-499	Lerma points
51	Russel Cave, Ala.	6,300 ± 350	M-591	Archaic
		7,950 ± 200	L-344	Human bones
		7,970 ± 450	M-846	Early Archaic
		8,240 ± 400	M-589	Middle Archaic
		8,350 ± 180	I-396	Early Archaic
		8,350 ± 500	M-847	Early Archaic
		8,450 ± 180	I-397	Early Archaic
		8,560 ± 400	M-590	Transitional zone
		8,750 ± 500	M-845	Middle Woodland
		9,020 ± 350	M-766	Paleo-Indian
52	Bull Brook, Mass.	6,940 ± 800	M-809	
		8,720 ± 400	M-808	Fluted points
		8,940 ± 400	M-810	Fluted points
		9,300 ± 400	M-807	Fluted points
53	Mangus site, Mont.	8,600 ± 100	SI-101	Hearth on Agate Basin (?) level
		8,690 ± 100	SI-98	Hearth on Agate Basin (?) level
54	Palli Aike Cave, Chile	8,639 ± 450	C-485	Palli Aike artifacts
55	Ojo de Agua Cave, Mexico	8,540 ± 450	M-500	Infiernillo phase
56	Santa Marta Cave, Mexico	8,730 ± 400	M-980	Santa Marta complex
57	Lind Coulee, Wash.	8,700 ± 400	C-827	Occupational level
58	Horner site, Wyo.	7,880 ± 1,300	SI-74	Cody complex ·
		8,750 ± 120	UCLA-697A	Cody complex
		8,840 ± 120	UCLA-697B	Cody complex
59	Cougar Mountain Cave, Ore.	8,510 ± 250	UCLA-112	Occupational level
60	Eagle Cave, Tex.	8,540 ± 120	Tx-140	Barbed points
		8,760 ± 150	Tx-107	Barbed points
61	Baker Cave, Tex.	8,910 ± 140	Tx-128	Plainview-like point
		9,030 ± 230	Tx-129	Plainview-like point
62	Anagula Island, Alaska	7,660 ± 300	W-1180	Obsidian flakes
		8,425 ± 275	I-715	Obsidian flakes and blades
63	Simonsen site, Iowa	8,430 ± 520	I-79	Side-notched points
64	Pigeon Cliffs, N.M.	8,280 ± 1,000	W-636	Meserve point
65	Ferry site, Ill.	8,160 ± 400	M-892	Milling stones and banner stones
66	Portales Cave, Mexico	8,200 ± 450	M-498	Infiernillo artifacts
67	Intihuasi Cave, Argentina	7,970 ± 100	Y-228	Ayampitin artifacts
		8,060 ± 100	P-345	Ayampitin artifacts
68	James Allen site, Wyo.	7,900 ± 400	M-304	James Allen points
69	Lamb site, Colo.	7,870 ± 240	SI-45	Eden and Scottsbluff points
70	Oconto, Wis.	7,510 ± 600	C-837	Old Copper culture (solid-carbon date questioned, see Mason and Mason, 1961)
71	Ash Cave, Wash.	7,940 ± 150	UCLA-131	End of "old Cordilleran"
94	Onion Portage, Alaska	7,180 ± 95	P-1111	Kobuk complex charcoal
		7,900 ± 103	P-1087	Kobuk complex charcoal

TABLE 1. Carbon-14 Dates of Early Man, Listed by Site—continued

No.	Site	C^{14} dates in years B.P.	Lab. No.	Association
		7,920 ± 102	P-984	Kobuk complex charcoal
		8,100 ± 206	P-985	Kobuk complex charcoal
95	Wilson Butte Cave, Idaho	14,500 ± 500	M-1409	Bifacial foliate point
		15,000 ± 800	M-1410	Mashed bone
96	El Inga, Ecuador	4,000 ± 190	I-557	Relation to specific artifact types uncertain
		5,500 ± 200	I-558	Same
		3,919 ± 121	NZ-1070/1	Same
		7,928 ± 132	NZ-1070/3	Same
		9,030 ± 144	NZ-1070/2	Same
97	Gordon Creek, Colo.	9,700 ± 250	GXO-530	Human bone collagen from burial
98	Horn Shelter, Tex.	9,290 ± 360	SM-689	
		9,500 ± 300	SM-761	Shell
		10,800 ± 300	SM-762	Shell
99	Packard site, Okla.	9,406 ± 193	NZ	Hearth 10 ft below surface
100	St. Albans, W. Va.	8,160 ± 100	Y-1540	Knauha level, charcoal
		8,250 ± 100	Y-1539	LeCroy level, charcoal
		8,915 ± 160	Y-1538	Kirk level, charcoal
101	Taima-taima, Venezuela	13,010 ± 280	IVIC-191-1	Bone with "pebble tools"
		14,440 ± 435	IVIC-191-2	Bone with "pebble tools"
102	Pine Spring, Wyo.	9,695 ± 195	GX-354	Bison bone and Plano artifact
		11,830 ± 410	GX-355	Bone date, rejected
103	Harris site, Calif.	7,620 ± 360	A-723	Above San Dieguito level
		8,490 ± 400	A-724	San Dieguito level
		8,490 ± 400	A-725	San Dieguito level
		9,030 ± 350	A-722A	Uncertain
104	Lake Le Conte, Calif.	9,630 ± 300	LJ-528	Shell in beach with artifacts

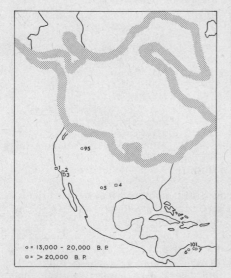

FIG. 1A

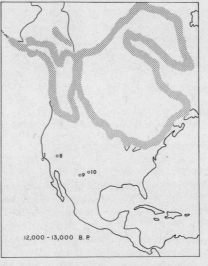

FIG. 1B

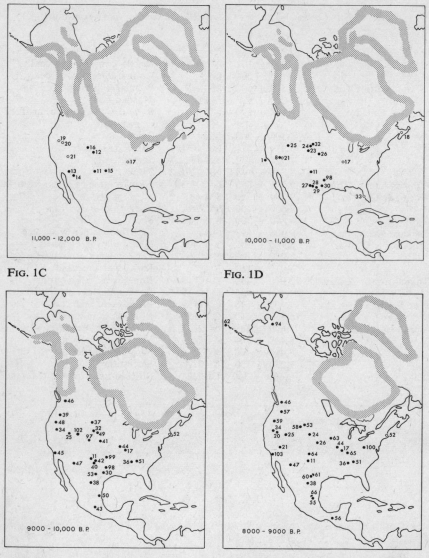

Fig. 1C

Fig. 1D

Fig. 1E

Fig. 1F

FIG. 1. A–F, Sequence of maps showing the distribution of carbon-14 dated sites in North America with respect to sea levels (Curray, 1965; Emery, 1965; Hopkins, 1959) and ice borders during deglaciation (Armstrong et al., 1964; Christiansen, 1965; Craig and Fyles, 1960; Easterbrook, 1966; Falconer et al., 1965; Geological Association of Canada, 1958; Hughes, 1965; Parry and MacPherson, 1965; Porter, 1964; Westgate, 1965; and Wright and Frey, 1965). There is reasonable probability that glacial boundaries were within the stippled zones sometime during each period shown. Cordilleran ice is not shown after 9000 B.P. because of even greater uncertainties regarding ice margins during deglaciation.

FIG. 2. The location of C^{14}-dated sites in South America between 8,000 and 11,000 years old.

in which the dashed areas represent the questionable site associations mentioned.

There are four C^{14}-dated sites older than 20,000 B.P. and allegedly associated with Early Man in the New World (Fig. 1A): Santa Rosa Island (site 1), Texas Street (site 2), and Scripps Service Yard (site 3) all in California; and Lewisville (site 4), Texas. In each case the C^{14} dates seem acceptable geologically, but clear evidence for their contemporaneity with man is lacking. These sites are among the most controversial issues facing American archaeology today. The Tule Springs site in Nevada would have remained in this category had it not been for the intensive interdisciplinary investigations that showed earlier C^{14} dates to be unrelated to Early Man (Shutler, 1965).

Sandia Cave, New Mexico, might also be included in the >20,000 B.P. group were it not for the questionable meaning of published C^{14} dates (Johnson, 1957; Hibben, 1957; Krieger, 1957). New geochronological investigations by George Agogino and myself are in progress, but all that has been determined so far by C^{14} dating is that Sandia artifacts are probably between 9,000 and 13,000 years old. Unfortunately, it is doubtful that further work in the cave will be able to fix the true C^{14} age more closely.

Although not directly associated with a C^{14} date, an object believed to be an artifact was found in a stratum of Altonian age in Illinois (Munson and Frye, 1965). The geological correlation of the site to Altonian sediments dated 35,000 to 40,000 years old appears to be unquestionable, but the artifactual nature of the object is not conclusively established.

The period 13,000 to 20,000 B.P. (Fig. 1A) is represented by dates that are as controversially related to man as those just mentioned. Santa Rosa Island has provided dates that seem geologically correct, but clear evidence for man's presence during this period is lacking. At the Scharbauer site (no. 5) near Midland, Texas, a fragment of a horse bone with anomalous scratches was found in the "white sand" from which snail shells were dated 13,400 ± 1200 B.P. (L-304C) (Wendorf and Krieger, 1959). Correlation of the white sand with the uppermost part of the Tehoka formation would be compatible with the date, but the origin of the scratches remains uncertain.

Two sites, Mauco (no. 7) and Rancho Peludo (no. 6) in Venezuela, have yielded C^{14} dates between 13,000 and 20,000 B.P., but none of the dates is clearly related to artifacts. In earlier reports the dates from Mauco were only tenuously related to artifacts, as they were obtained on bone from "mucky" spring deposits containing bones of extinct animals, projectile points, and bits of glass bottles (Rouse and Cruxent, 1963). In later reports (Rouse, 1964), the relation of the dates to human activity is assumed, because one of the dated bones was burned. The burning suggests but does not prove human activity. How clear is the evidence of burning? It has been shown that the black-to-brown coloring of some fossil bone is due to manganese replacement and not to pyrolysis (Brooks, 1963). A report of the behavior of the sample during laboratory pretreatment would be helpful in solving this question, as manganese oxides would dissolve in hydrochloric acid, whereas a carbon residue would remain if the bone had been burned.

The date of 13,915 ± 200 B.P. (Y-1108-IV) at Rancho Peludo was entirely unexpected by the investigators and is attributed to a hypothetical Paleo-Indian horizon below the ceramic-bearing horizons that were being dated (Rouse and Cruxent, 1963).

The most plausible Early Man dates in this range are from Wilson Butte Cave in Idaho (Gruhn, 1965) and the Taima-taima site in Venezuela (Tamers, 1966, p. 206). Two successive levels in Wilson Butte Cave in Idaho dated 14,500 ± 500 B.P. and 15,000 ± 800 B.P. These are on bone which, more often than not, yields dates that are too young. If there were ample conclusive evidence of bifacially flaked projectile points in the New World, the Wilson Butte Cave dates probably would be accepted without question; but because of the paucity of such dates and the implications of the artifacts, it may be useful to examine alternative possibilities.

As yet we do not understand the source of carbon contaminants in bone, but I have found in preliminary studies that absorbed(?) humic acids can represent more than 50 per cent of the organic carbon remaining in fossil bone. This source of contamination is not removed by standard techniques of recovering bone collagen. Humic acid contaminants are commonly derived from younger soils, but it is quite possible that humic acids older than the bone could be absorbed from ground water. At Wilson Butte Cave this possibility would have to be evaluated on geochemical grounds.

Humic ACID contaminants [handwritten marginal note]

Another possibility is that bones from stratum E were redeposited into the overlaying stratum C. Statistically the two dates could have been two counts on the same sample. Many rodent bones occur in these units and rodents are notorious for mixing strata. The abundance of instrusive rodent burrows was noted by Gruhn (1961, p. 24). It is also possible for such disturbed deposits to become compacted so that the evidence for disturbance is not obvious.

If such possibilities can be precluded at Wilson Butte Cave, it contained the earliest bifacial projectile point complex in the New World. On the other hand, the top of stratum C was dated 8,850 ± 300 B.P. (M-1087), which would mean that deposition without a significant stratigraphic break occurred through a period of 7,500 years, spanning Cary, Mankato, Two Creeks, and Valders time.

Dates of 13,010 ± 280 and 14,440 ± 435 B.P. were obtained on the organic portion of bones from Taima-taima site (no. 101), which is only 8 km from the Muaco site. The artifacts reportedly associated with the bones are "pebble tools" (Tamers, 1966, p. 206). If the dates are accurate and the associations valid, Taima-taima is the oldest site in South America, coeval with the early occupation at Wilson Butte Cave in North America. If both sites are valid, cultural diversification had already occurred in the New World 14,000 years ago. The implications of these dates are so important that proper evaluation at Taima-taima must await a scientific report describing the circumstances.

Between 12,000 and 13,000 years ago there are four C^{14}-dated sites that have been attributed to Early Man (Fig. 1B). At Santa Rosa Island clear-cut evidence for human occupation is again lacking for this time period. At Tule Springs, Nevada (site 8), a small polished bone object and other possible bone tools were found in a deposit that was repeatedly C^{14} dated at between 12,000 and 13,000 B.P. However, the artifactual nature of the bone objects has not gone unquestioned (Shutler, 1965).

Carbon-14 dates from Hermit's Cave (site 9) in New Mexico do not relate to human occupation of the cave, according to Edward Ferdon (personal communication), who conducted the archaeological excavations (Ferdon, 1946). On the other hand, the one date (W-495) that was reportedly on charcoal from a hearth invites further explanation.

At the Lubbock Reservoir in Texas (site 10), stratum 1 has been dated between 12,000 and 13,000 B.P. and has been correlated with the Clovis-point-bearing "gray sand" at Blackwater No. 1 in New Mexico (Sellards, 1952). However, on the basis of both extensive geological investigations in Blackwater Draw and a review of previous archaeological reports, I have suggested that the gray sand is pre-Clovis in age and contains intrusive artifacts (Haynes, 1966).

If any or all of the mentioned sites are truly archaeological, they indicate rather ill-defined cultural records. If the northern part of North America was covered by the Laurentide and Cordilleran ice sheets from 12,000 to at least 25,000 years ago, early migrants from Siberia might have entered the New World before that time. Of course, other sites may represent these early peoples, but they have not been dated by carbon-14 (Kreiger, 1964). The undated Valsequillo site in Mexico (Irwin-Williams, 1967) yields evidence of a unifacial projectile-point culture associated with extinct fauna.

In the period 11,000 to 12,000 B.P. (Fig. 1C), we find the first undisputed evidence of projectile points in the New World. Without exception they are Clovis fluted points, found from coast to coast in North America and from Canada to Mexico. Of the five C^{14}-dated Clovis levels, all are between 11,000 and 11,500 years old when the averages of multiple-dated horizons are considered (Haynes, 1964). While no projectile points were found at the Union Pacific mammoth site in Wyoming (no. 16), a bifacial knife and an ovoid form were found associated with the mammoth that was dated 11,280 ± 350 B.P. (I-499) on unweathered tusk ivory.

Dates of 9,695 ± 195 (GX-354) and 11,830 ± 410 B.P. (GX-355) are reported from the lowest occupation horizon at the Pine Spring site in

Wyoming, but the older date is not considered accurate and has been discounted (Sharrock, 1966).

Other sites with C^{14} dates between 11,000 and 12,000 B.P. are rock shelters—Fishbone Cave, Nevada, site 19; Leonard Rock Shelter, Nevada, site 20; and Modoc Rock Shelter, Illinois, site 17—in which the associations of the C^{14} dates are questionable or ill-defined. Danger Cave is not included in this period, because the only C^{14} dates representing it are on animal dung, whereas cultural features (hearths) in the same level produced dates in the succeeding period.

In North America between 10,000 and 11,000 B.P. (Fig. 1D), we find diversification of cultures as represented by Folsom, Midland, Plainview, Agate Basin, and undoubtedly other types of projectile points (Fig. 3). The first three types mentioned could well have derived from the Clovis type, but the genetic relation, if any exists, of Agate Basin to these types is uncertain.

A recent discovery near Debert, Nova Scotia (site 18), has produced fluted points that are somewhat different from either Clovis or Folsom points (MacDonald, 1967). Geological estimation of a Valders age by Borns (1965, 1967) has been confirmed by C^{14} dates, thirteen of which average 10,585 ± 47 B.P. (Stuchenrath, 1966). This indicates that the

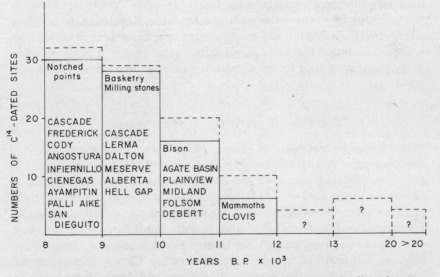

FIG. 3. Histogram of C^{14}-dated sites in the New World showing projectile-point types and other traits that appear within the thousand-year periods. Dashed bars represent the additional sites of questionable association shown by circles and squares in Figure 1.

Debert points are time-equivalent to Folsom points in the West and are post-Clovis.

By 9,000 B.P. (Fig. 1E) there was further diversification, with Cascade, Lerma, Dalton, Meserve, Alberta, and Hell Gap points centered in various geographic areas from Canada to South America (Fig. 2). The earliest C^{14} dates for basketry are in this period from 9,000 to 10,000 B.P. and the use of milling stones is in evidence at Danger Cave, Utah (site 25), and in Sulphur Springs Valley, Arizona (site 47). The Gordon Creek burial, dated 9,700 ± 250 B.P. (GX-530), is the oldest collagen date on a human skeleton in the New World (Anderson, 1966).

During the period of 8,000 to 9,000 B.P. (Fig. 1F), cultural complexes and point types from twenty-seven dated sites include Cascade, Frederick, Cody (Eden and Scottsbluff points), Angostura, San Dieguito, Kirk, Infiernillo, Cienegas, Ayampitin, and Palli Aike. The most significant archaeological event is the appearance of notched projectile points at several sites scattered across the United States. To several archaeologists, this marks the beginning to the Archaic period.

Both the number of C^{14}-dated sites and the number of projectile-point types in the New World increase with time, beginning with Clovis sites of the Llano complex about 11,500 years ago (Fig. 3). Candidates for earlier sites are few and far apart in both space and time; earlier peoples were few in number, unsuccessful in their New World environment, or possibly nonexistent in North America. Regarding the relationship between Early Man and extinct fauna, a focal point of this volume, it should be noted that the reliably dated associations of both man and megafauna other than bison fall within a narrow time range, perhaps no more than a thousand years.

CARBON-14 DATES AND STRATIGRAPHY

Multiple-component sites of Early Man with artifacts in stratified sequence provide the best insights into cultural change with time, but unfortunately such sites are rare. In the Southwest we are fortunate in having two such sites with radiocarbon control, and evaluation of these provides some useful information regarding the rate of cultural change, temporal overlap of cultures, and certain limitations of C^{14} dating.

At the Clovis type site, or Blackwater No. 1 locality in New Mexico (Fig. 4), cultural succession is represented by Clovis artifacts at the base of the sequence (unit C) through Folsom and Agate Basin (units D_1 and D_2) to Scottsbluff, Portales, and Frederick artifacts at the top (unit E). The preceding unit (B) contains abundant fauna, but the Clovis artifacts therein may be intrusive (Haynes and Agogino, 1966). It is

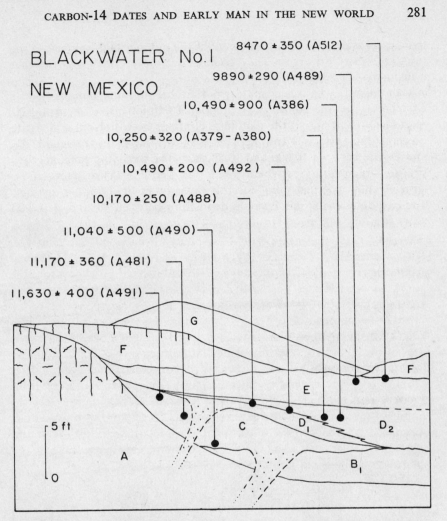

BLACKWATER No.1

NEW MEXICO

8470 ± 350 (A512)

9890 ± 290 (A489)

10,490 ± 900 (A386)

10,250 ± 320 (A379 - A380)

10,490 ± 200 (A492)

10,170 ± 250 (A488)

11,040 ± 500 (A490)

11,170 ± 360 (A481)

11,630 ± 400 (A491)

G

F

E

F

5 ft

C

D₁

D₂

A

B₁

0

FIG. 4. Generalized stratigraphic diagram of the Clovis type site or Blackwater No. 1 locality, New Mexico, showing the position of C¹⁴-dated samples with respect to stratigraphic units (lettered), paleosols (vertical hachuring), and spring-feeder conduits (stippled). No horizontal scale.

now known that the artifact-bearing sediments are in part springlaid and that several concentrations of artifacts and flakes occurred in the feeder conduits of the ancient springs (Haynes and Agogino, 1966). Within these sedimentary strata are thin (ca. 2 cm) layers of decomposed plants that were sampled to provide the dates shown in Figure 4. Unit C (Brown Sand Wedge) contains mammoth bones, Clovis artifacts, and material for three C¹⁴ dates. The reversed stratigraphic order of the three dates is insignificant, because all of the numbers are well within the range of statistical error. They could therefore record an event that lasted about 600 years.

Unit D_2 is a diatomaceous earth that has become famous for its content of Folsom points and remains of extinct bison. A springlaid sand facies (unit D_1) of the diatomite contains both Folsom and Agate Basin artifacts (Haynes and Agogino, 1966). Four dates from units D_1 and D_2 are between 10,000 and 10,500 B.P. The overlying carbonaceous silt (unit E) contains Scottsbluff, Frederick, and Portales artifacts, and a large Archaic notched point recently was found at its upper contact. The two dates from unit E reveal deposition between 8,000 and 10,000 years ago.

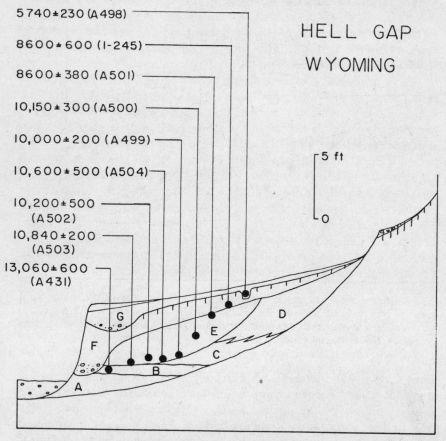

FIG. 5. Generalized stratigraphic diagram of the Hell Gap site, Wyoming, showing the position of C[14]-dated samples with respect to stratigraphic units (lettered) and paleosols (vertical hachuring). No horizontal scale.

At the Hell Gap site in Wyoming (no. 24) we find a very similar chronology (Fig. 5). The earliest dated unit (B) contains no evidence of man's presence. Although Clovis artifacts have not as yet been found at the site, mammoth bones and a Clovis point have been found separately in loess of unit D elsewhere in the area. Five C^{14} dates between 10,000 and 11,000 B.P. are on thin (ca. 1 in.) layers of carbonaceous matter (partially burned) that are contemporaneous with zones of human occupation in the lower part of unit E. These occupations are represented by Midland at the bottom (A-499 and A-504), Agate Basin, and Alberta–Hell Gap (A-500) at the top. Two dates (A-502 and A-503) are from culturally unidentified layers near the base of unit E. Although the mean values of the C^{14} dates are again not in proper stratigraphic sequence, all but the oldest one (A-503) are within the range of their statistical error. The top of unit E (with artifacts of the Cody complex) and the base of unit F (with Frederick artifacts) both provided the same C^{14} mean value of 8,600, consistent with dates on these complexes elsewhere.

The data at these two sites clearly show that cultural changes, or at least change in projectile points, have occurred within periods of about five hundred years or less, periods that are too short to be subdivided accurately by C^{14} dating at the present time. With larger and more reliable samples and longer laboratory counting runs, we might be able to distinguish between two hundred years at 10,000 to 11,000 B.P. Until such precision is obtained at Early Man sites, we must continue to rely on detailed, stratigraphically controlled excavation to determine cultural succession. Obviously, both procedures are necessary to provide the most reliable data.

A cursory examination of the C^{14} data from Early Man levels in caves indicates that, compared with stratified open sites, certain anomalies exist in both cultural succession and in dating. At the Modoc Rock Shelter in Illinois (site 17) the earliest occupational level has six C^{14} dates ranging in age from about 8,000 to 11,000 B.P., with both notched and lanceolate projectile points, but the relation of dates to the artifacts is not clear (Fowler, 1959). None of the three Gypsum Cave dates (site 21) is known to apply to a specific culture. A totally unacceptable date from Russell Cave in Alabama (site 51) is nearly 9,000 B.P. for Woodland occupation, but fortunately there are enough consistent dates to make the anomaly stand out.

During the initial examination of Pintwater Cave in Nevada, Richard Shutler and I collected a hearth-charcoal sample from less than six inches below the present surface. Basketmaker artifacts in the overlying silt and the fresh appearance of the hearth charcoal led us to expect a date of less than 2,000 B.P., but the analysis came out about 9,000 B.P. I offer the suggestion that a Basketmaker Indian robbed an ancient woodrat nest in the shelter to obtain kindling for his fire. It is now known that woodrat (*Neotoma*) nest in the same area contain wood and twigs of various ages, some exceeding 30,000 B.P. (Wells and Jorgensen, 1964).

From the data reviewed here it is clear that Early Man was well established in the New World between 11,000 and 12,000 B.P. and that in each succeeding millennium his population increased and his culture diversified. If Clovis points are relics of the first major migration to the New World, it is reasonable to suggest that other cultural groups followed to add to the increase in populations and cultures. Because of differences between Agate Basin and Folsom–Midland–Plainview points, it is possible that the former might represent a separate and post-Valders migration, while the latter represent diversification of the Clovis manufacturing techniques of the Llano Complex. Similar diversification in eastern North America is represented at the Debert and Bull Brook sites.

Evidence that Early Man was in the New World prior to 12,000 or 13,000 B.P. is inconclusive; at best his population was either small or relatively localized compared to Clovis times 11,000 to 12,000 B.P. The period 11,000 to 13,000 B.P. is critical for evaluating the peopling of the New World. If man was here before this time, he presumably reached the midcontinent sometime prior to the maximum of late-Wisconsin glaciation.

To date, the limitations of radiocarbon dating are insufficient for establishing cultural succession within periods of two hundred years in stratified open sites and within thousand-year periods in some cave sites. Detailed stratigraphic control is the only way to establish successions within these time periods.

Problems in dating some cave deposits may be attributed to (1) rodent disturbance concentrated by confinement, (2) prehistoric human disturbance, (3) depositional rates that are too low to provide the stratigraphic separation of cultures observed in alluvial sites, and (4) prehistoric use of ancient wood from "fossil" woodrat nests.

In spite of these difficulties the general trend of cultural expansion and diversification in the New World is now well established because of radiocarbon dating. With greater precision and better stratigraphic control, we can look forward to substantial increases in our knowledge of Early Man, paleoclimate, and extinction of Pleistocene fauna.

References

Anderson, D. C., 1966, The Gordon Creek burial: *Southwestern Lore*, v. *32*, p. 1–9

Armstrong, J. E., Crandell, D. R., Easterbrook, D. J. and Noble, J. B., 1964, Late Pleistocene stratigraphy and chronology in southwestern British Columbia and northwestern Washington: *Geol. Soc. Amer. Bull.*, v. *76*, p. 321–30

Borns, H. W., Jr., 1965, Late-glacial ice-wedge casts in northern Nova Scotia, Canada: *Science*, v. *148*, p. 1223–25

———— 1967, The Paleo-Indian's geography of Nova Scotia: VII INQUA Cong. (Boulder, 1965), Proc., v. *15* (Rome, Quaternaria, in press)

Brooks, R. H., 1963, An interpretation of polished, split bone from Tule Springs, Nevada: Soc. Amer. Archaeology, 28th ann. meeting (Boulder, 1963)

Bryan, K., 1950, Correlation with Glacial Chronology, *in* Haury, E. W., *Ventana Cave:* Albuquerque, Univ. New Mexico Press

Christiansen, E. A., 1965, Ice frontal positions in Saskatchewan: Research Council of Saskatchewan, Map 2

Craig, B. G., and Fyles, J. G., 1960, Pleistocene geology of Arctic Canada: Geol. Survey Canada Paper, 60–10, p. 1–21

Curray, J. R., 1965, Late Quaternary history, continental shelves of the United States, p. 723–35, *in* Wright and Frey, 1965

Easterbrook, D. J., 1966, Radiocarbon chronology of Late Pleistocene deposits in northwest Washington: *Science*, v. *152*, p. 764–67

Emery, K. O., 1965, Submerged shore deposits of the Atlantic continental shelf: VII INQUA (Boulder, 1965), Abstracts, p. 127–28

Falconer, G., Andrews, J. T., and Ives, J. D., 1965, Late-Wisconsin end moraines in northern Canada: *Science*, v. *147*, p. 608–10

Ferdon, E., 1946, An excavation of Hermit's Cave, New Mexico: School Amer. Res., Monog. no. *10*, 29 p.

Fowler, M. L., 1959, Summary report of Modoc Rock Shelter: Illinois State Mus. Rept. Investigation, no. *8*

Geological Association of Canada, 1958, Glacial map of Canada

Gruhn, R., 1961, The archaeology of Wilson Butte Cave, south-central Idaho: Idaho State Mus., Occasional Papers, no. *6*, 242 p.

———— 1965, Two early radiocarbon dates from the lower levels of Wilson Butte Cave, south-central Idaho: *Idaho State Mus. Tebiwa*, v. *8*, no. *2*, p. 57

Haynes, C. V., Jr., 1964, Fluted projectile points; Their age and dispersion: *Science*, v. *145*, p. 1400–13

———— 1966, Pleistocene and Recent stratigraphy of Blackwater Draw, New Mexico, and Rich Lake, Texas, *in* Wendorf, F., and Hester, J. J., *Assemblers, Paleoecology of the Llano Estacado*, v. *2:* Santa Fe, Ft. Burgwin Res. Center

Haynes, C. V., Jr., and Agogino, G. A., 1966, Prehistoric springs and geochronology of the Clovis site, New Mexico: *Amer. Antiq.*, v. *31*, p. 812–21

Hibben, F. C., 1957, Radiocarbon dates from Sandia Cave; *Science*, v. *125*, p. 235

Hopkins, D. M., 1959, Cenozoic history of the Bering land bridge: *Science*, v. *129*, p. 1519–27

Hughes, O. L., 1965, Surficial geology of part of the Cochrane District,, Ontario, Canada: Geol. Soc. Amer. Spec. Paper 84, p. 535–65

Irwin-Williams, Cynthia, 1967, Associations of Early Man with horse, camel, and mastodon at Hueyatlaco, Valsequillo (Puebla, Mexico) (this volume)

Johnson, F., 1957, Radiocarbon dates from Sandia Cave, correction: *Science* v. *125*, p. 234–35

Krieger, A. D., 1957, Sandia Cave in *Notes and News: Amer. Antiq.*, v. *22*, p. 435–36

———— 1964, Early Man in the New World, p. 23–84, *in* Jennings, J. D., and Norback, E., *Editors, Prehistoric Man in the New World:* Univ. Chicago Press

MacDonald, G. F., 1967, The technology and settlement pattern at a Paleo-Indian site at Debert, Nova Scotia: VII INQUA (Boulder, 1965), Proc., v. *15* (Rome, Quaternaria, in press)

Munson, P. J., and Frye, J. C., 1965, Artifact from deposits of Mid-Wisconsin age in Illinois: *Science*, v. *150*, p. 1722–23

Parry, J. T., and MacPherson, J. C., 1965, the St. Faustin–St. Narcisse Moraine and the Champlain sea: *Rev. Géo. Montréal*, v. *4*, 235–48

Porter, S. C., 1964, Late Pleistocene glacial chronology of north-central Brooks Range, Alaska: *Amer. Jour. Sci.*, v. *262*, p. 446–60

Rouse, I., 1964, Prehistory of the West Indies: *Science*, v. *144*, p. 499–513

Rouse, I., and Cruxent, J. M., 1963, Some recent radiocarbon dates for western Venezuela: *Amer. Antiq.*, v. *28*, p. 537–40

Sellards, E. H., 1952, *Early Man in America:* Austin, Univ. Texas Press, 211 p.

Sharrock, F. W., 1966, Prehistoric occupation patterns in southwest Wyoming and cultural relationships with the Great Basin and Plains culture areas: Univ. Utah, Anthropol. papers 77, 215 p.

Shutler, R., Jr., 1965, Tule Springs expedition: *Current Anthropol.*, v. *6*, p. 110

Stuchenrath, R., 1966, The Debert archaeological project; Radiocarbon dating: VII INQUA (Boulder, 1965), Proc., v. *15* (Rome, Quaternaria, in press)

Tamers, M. A., 1966, Instituto Venezolano de Investigaciones Cientificas, Natural Radiocarbon Measurements II: *Radiocarbon*, v. *8*, p. 204–12

Wells, P. V., and Jorgensen, L. D., 1964, Pleistocene woodrat middens and climatic change in Mohave Desert; A record of juniper woodlands: *Science*, v. *143*, p. 1171–74

Wendorf, F., and Krieger, A. D., 1959, New light on the Midland Discovery: *Amer. Antiq.*, v. *25*, p. 66–78

Westgate, J. A., 1965, The Pleistocene stratigraphy of the Foremost–Cypress Hills area, Alberta: Alberta Soc. Petroleum Geologists, *15th Amer. Field Conf. Guidebook*, p. 85–111

Wright, H. E., Jr., and Frey, D. G., *Editors*, 1965, *The Quaternary of the United States:* Princeton Univ. Press, 922 p.

	late Pleistocene		Recent
30%	M. titan	→	M. ganteus giganteus
25%	O. cooperi	→	O. robustus
15% 10%	M. siva	→	Macropus agilis
5%	Wallabia vishnu	→	W. bicolor
5%	Sarcophilus laniarius		S. harrisii
5% 2%	D. maculatus		D. maculatus

Keilor (Modern fauna)	Lake Victoria	Lake Menindee
17,000-18,000	18,000	
25,000-31,000 (late Pleist. fauna)		26,000

Keilor Terrace - Extinct megafauna directly associated with
man in charcoal.

1/2   Pleistocene → Recent   26 - 18000
                                               Charcoal

1) Small marsupials ——→ unchanged

2) "Medium" marsupials ——→ reduction in size }

3) "Large" marsupials → extinction   } 26-18000 yrs

   Larger than MICROPUS giganteus in
   absolute size.

1) Guilday - (Martin, 1967)

2) Density dependent - Kurten, Pleistocene
                          mammals in Europe Book

3) Inadequacy to maintain significant gene
   pool - inability to adapt to habitat shifts.

   Change 26000 - 18,000 - Megafauna
                                   extinction
   infer major climatic change - pluvial - dry
                               (wet cool - dry warm

Keilor Terrace - 31 - 25,000 BP.
Macropus rufa — 17 -18000 BP.
Macropus agilis - agile wallaby - presently in
Queensland — present ecology warmer summers
shift in rainfall to from winter rather than summer
cannot reproduce in cold winter climates.

ERNEST L. LUNDELIUS, JR.

Department of Geology
University of Texas
Austin, Texas

LATE-PLEISTOCENE AND
HOLOCENE FAUNAL HISTORY
OF CENTRAL TEXAS

Abstract

The late-Pleistocene and post-Pleistocene faunal sequence of central Texas consists of three types of assemblage. The late-Pleistocene fauna consists of (1) large extinct forms such as proboscidians (Mammuthus, Mammut), *sloths* (Paramylodon, Megalonyx), *glyptodonts,* Panthera atrox, *saber-toothed cats* (Smilodon, Dinobastis), Bison antiquus, B. occidentalis, Camelops, Tanupolama, *horses, peccaries* (Platygonus, Mylohyus): *(2) extant species no longer found in central Texas,* Sorex cinereus, Blarina brevicauda, Mustela erminea, Microtus ochrogaster, Synaptomys cooperi, Ondatra zibethica; *(3) species still occurring in central Texas such as* Odocoileus virginianus, Canis latrans, C. lupus, Spilogale putorius, Mephitis mephitis, Cryptotis parva, Lynx rufus, Lepus californicus. *This fauna persisted until approximately 8,000 years ago, when the majority of the large Pleistocene mammals (group 1) became extinct. Subsequently, the faunal assemblage contained the species of groups 2 and 3. The northern species gradually withdrew from central Texas, the last disappearing approximately 1,000 years ago.*

The Recent faunal composition was attained with the appearance of such species as Tayassu tajacu *and* Dasypus novemcinctus. *Remains of these two species are absent in all but the very late-Holocene deposits.*

The withdrawal of a number of extant northern species in the last 10,000 years suggests that the climate has become drier and/or warmer. A change to a climate with stronger seasonal contrast, marked by the hot, dry summers now characteristic of central Texas, may be sufficient to account for their disappearance.

One objective of vertebrate paleontology is the understanding of past faunas and their histories, based on an examination of the fossil record.

This provides the historical basis of present animal distributions. It also can be used for the interpretation of climatic history.

The late-Pleistocene and Holocene faunal history of central Texas is of considerable interest because of the geographic position of this area. It is immediately north of Mexico, which has been postulated as a refuge for warmth-adapted animals during the glacial stages. It also lies between now disjunct distributions of animals in the northeastern United States and the highlands of Mexico. This area should have been crossed by both groups of animals during their periods of dispersal, and its faunal history would be important in determining the times and rates of dispersal of animals across it.

REVIEW OF PAST WORK

Previous work on the faunal and climatic history of central Texas has been based mostly on distribution patterns of Recent organisms and has been aimed primarily at explaining these distributions by past climatic changes. Adams (1902) first proposed that distributional patterns could be explained by climatic changes in the southern United States during the Pleistocene. Deevey (1949) stated the idea clearly and reviewed the evidence. Blair (1958) reviewed in considerable detail the evidence available from Recent distribution records and from the fossil record and concluded that many species showing disjunct distributions on either side of central Texas must have been distributed across this area in the late-Pleistocene. At the time Blair's paper was written, the late-Pleistocene fossil record from central Texas was very poorly known.

GEOGRAPHIC AND CHRONOLOGIC SETTING

The faunal assemblages are primarily from the Edwards Plateau, a structural plain underlain by flat-lying limestones of early Cretaceous age. The central part of the Edwards Plateau has little relief, aside from the valleys of the major streams such as the Llano River. The eastern and southern edges are highly dissected by deep and humid canyons which provide ecological diversity and habitats for zoogeographic relicts.

The present climate is characterized by a westward decrease in rainfall. The climate in the eastern part is classified by Thornthwaite (1948) as C_1B_4 (dry subhumid, mesothermal, with an average annual potential evapotranspiration between 39.27 and 44.88 inches). The climate of the western part is classified as DB_3 (semiarid, mesothermal, with average annual potential evapotranspiration between 33.66 and 39.27 inches).

Most of the fossil localities are caves along or near the edge of the Edwards Plateau. The dissection of this area has resulted in the opening of many caves. Approximately half the localities are archaeological sites. The large percentage of cave localities introduces a bias against the fossilization of the larger animals. With some exceptions (e.g. Freisenhahn and Kincaid caves), larger animals are usually poorly represented in cave deposits as compared with alluvial deposits. The absence of some species from central Texas may thus be explained.

The time interval represented by the cave faunas covers the latter part of the Wisconsin glacial stage and post-Wisconsin time. This interval is of considerable interest because of the many drastic changes in the physical environment and the biota. It is a time during which a large number of mammal species became extinct and man is first definitely known from North America.

The dating of the faunas is based on radiocarbon analyses and on artifacts associated with the bones. Several radiocarbon dates (indicated in the table) were obtained on residual organic material from bone, but there is considerable difference of opinion regarding this reliability. Dates on the residual collagen of associated bone and charcoal from Texas at the University of Texas laboratory indicate that the bone dates are consistently too young and that the discrepancy increases with age (Tamers and Pearson, 1965). Consequently, the dates based on bone that are quoted here are used as minimum dates. The dating based on associated artifacts is at best only approximate.

The analyses are based entirely on the mammalian remains; study of the reptiles, amphibians, and birds in many deposits is incomplete.

LATE-PLEISTOCENE ASSEMBLAGES AND ENVIRONMENTS

The principal late-Pleistocene faunas discussed here are summarized in Table 1; their locations are shown in Figure 1. Several other localities have produced a limited amount of material.

Faunas of late-Wisconsin age in central Texas are made up of three groups of species. One is represented by extinct species such as *Mammut americanum, Mammuthus columbi, Smilodon* sp., *Camelops* sp., *Aenocyon dirus, Platygonus* sp., and others which are usually thought of as being characteristic of the Pleistocene. Most of these species disappeared about 8,000 years ago (Hester, 1960). A few, such as *Bison antiquus* and *Mammut americanum*, survived an additional 1,000 to 2,000 years (Hester, 1960; Skeels, 1962).

A second group contains species and subspecies that are extant but that either do not inhabit central Texas today or are present only as

TABLE 1. Species Distribution in Central Texas Fossil Localities

Species	Fern Cave Upper Zone	Fern Cave Middle Zone	Fern Cave Lower Zone	Rattlesnake Cave Zone 1	Rattlesnake Cave Zone 2	Rattlesnake Cave Zone 3	Rattlesnake Cave Zone 4 (oldest)	Kyle Site Toyah Focus Zone (674 ± 165–389 ± 130)	Kyle Site Austin Focus Zone (1,389 ± 150–979 ± 170)	Centipede Cave Upper Zone	Centipede Cave Middle Zone	Centipede Cave Lower Zone	Barton Road Cultural unit (1,015 ± 105)	Barton Road Noncultural unit (3,450 + 150)	Wunderlich Site (5,405 ± 300–4,170 ± 200)	Felton Cave (7,770 ± 130)‡	Black Soil unit	Red Clay unit	Longhorn Breccia	Miller's Cave Brown Clay unit (3,008 ± 410)	Miller's Cave Travertine unit (7,290 ± 260)‡	Montell Shelter	Zesch Cave	Levi Shelter Zone V	Levi Shelter Zone IV (6,750 ± 150–9,300 ± 160)	Levi Shelter Zone III	Levi Shelter Zone II (10,000 ± 175)	Levi Shelter Zone I (oldest)	Kincaid Shelter	Clamp Cave	Laubach Cave	Cave Without a Name (10,900 ± 190)‡	Friesenhahn Cave
Order Marsupialia																																	
Didelphis marsupialis Linnaeus				×									×	×																			×
Order Insectivora																																	
Sorex cinereus Kerr																																×	
Blarina brevicauda (Say)													×			×	×				×											×	×
Cryptotis parva (Say)				×									×	×		×		×			×											×	×
Notiosorex crawfordi (Coues)				×	×	×	×		×				×	×		×	×						×										×
Scalopus aquaticus Linnaeus									×									×			×		×					×				×	

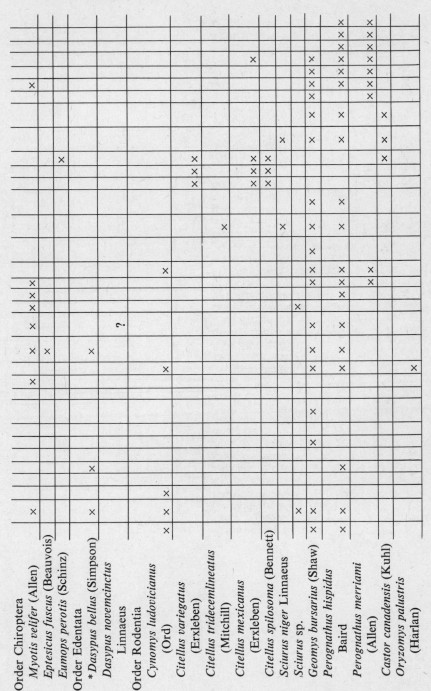

Order Chiroptera
Myotis velifer (Allen)
Eptesicus fuscus (Beauvois)
Eumops perotis (Schinz)
Order Edentata
*Dasypus bellus (Simpson)
Dasypus novemcinctus Linnaeus
Order Rodentia
Cynomys ludovicianus (Ord)
Citellus variegatus (Erxleben)
Citellus tridecemlineatus (Mitchill)
Citellus mexicanus (Erxleben)
Citellus spilosoma (Bennett)
Sciurus niger Linnaeus
Sciurus sp.
Geomys bursarius (Shaw)
Perognathus hispidus Baird
Perognathus merriami (Allen)
Castor canadensis (Kuhl)
Oryzomys palustris (Harlan)

TABLE 1.—continued

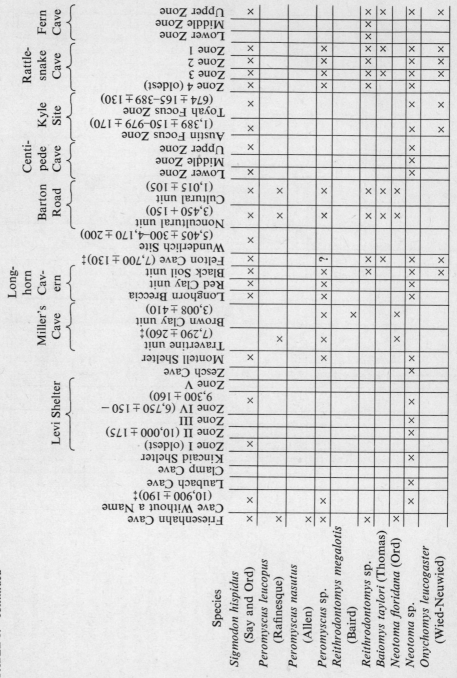

Species	Fern Cave Upper Zone	Fern Cave Middle Zone	Fern Cave Lower Zone	Rattlesnake Cave Zone 1	Rattlesnake Cave Zone 2	Rattlesnake Cave Zone 3	Rattlesnake Cave Zone 4 (oldest)	Kyle Site Toyah Focus Zone (674 ± 165–389 ± 130)	Kyle Site Austin Focus Zone (1,389 ± 150–979 ± 170)	Centipede Cave Upper Zone	Centipede Cave Middle Zone	Centipede Cave Lower Zone	Barton Road Cultural unit (1,015 ± 105)	Barton Road Noncultural unit (3,450 + 150)	Wunderlich Site (5,405 ± 300–4,170 ± 200)	Longhorn Cavern Felton Cave (7,700 ± 130)‡	Longhorn Cavern Black Soil unit	Longhorn Cavern Red Clay unit	Longhorn Cavern Longhorn Breccia	Miller's Cave Brown Clay unit (3,008 ± 410)	Miller's Cave Travertine unit (7,290 ± 260)‡	Montell Shelter	Zesch Cave	Levi Shelter Zone V	Levi Shelter Zone IV (6,750 ± 150–9,300 ± 160)	Levi Shelter Zone III	Levi Shelter Zone II (10,000 ± 175)	Levi Shelter Zone I (oldest)	Kincaid Shelter	Clamp Cave	Laubach Cave	Cave Without a Name (10,900 ± 190)‡	Friesenhahn Cave
Sigmodon hispidus (Say and Ord)	×			×	×	×	×	×		×		×	×	×	×	×	×	×	×			×			×			×				×	×
Peromyscus leucopus (Rafinesque)													×	×							×												×
Peromyscus nasutus (Allen)																																	×
Peromyscus sp.				×	×	×	×						×	×		?	×	×	×	×	×	×										×	×
Reithrodontomys megalotis (Baird)													×	×		×				×													
Reithrodontomys sp.		×	×	×	×	×	×						×	×		×	×																
Baiomys taylori (Thomas)	×			×	×	×							×	×						×	×												×
Neotoma floridana (Ord)																																	×
Neotoma sp.	×			×	×	×	×	×	×	×	×	×			×	×	×	×	×			×	×		×	×	×	×			×	×	
Onychomys leucogaster (Wied-Neuwied)	×			×	×	×		×	×							×	×																

Ondatra zibethica Linnaeus												×	×																				
Microtus ochrogaster (Wagner)													×	×																			
Microtus pennsylvanicus (Ord)		×																															
Pitymys pinetorum (Le Conte)																									×	×							
Microtus sp.	×	×				×					×	×						×		×	×							×	×				
Synaptomys cooperi Baird		×											×		×	×																	
Mus musculus Linnaeus																	×												×	×			
Erethizon dorsatum Linnaeus				×																				×									
Order Carnivora																																	
Canis latrans (Say)	×	×	×	×			×		×				×		×				×						×								
†*Aenocyon dirus* (Leidy)	×			×	×	×																											
Vulpes velox (Say)											×																						
Urocyon cinereoargenteus (Schreber)																						×		×									
†*Arctodus pristinus* Leidy	×																																
Ursus americanus Pallas	×	×		×							×		×		×																		
Bassariscus astutus (Lichtenstein)																×					×	×	×	×									
Procyon lotor simus Gidley		×	×						×										×						×	×							
Procyon lotor fuscipes Mearns																				×	×			×									
Mustela erminea (Richardson)		×																															
Taxidea taxus (Schreber)				×							×														×	×							

TABLE 1.—continued

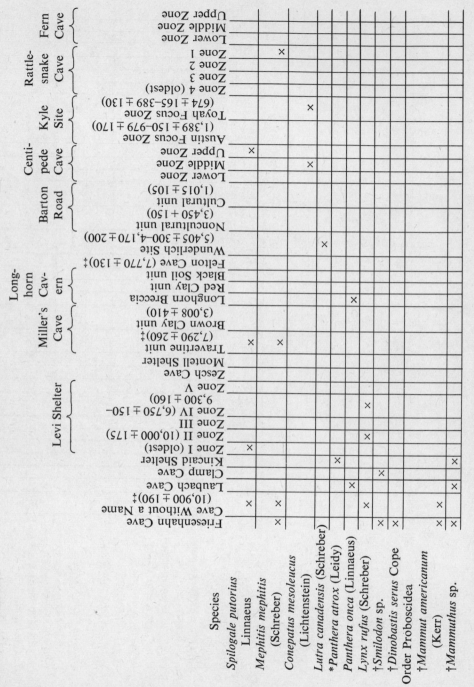

Site / Zone	Spilogale putorius Linnaeus	Mephitis mephitis (Schreber)	Conepatus mesoleucus (Lichtenstein)	Lutra canadensis (Schreber)	*Panthera atrox (Leidy)	Panthera onca (Linnaeus)	Lynx rufus (Schreber)	†Smilodon sp.	†Dinobastis serus Cope	Order Proboscidea	†Mammut americanum (Kerr)	†Mammuthus sp.
Fern Cave, Upper Zone												
Fern Cave, Middle Zone												
Fern Cave, Lower Zone												
Rattlesnake Cave, Zone 1					×							
Rattlesnake Cave, Zone 2												
Rattlesnake Cave, Zone 3												
Rattlesnake Cave, Zone 4 (oldest)												
Kyle Site, Toyah Focus Zone (674 ± 165–389 ± 130)				×								
Kyle Site, Austin Focus Zone (1,389 ± 150–979 ± 170)												
Centipede Cave, Upper Zone	×											
Centipede Cave, Middle Zone				×								
Centipede Cave, Lower Zone												
Barton Road, Cultural unit (1,015 ± 105)												
Barton Road, Noncultural unit (3,450 + 150)												
Wunderlich Site (5,405 ± 300–4,170 ± 200)				×								
Felton Cave (7,770 ± 130)‡												
Longhorn Cavern, Black Soil unit												
Longhorn Cavern, Red Clay unit												
Longhorn Breccia								×				
Miller's Cave, Brown Clay unit (3,008 ± 410)												
Miller's Cave, Travertine unit (7,290 ± 260)‡	×	×										
Montell Shelter												
Zesch Cave												
Levi Shelter, Zone V												
Levi Shelter, Zone IV (6,750 ± 150–9,300 ± 160)							×					
Levi Shelter, Zone III												
Levi Shelter, Zone II (10,000 ± 175)							×					
Levi Shelter, Zone I (oldest)	×											
Kincaid Shelter						×						×
Clamp Cave							×					
Laubach Cave						×						×
Cave Without a Name (10,900 ± 190)‡	×	×					×				×	
Friesenhahn Cave		×					×	×	×		×	×

Order Lagomorpha
Lepus californicus Gray
Sylvilagus floridanus (Allen)
Sylvilagus auduboni (Baird)
Sylvilagus sp.
Order Artiodactyla
†Mylohyus nasutus (Leidy)
†Platygonus sp.
†Camelops sp.
†Tanupolama sp.
Odocoileus virginianus (Zimmerman)
†Capromeryx sp.
Antilocaprid
Bison sp.
†Ovibovine
Order Perissodactyla
*Tapirus veroensis Sellards
*Tapirus sp.
*Equus sp.
*Hemionus lambei (Hay)

† Genus extinct.
* Species or subspecies extinct.
‡ Date based on bone.
Source for following caves are: Levi Shelter, Alexander (1963); Wunderlich, Johnson (1962); Centipede Cave, Epstein (1963); Kyle Site, Jelks (1962).

relics. These are: *Sorex cinereus, Blarina brevicauda, Mustela erminea, Procyon lotor simus, Peromyscus nasutus, Cynomys ludovicianus, Synaptomys cooperi, Pitymys pinetorum, Microtus pennsylvanicus, M. ochrogaster*, and *Ondatra zibethica*, which are now found either to the north and east or northwest, in generally wetter and cooler climates.

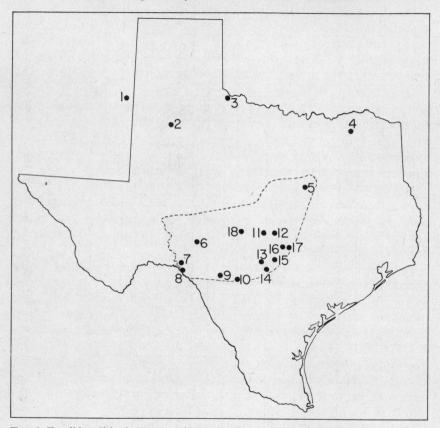

FIG. 1. Fossil localities in Texas and New Mexico. 1, Blackwater Draw, New Mexico; 2, Lubbock Reservoir site, Lubbock Co.; 3, Howard Ranch, Hardeman Co.; 4, Ben Franklin, Delta Co.; 5, Kyle site, Hill Co; 6, Felton Cave, Sutton Co.; 7, Fern Cave, Val Verde Co.; 8, Centipede Cave, Val Verde Co.; 9, Montell Shelter, Uvalde Co.; 10, Kincaid Shelter, Uvalde Co.; 11, Miller's Cave, Llano Co.; 12, Longhorn Cavern, Burnet Co.; 13, Cave Without a Name, Kendall Co.; 14, Friesenhahn Cave, Bexar Co.; 15, Wunderlich site, Comal Co.; 16, Levi Shelter, Travis Co.; 17, Barton Springs Road site, Travis Co.; 18, Zesch Cave, Mason Co.

The third group of species is composed of the Recent local fauna. Almost all Recent species of central Texas are found in the late-Wisconsin faunas. Major exceptions are the nine-banded armadillo *Dasypus novemcinctus* and the collared peccary *Tayassu tajacu*. These seem to be ecological equivalents of *Dasypus bellus* and *Platygonus* sp. respectively. The latter were not immediately replaced after their disappearance at the end of the Wisconsin, for the Recent species are not represented in any but very late archaeological sites.

The use of late-Wisconsin extinct species for environmental interpretation is subject to several difficulties. It is obviously impossible to obtain experimental or field data regarding their tolerances and habitat preferences. Data obtained from their closest living relatives may be questioned or rejected because of specific differences in habitat requirements.

Many of the extinct species were very widely distributed in North America during the Pleistocene. This might be interpreted as indicating either wide tolerances for different conditions or else a relatively uniform environment over large areas of North America at the time. Some of these wide distributions may eventually prove to reflect improper taxonomic treatment or imprecise dating of the fossil faunas.

The extent of local differentiation within these species is largely unknown, but Simpson (1941) indicated that the large Pleistocene cats show geographic differences. It is likely that when they are better known, many fossil species will show geographic differentiation comparable to that of many living mammals.

Some fossil species are absent or show differences in relative abundance in different areas for reasons that are not always apparent but are probably a function of local ecologic conditions. In general, the mastodon *Mammut americanum* was much more common on the Gulf Coastal Plain than on the High Plains or the Edwards Plateau (Slaughter et al., 1962), where *Mammuthus* was the common proboscidian. The mammoth was common over most of Texas, including the Gulf Coastal Plain. This change in the relative abundance of mastodon and mammoth suggests differences in the environment of these areas. The most obvious difference in the two animals is in the teeth. The complex high-crowned teeth of the mammoth would be better adapted to a diet of harsh vegetation such as grass than the simpler, low-crowned teeth of the mastodon. This indicates more forest on the Coastal Plain than on the High Plains. The mastodon was not completely absent from the High Plains or the Edwards Plateau, but it was probably restricted to locally favorable areas such as stream valleys.

The peccary *Mylohyus nasutus* and an ovibovine were also part of the Wisconsin fauna of central Texas. The distribution of the former is principally in the eastern part of North America where it is associated with other forest species. The Pleistocene ovibovines are distributed across the northern part of North America. In independent studies

Hibbard (1951), Kitts (1953), and Semken et al. (1964) have concluded that *Symbos* and probably *Bootherium* were forest forms.

Glyptodonts, chlamytheres, and the large Pleistocene beaver *Castoroides* have not been found on the Edwards Plateau. *Castoroides* (Fig. 2) has been found in Texas only in the northeastern corner of the state (Slaughter, 1963). This species is widely distributed in late-Pleistocene deposits in the eastern part of the United States with one record from Oregon (Hay, 1923, 1927). The known distribution of the giant beaver suggests that it occupied forested areas with a cool to cold humid climate.

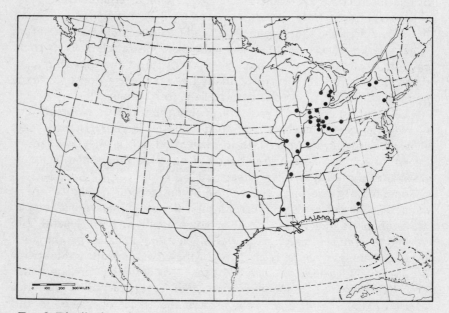

FIG. 2. Distribution of *Castoroides ohioensis*.

Both the glyptodonts and the chlamytheres (Fig. 3) were restricted to the southern part of the United States throughout the Pleistocene (Melton, 1964; James, 1957). In Texas both types were restricted to the southern part of the state in the late Wisconsin. Both are absent from the Ben Franklin and Howard Ranch local faunas (Slaughter, 1963; Dalquest, 1965). Both are present in the Berclair Terrace deposits in Bee County in southern Texas (Sellards, 1940). Evans and Meade (1945) record a questionable occurrence of a chlamythere from Lubbock County and a glyptodont from Andrews County, both on the High Plains. These are the only two records of these species out of twenty localities of Wisconsin age recorded by Evans and Meade on the High

Plains. Andrews County is in the southwestern part of the High Plains and may have been ecologically related to the southwest. The record of the chlamythere from Lubbock County is questioned by Evans and Meade. The restriction of these two groups to the southern United States suggests that they were intolerant of severe cold. Admittedly their absence from the Edwards Plateau may be related to the poor showing of large animals in cave deposits.

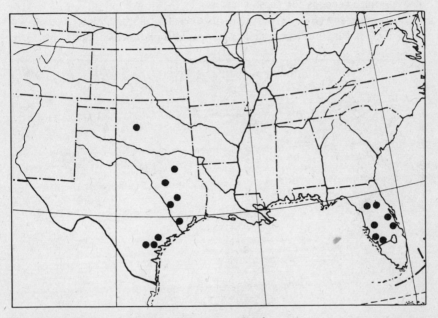

FIG. 3. Distribution of *Chlamytherium septentrionale*.

The following genera and species are not known from faunas of Wisconsin age in central Texas: *Preptoceros, Euceratherium, Nothrotherium*, and *Sangamona*. The first three are found in scattered localities in the Southwest (Hibbard, 1958). The last is known only from Burnet Cave, New Mexico.

The antilocaprids are uncommon in the Pleistocene faunas of central Texas. *Capromeryx* is present only at Kincaid Shelter in central Texas and is known from adjacent areas both northwest on the Llano Estacado (Lundelius, in press; Slaughter, in press; Green, 1961) and to the south on the southern Gulf Coastal Plain (Sellards, 1940).

There is some indication from scanty material that the species of *Camelops* from the Edwards Plateau is a large form that is similar to the High Plains species from Blackwater Draw.

Species that are still extant but that do not now live in central Texas are a more reliable group upon which to base environmental interpretations. There is little direct information on animal tolerances under extreme conditions; thus the present distributions are used as approximate indicators of their ecological requirements. Several of these species are known from late-Pleistocene deposits in Mexico or have isolated living populations in Mexico. The central-Texas occurrences provide intermediate records for these species.

Sorex cinereus (Fig. 4) is found in several late-Wisconsin deposits on the Edwards Plateau, in north Texas, and at San Josecito Cave, Mexico.

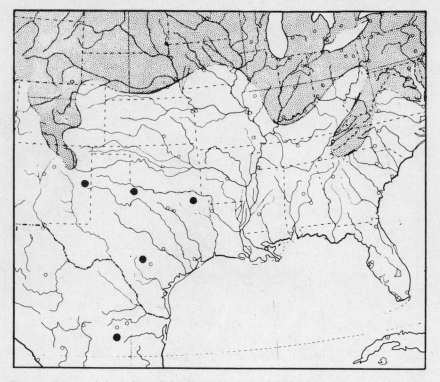

FIG. 4. Recent (stippled) and fossil (solid circles) distribution of *Sorex cinereus*.

Its present distribution is throughout northern North America, with southern extensions at higher elevations into New Mexico and Tennessee and North Carolina. All these areas are cooler and wetter than the present climate of central Texas.

Blarina brevicauda (Fig. 5) is abundant in most deposits of Wisconsin age in central Texas. Its present distribution is the eastern United States;

in eastern Texas it extends from Sour Lake, Hardin County, northward to Cherokee County, and an isolated population is reported from Aransas County. All the material of this species found to date falls in the size range of the southern subspecies, which is smaller than the northern (Hibbard, 1963).

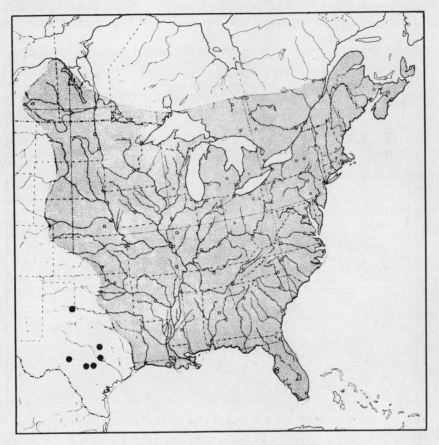

FIG. 5. Recent (stippled) and fossil (solid circles) distribution of *Blarina brevicauda*.

Microtus pennsylvanicus (Fig. 6) is known from only two specimens found at Cave Without a Name. The scarcity of material suggests that this species was quite rare; its Recent distribution extends from the northern edge of Hudson Bay to a line running from Washington to Georgia, with southward extensions into Utah and New Mexico (Hall and Kelson, 1959). Other Wisconsin records are Hardeman County (Dalquest, 1965); Brown Sand Wedge at Blackwater Draw, New Mexico

(Slaughter, unpublished); Lubbock Reservoir site and the Ben Franklin local fauna, Delta County, Texas (Slaughter, 1963). The Cave Without a Name is its southernmost known occurrence.

Pitymys pinetorum (Fig. 7), an eastern forest form with relict populations in Kerr County, Texas (Bryant, 1941), and Wichita Mountains, Oklahoma (Blair, 1958), has a close relative, *P. quasiater*, in eastern Mexico. *Pitymys* is present in several late- and post-Wisconsin faunas on

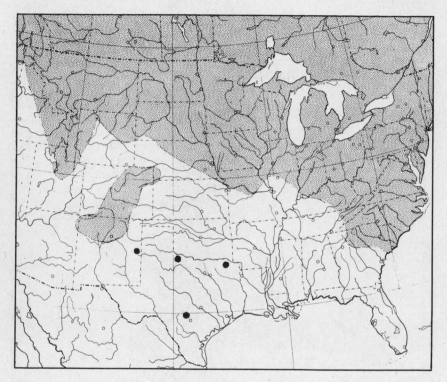

FIG. 6. Recent (stippled) and fossil (solid circles) distribution of *Microtus pennsylvanicus.*

the Edwards Plateau. Its present distribution in Texas is confined to the eastern forest. The fossil records indicate a much wider distribution during the Pleistocene. Because of the fragmentary nature of much of the material in many localities, *Pitymys* cannot be reliably distinguished from another vole, *Microtus ochrogaster* (Fig. 8). Both are known to have been present on the Edwards Plateau during the late Wisconsin, and considerable ecological information might be gained if they could be reliably distinguished. The central-Texas fossil records of *M. ochrogaster*

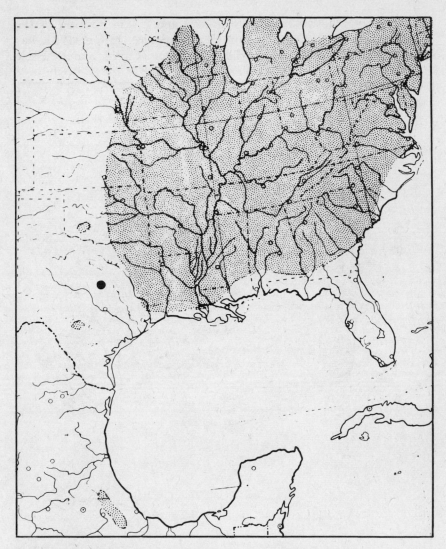

FIG. 7. Recent (stippled) and fossil (solid circles) distribution of *Pitymys pinetorum* (in the U.S.) and *P. quasiater* (in Mexico).

are well south of the southern edge of its Recent distribution in central Oklahoma.

Synaptomys cooperi (Fig. 9) has a wide distribution in Wisconsin faunas of central Texas and is also present in San Josecito Cave, Nuevo Leon, Mexico. Its recent distribution extends from north of the Great Lakes to southern Kansas and northeastern Arkansas. Hibbard (1963) has observed a north–south cline in size and in the position of the posterior end of the lower incisor in this species. Patton (1963) has shown that the *Synaptomys* material from Miller's Cave is about the size

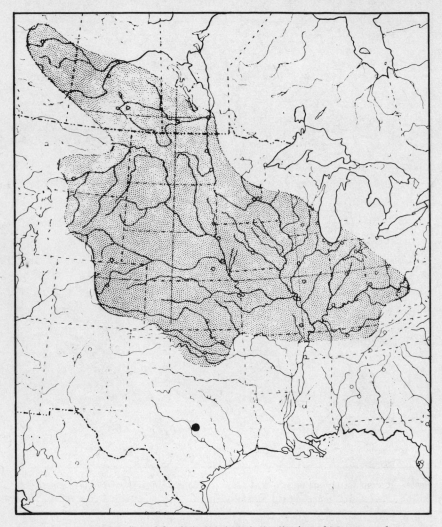

FIG. 8. Recent (stippled) and fossil (solid circles) distribution of *Microtus ochrogaster*.

of living specimens from southwestern Kansas and is smaller than the more southern *Synaptomys australis* from Florida. The position of the posterior end of the lower incisor is also similar to that of the living Kansas form.

Another northern element in the central-Texas fauna is *Mustela erminea* (Fig. 10), whose present distribution is from the Arctic to the northern United States, extending southward at higher elevations.

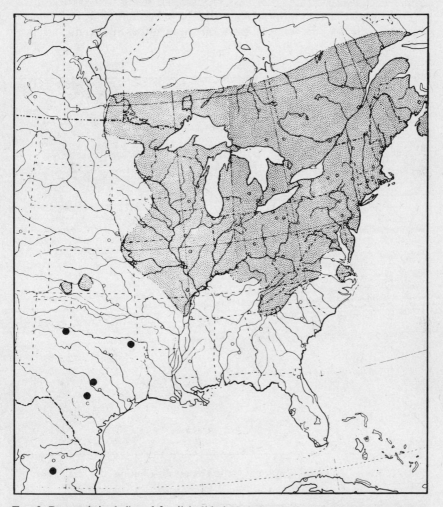

FIG. 9. Recent (stippled) and fossil (solid circles) distribution of *Synaptomys cooperi*.

The raccoon *Procyon lotor* is a member of both late-Wisconsin and Recent faunas, but the successive populations show morphological differences. The Wisconsin and some of the later populations are indistinguishable from the subspecies *P. l. simus* known from the late Pleistocene and Recent of the Pacific Northwest (Wright and Lundelius, 1963).

This form of *Procyon* is also known from the Brown Sand Wedge at Blackwater Draw, New Mexico, which is late Wisconsin in age (Slaughter, unpublished). Specimens of *P. l. simus* have much more massive mandibles and skulls than representatives of late-Pleistocene or Recent populations from the eastern United States that have been studied; *P. l. simus* is characteristic of the western part of the United States during late-Wisconsin and early-Holocene time.

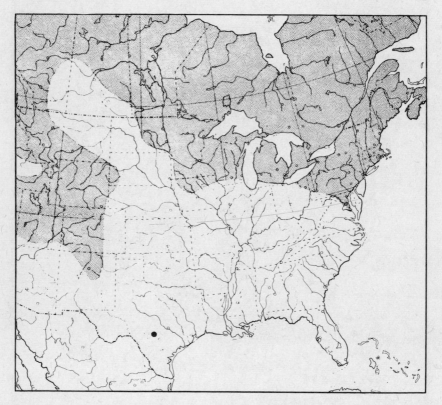

Fɪɢ. 10. Recent (stippled) and fossil (solid circles) distribution of *Mustela erminea*.

Peromyscus nasutus is known from undated but certainly late-Wisconsin deposits in Friesenhahn Cave (Tamsitt, 1957). It has been questionably identified from Cave Without a Name in Kendall County. Its present range is northwest of central Texas in New Mexico, Colorado, Utah, and Arizona (Fig. 11).

The prairie dog *Cynomys ludovicianus* is present in deposits of Wisconsin age at Friesenhahn Cave, Bexar County; Cave Without a Name,

Kendall County; and Laubach Cave, Williamson County. The southern-most Recent record of this species is Mason County which is on the northwest edge of the Edwards Plateau. It is widely distributed through-out the Great Plains, west Texas, and New Mexico.

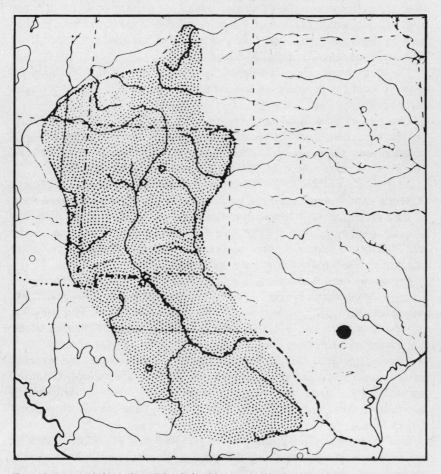

Fig. 11. Recent (stippled) and fossil (solid circles) distribution of *Peromyscus nasutus*.

The muskrat *Ondatra zibethica* occurs in Montell shelter in Uvalde County. It is also known from several early-Holocene deposits in central Texas and was probably more widely distributed in the Wisconsin than records indicate. It is present in late-Wisconsin deposits at Blackwater Draw, New Mexico, and Lubbock County, Texas. This species requires dependable year-round water (Hollister, 1911). At present it is confined in Texas to the Gulf Coast, the Canadian River in the Texas Panhandle, and the Rio Grande in Trans-Pecos Texas.

The eastern mole *Scalopus aquaticus* and the plains pocket gopher *Geomys bursarius* are widely distributed in both late-Wisconsin and post-Wisconsin deposits on the Edwards Plateau. At present both occur in the area only on small patches of suitable soil. Both species extend farther west on the High Plains, and *Scalopus aquaticus* has isolated populations in west Texas and Coahuila, Mexico.

The least shrew, *Cryptotis parva*, is another species that is found in many Wisconsin deposits on the Edwards Plateau but does not occur there today.

Two species of mammals which had range extensions southward during the Wisconsin have not been found in central Texas. These are *Sorex palustris* and *Thomomys talpoides*, which are associated with a date of $16,775 \pm 565$ years B.P. in Hardeman County (Dalquest, 1965).

As indicated above, the Pleistocene mammalian fauna of North America shows geographic differences. In Texas these differences seem to be a reflection of a temperature gradient in a north–south direction and an east–west change from forest to grassland or savanna. Some late-Pleistocene faunas contain species whose present ranges do not overlap or are even separated by long distances.

Species whose distributions are far north of central Texas are *Sorex cinereus*, *Synaptomys cooperi*, *Mustela erminea*, *Microtus ochrogaster*, *M. pennsylvanicus*, and one subspecies, *Procyon lotor simus*. The presence of these forms in a late-Wisconsin fauna is not unexpected, in view of the southward movement of the climatic zones during a glacial stage.

Another group of mammals has Recent distributions that are centered in the southeast: *Didelphis virginianus*, *Blarina brevicauda*, *Cryptotis parva*, *Scalopus aquaticus*, *Oryzomys palustris*, and *Sigmodon hispidus*. In addition to these there is a third group that is widespread through the Great Plains.

The problem of accounting for the presence in Pleistocene faunas of species whose ranges are today completely allopatric and whose habitat requirements seem to be mutually exclusive has been discussed by Hibbard (1960) and Dalquest (1965). Both have concluded that if climatic extremes are replaced by a more uniform climate the coexistence of these divergent types becomes feasible. This hypothesis is based on the assumption that the northern species are limited to the south by the summer temperature maxima and the southern species are limited to the north by the winter minima. The distribution maps of these species and the climatic maps of Visher (1945) support this idea. Several species extend farther south in the Rocky Mountains and Appalachian Mountains than they do in the central part of the continent. Visher's maps show that both higher rainfall and lower temperature extend farther south in these regions than in the central United States. Slaughter (1957) has used the interpretation of a more uniform, equable Pleistocene climate as causing Pleistocene megafaunal extinction when continentality developed.

It should be pointed out that most of those species that are usually thought of as requiring relatively mild winters extend northward into areas where they must tolerate rather low winter temperatures. *Didelphis* extends to central Wisconsin, central Michigan, and southern Ontario; *Sigmodon hispidus* to northeast Kansas and southeast Virginia; *Oryzomys palustris* to east-central Kansas, southern Illinois, and up the east coast to southeastern Pennsylvania; *Cryptotis parva* north to central South Dakota, central Wisconsin, central Michigan, and central New York; *Blarina brevicauda* extends into Canada; and *Scalopus aquaticus* extends to central Minnesota, Wisconsin, and Michigan and up the east coast to Massachusetts.

All these species except *Sigmodon hispidus* are in areas of rather high rainfall. The western limit of most of them lies somewhere on the Great Plains or east of the Great Plains. It seems likely that their distributions are more readily controlled by moisture and soil conditions than by low temperatures, and that their presence in central Texas during the Wisconsin, associated with the northern elements, reflects increased rainfall and reduced summer evaporation rather than change in winter temperatures. Although it seems almost certain that central Texas did not have arctic winters, there seems to be no reason why the minima could not have been as low or perhaps lower than at present.

The presence of the northern species in the Wisconsin faunas of central Texas almost certainly indicates greater and more evenly distributed moisture and lower summer maxima. Several species now have their southernmost occurrence in either the Rocky Mountains or the Appalachian Mountains or both. In the Appalachians there is little difference in rainfall from the lower elevations, but the temperatures are lower (Visher, 1945).

Several late-Wisconsin faunas contain species indicating the presence of diverse ecological situations in the areas surrounding the sites from which the fossils were recovered. Friesenhahn Cave, Cave Without a Name, and Longhorn Cavern contain *Perognathus hispidus* as well as one or more species of microtines (*Synaptomys cooperi*, *Microtus pennsylvanicus*, *Pitymys pinetorum*) and shrews (*Blarina brevicauda*, *Sorex cinereus*). The habitat preferences of these species are very different. *Perognathus hispidus* inhabits open grassland areas with scattered stands of herbaceous vegetation. *Sorex cinereus*, *Blarina brevicauda*, and *Pitymys pinetorum* occupy areas of varying temperature and humidity but are generally forest forms.

The overlap of species of such divergent habitats can be attributed to the topography of the margins of the Edwards Plateau. This area today has numerous deep canyons that are much cooler and more humid during the summer than the adjacent divides. Although the exact relief at the time is unknown, the situation during the Wisconsin was probably much the same.

I suggest that the species now occupying open situations were living in the upland areas during the Wisconsin. The species living today in forests and/or bogs then occupied the canyons. Predators such as owls, using caves as shelters, would hunt in both areas and bring remains of animals from both habitats into the caves.

The fauna of central Texas as well as the rest of North America underwent a drastic change, 8,000 to 10,000 years ago. Hester (1960) adopted a terminal date of 8,000 years for most of the megafauna but Martin (1957) believes that over 10,000 years is more likely. Radiocarbon dates from Cave Without a Name and Levi Shelter indicate the presence of extinct forms approximately 10,000 years ago. Younger radiocarbon dates from Felton Cave and Wunderlich Site are not associated with the extinct fauna. These dates support Martin's thesis of a terminal date of approximately 10,000 years, but no faunas from central Texas are known from the critical interval of 8,000 to 10,000 years ago. In addition to the loss of the extinct species, four extant species, *Sorex cinereus*, *Cynomys ludovicianus*, *Microtus pennsylvanicus*, and *Mustela erminea* withdrew from central Texas during this interval. More data are needed to demonstrate how closely their disappearance coincides with the extinction of the megafauna.

POST-PLEISTOCENE ASSEMBLAGES

After the extinction of the large mammals about 8,000 to 10,000 years ago, the fauna of central Texas consisted of the Recent fauna plus several species that are now found north and east of central Texas in regions of cooler and more humid climates.

With the exception of *Procyon lotor simus*, the species that are today distributed farthest north disappeared earliest from central Texas. *Sorex cinereus*, *Microtus pennsylvanicus*, and *Mustela erminea* are last known from Cave Without a Name (10,900 B.P.). They seem to have disappeared from central Texas with the large extinct fauna. *Synaptomys cooperi* and *Peromyscus nasutus* are last known from Miller's Cave

($\sim$ 7,300 B.P.) and Felton Cave (7,800 B.P.) respectively; *Microtus ochrogaster* was present at Miller's Cave 3,000 years ago, *Scalopus aquaticus* at Austin (Barton Road site) 1,000 years ago, and *Procyon lotor simus* at the Kyle site 800 years ago.

The present data indicate that *Blarina brevicauda* disappeared earlier from the western part of the Edwards Plateau than from the eastern part. It is present in almost all faunas in the eastern part of the Plateau until about 1,000 years ago. It is last known in the western part of the Plateau from Felton Cave 7,800 years ago. It has not been found in the abundant small-mammal faunas of either Centipede or Damp caves in Val Verde County. This absence is especially significant because both caves are located along the canyon of the Rio Grande, which one might expect would have remained humid after the uplands had become quite arid and would provide a refuge for such species as *Blarina brevicauda.*

I believe that the order in which these species disappeared is an indication of a gradually progressive warming and drying of the climate from west to east across the Edwards Plateau. Frank (1965) has shown that the clay minerals making up the cave sediments indicate a shift to drier conditions had begun earlier in the western part of the Edwards plateau.

The restriction of the range of *Geomys bursarius* has taken place during post-Wisconsin time. There seems to be no pattern, geographic or chronologic, to the changes in its distribution. This, plus the fact that its present distribution in central Texas is spotty and related to locally favorable soil conditions, suggests that soil erosion is responsible for the range restriction. The soil cover over most of the Edwards Plateau is very thin today, and many species are not found in this region apparently for this reason. The presence of *Geomys bursarius* in very young deposits in areas in which it does not now occur indicates that this process has been operative in very recent times and probably is still going on.

On the presence of *Mus musculus*, the black fill unit in Longhorn Cavern (Semken, 1961), has been dated as younger than 1700 A.D. It contains *Geomys bursarius*, which is not found today in Burnet County. The nearest known occurrence is in Llano County approximately forty miles away. *G. bursarius* is also found in Rattlesnake Cave in Kinney County, again associated with *Mus musculus* in the top twelve inches of a black soil unit. The nearest modern occurrence is in Dimmit County, seventy-five miles south. In both instances *Geomys* remains have been recovered from a black soil unit that was derived from a black soil similar to that found on the Edwards Plateau today. Other older, but still late post-Pleistocene, occurrences of *Geomys bursarius* in areas in which it is not now found are Barton Road site in Austin, Travis County (1000 A.D.), and Kyle site, Hill County (1000 A.D.).

The mole *Scalopus aquaticus* appears to present much the same picture as *Geomys bursarius*, although the record is not so good. Its past distribution was more extensive than at present.

The prairie dog *Cynomys ludovicianus* is another species that had a more extensive distribution in central Texas during the Wisconsin. Although it is known historically from Mason, Sutton, and Schleicher counties, it has not been found in any of the post-Pleistocene faunas from other parts of the Plateau.

These species are fossorial to a large degree and demand a reasonable thickness of soil. Their scarcity on the Edwards Plateau today appears to be caused by the generally thin soil in this region. The removal of soil from the Plateau undoubtedly began with the entrenchment of the streams during the Pleistocene and was probably accelerated by the post-Wisconsin climatic change to drier conditions. The last stage in the removal of topsoil from this region has taken place since about 1800 by means of extensive overgrazing which has destroyed much of the vegetation (Semken, 1961). *Geomys bursarius* and *Scalopus aquaticus* are found today in areas of deep sandy soil derived from either granite or sandy formations of Paleozoic or Lower Cretaceous age.

The muskrat *Ondatra zibethica* was not widely distributed in central Texas either during or after the Wisconsin. It is known from a Wisconsin fauna from the Montell Shelter, Uvalde County, from the older unit in Miller's Cave dated at least 7,300 years B.P. (Patton, 1961), and from a relatively recent deposit in Damp Cave, Val Verde County (Lundelius, *in* Epstein, 1960). This species is an inhabitant of marshes and stream and lake borders and spends most of its time in the water (Hollister, 1911). It seems likely that its presence in a fauna indicates the existence of permanent water in the immediate area. All three occurrences mentioned above are close to streams that either have permanent water now or that could have had it in the past.

There is no indication in any of these faunas that the climate during the interval 4,000–6,000 years ago was any drier or warmer than at present. In fact the occurrences of the otter *Lutra canadensis* in the Wunderlich site in Comal County 5,000 years ago indicates more humid conditions. The faunal material from Centipede Cave in Val Verde County 5,000 years ago shows no indication of conditions drier than the present. This is in agreement with the conclusions of Martin (1963) and Bryant (1966) that the Altithermal in Arizona, New Mexico, and southern Texas was actually wetter than at present.

The modern faunal composition of central Texas was attained about 1,000 A.D. It is probable that *Pitymys pinetorum* was more widely distributed than at present, and that the raccoon was *Procyon lotor simus*, but both taxa were rapidly disappearing. Three species, *Dasypus novemcinctus*, *Thomomys bottae*, and *Tayassu tajacu*, which are members of the present fauna, are apparently very late arrivals in central Texas.[1] None of them has been found in any except the most recent deposits. There is no record of *Tayassu tajacu* in any archaeological sites in central Texas.

The armadillo *Dasypus novemcinctus* has long been cited as an animal whose range is expanding at a rapid rate. The absence of this animal in all but very late archaeological sites is consistent with its recent expansion.

The botta pocket gopher *Thomomys bottae* is also unknown in any of the Pleistocene or post-Pleistocene deposits. Its present distribution is mainly west of the Edwards Plateau, but there are local populations in the western part of the Plateau in Uvalde, Edwards, Sutton, and Val Verde Counties. This species probably is a recent invader of the Edwards Plateau and reflects the postglacial trend toward aridity. It is capable of living in areas with many different types of soil and does not require as thick a soil as does *Geomys bursarius*.

CLIMATIC INTERPRETATION

The Pleistocene faunas of central Texas contribute to the interpretation of the distributions of vertebrates in the southern United States and Mexico. Adams (1902), Deevey (1949), and Blair (1958) have argued that drastic changes in the environment of the southern United States accompanied the peaks of glacial advance into the northern United States. The results were the withdrawal of warmth-adapted animals into Florida and Mexico and the extension of many cold-adapted forest species into the southern United States and at least the higher elevations in Mexico. Withdrawal of the glaciers and climatic warming resulted in the northward retreat of the cold-adapted forms. Relicts were left in locally favorable situations and the warmth-adapted species expanded from the refuges.

Braun (1955) has opposed this hypothesis and argued that there was little change in the southern United States during the glacial stages. Braun also claimed that the forests of the mountains of Mexico and the eastern United States were last connected in mid-Tertiary time.

Martin and Harrell (1957) have put forward essentially the same argument as Braun. They maintain that there is no evidence for a forest

1. The record of *Dasypus novemcinctus* from the brown clay unit of Miller's Cave (Patton 1963) could represent a Recent specimen rather than a 3,000-year old record. The brown clay unit is thin (12 to 18 in.) and forms the present surface over part of the cave, and the 3,000-year date is based on charcoal from the base.

corridor across Texas and northeastern Mexico during the Pleistocene and that those species represented by relics in Mexico today were dispersed through a savanna.

The fossil faunas from central Texas indicate that a number of northern, cold-adapted species inhabited this area in the past. The radiocarbon dates of some of these demonstrate that their occurrence in central Texas coincides with the last glacial stage. All the species of eastern North American forest mammals that have relics in Mexico or have been found in late-Pleistocene deposits in Mexico, with the exception of *Glaucomys volans*, are members of the late-Pleistocene faunas of central Texas.

The suggestion of Martin and Harrell (1957), that the species now showing disjunct distributions in Mexico and the eastern United States dispersed through Texas and northeastern Mexico under savanna conditions, can be adequately checked only when pollen analyses have been done for this area. The presence of *Blarina brevicauda*, *Sorex cinereus*, and *Pitymys pinetorum* indicates forest conditions. Although *Glaucomys volans* has not been found in Wisconsin deposits in central Texas, its disjunct distribution (eastern U.S. and central Mexico) indicates the presence of forest conditions in intermediate areas. As noted above, the presence of *Perognathus hispidus* and *P. merriami* indicates that there were at least local open areas. These situations are believed to have existed on the stream divides, with forest occupying the valleys and possibly the valley slopes.

THE EXTINCTION PROBLEM

Many theories have been advanced to account for the late-Pleistocene extinction. There has been much disagreement over the relative importance of man as a cause. Many biologists believe that Early Man as a hunter would bear the same relation to his prey (in this case the large mammals) that other predators have to their prey; i.e. that the size of early human populations would be as much determined by those of their prey as vice versa. This may not have been the case, if man's ability to diversify his diet is considered. It is difficult to understand why such other animals as deer, elk, bison, antelope, moose, caribou, etc. which were utilized as food did not also become extinct if pressure of human predation was the only cause.

Pleistocene extinction in central Texas appears to coincide with that of other areas in North America. Among the faunas considered here, all those containing extinct species are older than 10,000 years. This supports Martin's contention (1967) that the major extinction occurred before 10,000 years ago, rather than 8,000 as proposed by Hester (1960). It should be pointed out that none of the faunas in this area dates from the critical period 8,000 to 10,000 years ago. All faunas younger than 7,800 years contain only Recent species.[2]

The faunal sequence from central Texas implies that climatic changes were underway when the extinction occurred. The nature of the faunal changes also suggests the nature of the climatic change. The late-Wisconsin faunas contained extant species whose distributions do not overlap today, so the late-Wisconsin climate of central Texas may have been more equable than the present. The changes in distribution of the extant forms are almost entirely concerned with the disappearance from central Texas of species that are found today to the north, northeast, or northwest in cooler and/or more humid areas, indicating that they are at present limited in southward distribution by the summer extremes of temperatures and aridity.[3] The overall indications are that the climatic change at the end of the Wisconsin involved an increase in seasonality of rainfall and temperature. Two extant species, *Mustela erminea* and *Microtus pennsylvanicus*, disappeared from central Texas with the extinct fauna. This could be interpreted to indicate that some change in climate was underway at this time.

The suggestions that climatic change can be ruled out as an agent of extinction because feral burros now occupy the driest parts of North America (Martin, 1963), and that the native American camels, llamas, and horses were even less susceptible to climatic changes than the mammoth, seem unsupportable. These feral burros are derived from Old World desert animals and are not the same species of horse (or ass) as the North American Pleistocene form. The native American camels, llamas, etc. are now extinct and did not necessarily have the same tolerance or adaptiveness as the modern forms.

Guilday (1967) has argued that desiccation was the major factor in the late-Wisconsin extinction: it destroyed some habitats and forced many species of herbivores into competition. Mehringer (1967) has rejected this hypothesis on the grounds that the pollen evidence reveals no decrease in habitat diversity. He does say that coniferous forest and woodland were more widespread than at present. Hafsten (1961) found conifer pollen widespread on the southern High Plains during the Wisconsin. The degree of this diversity created by the presence of conifers, etc. in areas

2. The record of *Dasypus bellus* in Miller's Cave, 7,300 years ago, is probably based on an inaccurate date.

3. I do not suggest that the animals are necessarily controlled directly by these climatic factors; it is more likely that the essential habitats are destroyed by the summer extremes.

which are now predominantly grassland is not measured only by the number of kinds of additional plant assemblages but also by the ecotones or border zones between the different plant assemblages. This is well summarized with examples by Allee et al. (1949). This is possibly the kind of diversity that was present during the Pleistocene and that was destroyed by postglacial climatic change.

Climatic change as the major cause of extinction is not so apparent in eastern North America (Russell, 1885; Martin, 1963), which does not seem to have undergone the large amount of faunal change seen in the West. There have been changes in the ranges of the extant fauna and flora in the East (see Guilday, 1962, 1964; Potzger and Tharp, 1947). Climatic changes involving a change in seasonality may have taken place there too, resulting in a reduction of the ecological diversity, diversity of niches, and the extinction of many species.

The post-Wisconsin faunal history of central Texas is one of extinction of the giant Pleistocene species, northward withdrawal of some species, and the recent addition of new forms. This is interpreted as showing a gradual drying of the climate and an increase in seasonality. There is no indication of drier conditions during the Altithermal. Several fossorial species suffered distributional restrictions caused by the removal of soil.

Two Recent species, *Sorex cinereus* and *Mustela erminea*, which disappeared from central Texas at the same time as the extinct species, have the northernmost distribution of the extant species found in the Pleistocene and post-Pleistocene faunas of this area. Their disappearance from central Texas indicates the beginning of the climate change that took place at the end of the Wisconsin. The coincidence of the disappearance from central Texas of these northern species and the extinct species indicates that the widespread extinction coincided with the beginning of the climatic change at the end of the Pleistocene.

Thanks are due to the following people who provided assistance and encouragement in the preparation of this paper: W. F. Blair and G. G. Raun for information on many Recent species; Dee Ann Story for information concerning the archaeological sites; Clayton Ray and Paul S. Martin for comments and criticisms; R. M. Frank for data on many caves and assistance in collecting and preparing material; W. A. Akersten and C. Seewald for sorting much of the material. This work·was supported in part by grants from the University Research Institute and the Geology Foundation, The University of Texas.

References

Adams, C. C., 1902, Southeastern United States as a center of geographical distribution of flora and fauna: *Biol. Bull.*, v. *3*, p. 115–31

Alexander, H. L., Jr., 1963, The Levi Site: A paleo-Indian campsite in central Texas: *Amer. Antiq.*, v. *28*, p. 510–28

Allee, W. C., Emerson, A. E., Park, Orlando, Park, Thomas, and Schmidt, K. P., 1949, *Principles of animal ecology:* Philadalephia and London, Saunders

Blair, W. F., 1958, Distributional patterns of vertebrates in the southern United States in relation to past and present environments, p. 433–568, *in* Hubbs, C. L., *Editor, Zoogeography:* Amer. Assoc. Sci., Publ. 51

Braun, E. Lucy, 1955, Phytogeography of unglaciated eastern United States and its interpretation: *Botan. Rev.*, v. *21*, p. 277–375

Bryant, M. D., 1941, A far southwestern occurrence of *Pitymys* in Texas: *Jour. Mamm.*, v. *22*, p. 202

Bryant, V. M., Jr., 1966, Pollen Analysis of the Devil's Mouth site, p. 129–64, *in* Story, D. A., and Bryant, V. M. J., Jr., *Assemblers, Paleoecology of the Amistad Reservoir Area:* National Science Foundation Final report

Dalquest, W. W., 1965, New Pleistocene formation and local fauna from Hardemann County, Texas: *Jour. Paleont.*, v. *39*, p. 63–79

Deevey, E. S., Jr., 1949, Biogeography of the Pleistocene: *Geol. Soc. Amer., Bull.*, v. *60*, p. 1315–1416

Dice, L. R., 1922, Some factors affecting the distribution of the Prairie vole, forest deer mouse and prairie deer mouse: *Ecology*, v. *3*, p. 29–47

Epstein, J. F., 1963, Centipede and Damp Cave; Excavations in Val Verde County, Texas, 1958, with appendices by Thomas W. McKern and Ernest Lundelius: *Texas Archeol. Soc. Bull.*, v. *33*, p. 1–29

Evans, G. L., and Meade, G. E., 1945, Quaternary of the Texas High Plains: Univ. Texas Publ. 4401, p. 485–507

Frank, R. M., 1965, Petrologic study of sediments from selected Texas caves: Univ. Texas, unpubl. master's thesis

Guilday, J. E., 1962, The Pleistocene local fauna of the Natural Chimneys, Augusta County, Virginia: *Ann. Carnegie Mus.*, v. *36*, no. 9, p. 87–122

——— 1967, Differential extinction during late-Pleistocene and Recent times (this volume)

Guilday, J. E., Martin, P. S., and McCrady, A. D., 1964, New Paris No. 4; A late Pleistocene cave deposit in Bedford County, Pennsylvania: *Nat. Speleol. Soc. Bull.*, v. *26*, p. 121–94

Hafsten, Ulf, 1961, Pleistocene development of vegetation and climate in the southern High Plains as evidence by pollen analysis, p. 41–59, *in* Wendorf, F., *Compiler, Paleoecology of the Llano Estacado:* Santa Fe, Museum of New Mexico

Hall, E. Raymond, and Kelson, Keith, 1959, *The mammals of North America:* New York, Ronald Press

Hay, O. P., 1923, The Pleistocene of North America and its vertebrated animals from the states east of the Mississippi River and from the Canadian provinces east of longitude 95: Carnegie Inst. Washington, Publ. 322, 499 p.

——— 1927, The Pleistocene of the western region of North America and its vertebrated animals: Carnegie Inst. Washington, Publ. 322B, p. 1–346

Hester, J. J., 1960, Late Pleistocene extinction and radiocarbon dating: *Amer. Antiq.*, v. *26*, p. 58–77

Hibbard, C. W., 1951, Animal life in Michigan during the Ice Age: *Michigan Alumnus Quart. Rev.*, v. *57*, p. 200–08

——— 1958, Summary of North American Pleistocene mammalian faunas: Michigan Acad. Sci., Arts Lett., Papers, v. *43*, p. 3–32

——— 1960, An interpretation of Pliocene and Pleistocene climates in North America: Michigan Acad. Sci., Arts Lett., 62nd Ann. Rept., (1959–60), p. 5–30

——— 1963, A late Illinoian fauna from Kansas and its climatic significance: Michigan Acad. Sci., Arts Lett., Papers, v. *48*, p. 187–221

Hollister, H., 1911, A systematic synopsis of the muskrats: *North Amer. Fauna*, v. *56*, 256 p.

James, G. T., 1957, An edentate from the Pleistocene of Texas: *Jour. Paleont.*, v. *31*, p. 796–808

Jelks, E. B., 1962, The Kyle site; A stratified central Texas aspect site in Hill County, Texas: Univ. Texas, Dept. Anthropology, Archaeology ser. 5

Kitts, D. B., 1953, A Pleistocene musk-ox from New York and the distribution of the musk-oxen: *Amer. Mus. Nat. Hist. Novitates*, no. *1607*, p. 1–8

Lundelius, E. L., Jr., 1967, Vertebrate remains from the Grey Sand, Blackwater Draw, New Mexico: Santa Fe, Fort Burgwin Research Center (in press)

Martin, P. S., and Harrell, B. E., 1957, The Pleistocene history of temperate biotas in Mexico and eastern United States: *Ecology*, v. *38*, p. 468–80

Martin, P. S., 1963, *The last 10,000 years; A fossil pollen record of the American southwest:* Tucson, Univ. Arizona Press, p. 1–87

——— 1967, Prehistoric overkill (this volume)

Mehringer, P. J., Jr., 1967, The environment of extinction of the late-Pleistocene megafauna in the arid southwestern United States (this volume)

Melton, W. G., Jr., 1964, *Glyptodon fredericensis* (Meade) from the Seymour formation of Knox County, Texas: Michigan Acad. Sci., Arts Lett., Papers, v. *49*, p. 129–46

Patton, T. H., 1963, Fossil vertebrates from Miller's Cave, Llano County, Texas: *Texas Memorial Mus. Bull.*, v. *7*, p. 7–41

Russell, I. C., 1885, Geological history of Lake Lahontan: *U.S. Geol. Survey Mono.* v. *11*, p. 1–288

Sellards, E. H., 1940, Pleistocene artifacts and associated fossils from Bee County, Texas (with notes on artifacts by T. N. Campbell, and notes on Terrace deposits by Glen L. Evans): *Geol. Soc. America Bull.*, v. *51*, p. 1627–58

Semken, H. A., Jr., 1961, Fossil vertebrates from Longhorn Cavern, Burnet County, Texas: *Texas Jour. Sci.*, v. *13*, p. 290–310

Semken, H. A., Miller, Barry, and Stevens, J. B., 1964, Late Wisconsin woodland musk oxen in association with pollen and invertebrates from Michigan: *Jour. Paleont.*, v. *38*, no. *5*, p. 823–35

Simpson, G. G., 1941, Large Pleistocene felines of North America: *Amer. Mus. Nat. Hist. Novitates*, no. *1136*, p. 1–27

Skeels, Margaret Anne, 1962, The mastodons and mammoths of Michigan: Michigan Acad. Sci., Arts Lett., Papers, v. *17*, p. 101–33

Slaughter, B. H., 1967, Animal ranges as a clue to late-Pleistocene extinction (this volume)

—— 1967, An ecological interpretation of the Brown Sand Wedge Local Fauna, Blackwater Draw, New Mexico: and a hypothesis concerning late Pleistocene extinction: Santa Fe, Fort Burgwin Research Center (in press)

Slaughter, B. H., Crook, W. E., Jr., Harris, R. K., Allen, D. C., and Seifert, Martin, 1962, The Hill-Schuler local faunas of the upper Trinity River, Dallas and Denton Counties, Texas: Univ. Texas, Bur. Econ. Geology, Rept. Invest. 48

Slaughter, B. H., and Hoover, B. R., 1963, Sulphur River formation and the Pleistocene mammals of the Ben Franklin local fauna: *Sou. Meth. Univ. Grad. Res. Cent. Jour.*, v. *31*, p. 132–51

Tamers, M. A., and Pearson, F. J., Jr., 1965, Validity of radiocarbon dates on bone: *Nature*, v. *208*, no. 5015, p. 1053–1055

Thornthwaite, C. W., 1948, An approach toward a rational classification of climate: *Geog. Rev.*, v. *38*, p. 55–94

Tamsitt, R., Jr., 1957, *Peromyscus* from the late Pleistocene of Texas: *Texas Jour. Sci.*, v. *9*, p. 355–63

Visher, S. S., 1945, Climatic maps of geologic interest: *Geol. Soc. Amer. Bull.*, v. *56*, p. 713–36

Wright, T., and Lundelius, Ernest, Jr., 1963, Post-Pleistocene raccoons from central Texas and their zoogeographic significance: Texas Memorial Museum, The Pearce-Sellards Series, no. 2, p. 5–21

GERALD E. SCHULTZ

Department of Geology
West Texas State University
Canyon, Texas

FOUR SUPERIMPOSED LATE-PLEISTOCENE VERTEBRATE FAUNAS FROM SOUTHWEST KANSAS

Abstract

Subsurface solution of salt and anhydrite from Permian beds has caused a series of sinkholes and collapse basins in overlying Pleistocene sediments in southwestern Kansas and northwestern Oklahoma in the region drained by the Cimarron River. Recent studies of a multiple collapse sink a mile north of the Cimarron River in southern Meade County, Kansas, have demonstrated the unique occurrence of four superimposed vertebrate faunas, which are assigned to the Illinoian, Sangamon, and late Wisconsin on the basis of stratigraphic position, faunal composition, and pollen evidence.

The Adams local fauna (Illinoian) contains fragmentary remains of large mammals and occurs in stream deposits of the ancestral Cimarron River that lie topographically below the present High Plains surface and indicate a period of downcutting during late-Pleistocene time. In the Butler Spring local fauna (late Illinoian), the occurrence of Sorex cinereus, Citellus richardsoni, *and* Microtus pennsylvanicus, *whose present ranges do not extend so far south as Meade County, Kansas, suggests summers that were cooler and more moist than those in the area today. The Cragin Quarry local fauna (Sangamon) occurs in overlying silts and sands of sinkhole origin and contains large extinct mammals and extant microvertebrates. A massive caliche separates this fauna from the Robert local fauna (late Wisconsin), which contains several northern mammalian species and yielded a radiocarbon date of 11,100 ±390 years* B.P.

Mollusks and pollen from several faunal horizons support the paleoecological interpretations derived from a study of the vertebrates.

In parts of southwestern Kansas and northwestern Oklahoma, circulating groundwater has removed considerable quantities of salt and anhydrite

from the subsurface Permian strata, causing the collapse of overlying
Pliocene–Pleistocene beds (Frye and Schoff, 1942). In some cases the
collapse has been slight and sudden, such as that which formed the
Meade "salt" sink (Mudge, 1879) south of Meade, Kansas, in 1879.
Other large basins have been formed over a greater period of time by

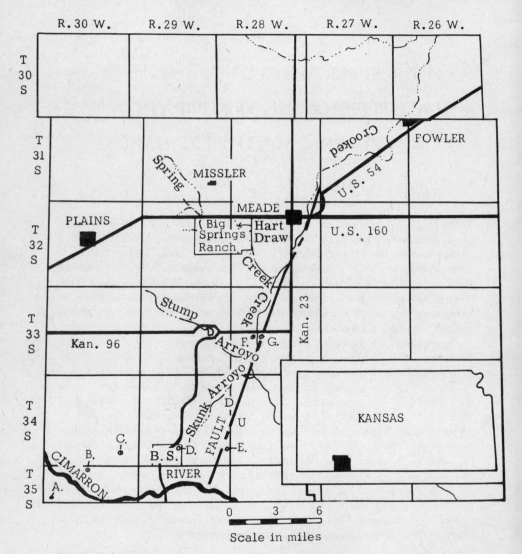

FIG. 1. Index map of Meade County, Kansas. (B.S. denotes Butler Spring area on
the XI Ranch. Other lettered localities denote the locations of measured sections
used in a geological study of this region.)

the slow sagging of the overlying sediments or by the coalescence of several sinks, resulting from a series of collapses. Many of the basins so formed were subsequently filled with Pleistocene sediments, some of which were laid down in small lakes or ponds. Some of these deposits contain fossil remains of animals that lived in and around the lakes or in nearby streams.

One such basin approximately a mile in diameter is situated on the XI Ranch just north of the Cimarron River in southern Meade County, Kansas (Fig. 1). This basin, known as the Butler Spring area, contains sediments of late-Pleistocene age, and both vertebrate and invertebrate fossils have been recovered here (Hibbard, 1939; 1943, p. 190; 1949a, p. 83). Early geological studies of the area were made by Smith (1940, p. 102). Recent studies of the area by Hibbard and Taylor (1960) and Schultz (1965) have demonstrated the presence of four distinct super-imposed late-Pleistocene local faunas in these deposits (Fig. 2): the faunas Adams (Illinoian, cold), Butler Spring (late Illinoian, cool), Cragin Quarry (Sangamon, warm), and Robert (late Wisconsin, cool). Although superimposed vertebrate faunas are not particularly uncommon, a total of four such faunas in stratigraphic sequence reflecting shifts in Pleistocene climates may be unique.

While this paper neither presents new theories on the causes of late-Pleistocene extinction nor champions any present theories, its inclusion in the extinction volume may be appropriate as a detailed example of the late-Pleistocene association of small extant species of vertebrates with the large extinct mammalian species. In the following faunal lists, extinct genera are designated by a dagger (†) and extinct species by an asterisk (*).

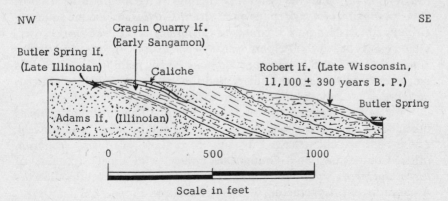

FIG. 2. Cross section through Butler Spring area, showing superposition of four late Pleistocene faunas.

ADAMS LOCAL FAUNA

Description. The oldest of the four faunas occurs in coarse yellow cross-bedded sands and gravels, which probably represent stream deposits of the ancestral Cimarron River laid down during Illinoian time. These beds are channeled into the High Plains surface, which is locally capped by the Crooked Creek Formation, considered to be of Kansan–Yarmouth age (Hibbard and Taylor, 1960, p. 19–20). The fossils from the yellow sands constitute the Adams local fauna. The molluskan fauna collected from Butler Spring localities 2 and 3 (Hibbard and Taylor, 1960, p. 46) includes

Quadrula quadrula (Rafinesque)
Valvata tricarinata (Say)
Stagnicola caperata (Say)
Fossaria dalli (Baker)
Gyraulus parvus (Say)
Promenetus kansasensis (Baker)
Physa anatina Lea
Gastrocopta cristata (Pilsbry and Vanatta)
G. procera (Gould)
G. tappaniana (Adams)

Pupilla blandi Morse
P. muscorum (Linnaeus)
P. sinistra Franzen
Vertigo ovata Say
V. milium (Gould)
Vallonia cyclophorella Sterki
V. gracilicosta Reinhardt
  cf. *Succinea*
Discus cronkhitei (Newcomb)
Helicodiscus singleyanus (Pilsbry)
Hawaiia minuscula (Binney)

The vertebrate fauna includes *Ambystoma tigrinum* (Green), tiger salamander (Tihen, 1955, p. 240); *Bufo* cf. *B. cognatus* Say, Plains toad (Tihen, 1962, p. 29); *Emydoidea twentei* (Taylor), Plains semibox turtle (Taylor, 1943); and the following mammals (Hibbard and Taylor, 1960, p. 41): †*Megalonyx* sp., ground sloth; *Citellus* sp., ground squirrel; *Geomys* sp., eastern pocket gopher; †*Castoroides* cf. *C. ohioensis* Foster, Ohio giant beaver; *Microtus pennsylvanicus* (Ord), meadow vole; *Canis* cf. *C. latrans* Say, coyote; †*Mammuthus* sp., mammoth; †*Camelops* sp., camel; *Equus scotti* Gidley, Scott horse; and *Equus* sp. Microvertebrates are not abundant in the fauna.

Kapp (1965) reported the results of pollen studies made of samples taken from clay and silt lenses within the yellow sands and gravels. These beds are characterized by high percentages of *Pinus, Artemisia,* other Compositae, and Gramineae pollen, with small amounts of *Picea, Juniperus* sim., deciduous trees, Ambrosineae, Chenopodiaceae-Amaranthaceae, Cyperaceae, *Myriophyllum,* and *Typha.*

Paleoecology and correlation. The Adams local fauna was taken from yellow cross-bedded sands and gravels, which are believed to represent deposits of the ancestral Cimarron River. The occurrence of the large freshwater mussel *Quadrula* indicates the existence of a medium-sized perennial stream considerably larger than the present Cimarron River. The stream deposits are channeled into the High Plains surface. Several of the molluskan species in the fauna are northern forms not found in Kansas today. These would indicate much cooler summers than those in the area now. The presence of a neotenic tiger salamander, *Ambystoma tigrinum* (Green), is of special interest. According to Tihen (1958, p. 5),

> Within a limited area of southwestern Kansas, it appears that the changing environments of the glacial and interglacial stages during the Pleistocene were accompanied by corresponding changes in the mode of living of populations of *tigrinum* inhabiting that area. Populations from deposits associable with major glacial advances were apparently neotenic, while those associable with the interglacial stages underwent normal metamorphosis (cf. Tihen, 1955).

Although most of the mammals in the Adams local fauna are large forms and not useful as climatic indicators, the presence of *Microtus pennsylvanicus* (Ord) is of note. The meadow vole is a northern form whose present range extends only as far south as extreme north-central Kansas (Burt and Grossenheider, 1964, p. 192; Fleharty and Andersen, 1964). It is known only from faunas younger than the Cudahy (late Kansan). Its presence in the Adams local fauna indicates summers markedly cooler than those of today.

The occurrence of spruce and pine pollen suggests a glacial age. On the basis of faunal, pollen, and stratigraphic evidence, the Adams biota is at present assigned to the maximum Illinoian (Schultz, 1965, p. 236).

BUTLER SPRING LOCAL FAUNA

Description. Overlying the yellow sands containing the Adams biota is the two- to three-foot bed of greenish clayey sand containing rich molluscan and microvertebrate remains known as the Butler Spring local fauna. This formerly included the fossils from the underlying yellow sands as well until these were shown to constitute the distinct Adams local fauna just discussed. Although a few specimens were obtained by surface collecting, most of the fossils from the now restricted Butler Spring local fauna horizon were obtained by use of the washing technique developed by Hibbard (1949b). The molluscan fauna collected from Butler Spring locality 1 was reported by Herrington and Taylor (1958) and Hibbard and Taylor (1960) and includes the following 53 species:

Anodonta grandis Say?
Sphaerium striatinum (Lamarck)
S. transversum Say
S. indet.
Pisidium casertanum (Poli)
P. compressum Prime
P. nitidum Jenyns
P. walkeri Sterki?
Valvata tricarinata (Say)
Probythinella lacustris (Baker)
Carychium exiguum (Say)
Stagnicola reflexa (Say)
S. caperata (Say)
Fossaria obrussa (Say)
F. dalli (Baker)
Lymnaea stagnalis jugularis Say
Anisus pattersoni (Baker)
Gyraulus circumstriatus (Tryon)
G. parvus (Say)
Helisoma anceps (Menke)
H. trivolvis (Say)
Promenetus kansasensis (Baker)
P. umbilicatellus (Cockerell)
Ferrisia meekiana (Stimpson)
Laevapex kirklandi (Walker)
Physa anatina Lea
P. gyrina Say

P. skinneri Taylor
Aplexa hypnorum (Linnaeus)
Gastrocopta armifera (Say)
G. contracta (Say)
G. holzingeri (Sterki)
G. tappaniana (Adams)
G. cristata (Pilsbry and Vanatta)
G. procera (Gould)
Pupoides albilabris (Adams)
P. inornatus Vanatta
Pupilla muscorum (Linnaeus)
P. blandi Morse
P. sinistra Franzen
Vertigo ovata Say
V. milium (Gould)
V. gouldi (Binney)
Vallonia cyclophorella Sterki
V. gracilicosta Reinhardt
V. parvula Sterki
cf. *Succinea* sp.
Oxyloma retusa (Lea)
Discus cronkhitei (Newcomb)
Helicodiscus singleyanus (Pilsbry)
Hawaiia minuscula (Binney)
Zonitoides arboreus (Say)
Stenotrema leai (Binney)

Freshwater ostracodes from the Butler Spring local fauna were identified by Edwin O. Gutentag of the U.S. Geological Survey and were reported by Schultz (1965, p. 240). The list includes *Cypridopsis vidua* (O. F. Müller), *Candona nyensis* Gutentag and Benson, *C. renoensis* Gutentag and Benson, *C. crogmaniana* Turner, *C. acuta* Hoff, *C. truncata* Furtos, *C. caudata* Kaufmann, *Cyprideis littoralis* Brady, *Cypricercus tuberculatus* (Sharpe), and *Physocypria* sp.

Fish remains from the fauna were reported by C. L. Smith (1958) and Schultz (1965) and include *Lepisosteus osseus* (Linnaeus), longnose gar; *L. platostomus*? Rafinesque, a shortnose gar; *Catostomus commersoni* (Lacépède), white sucker; *Ictalurus punctatus* (Rafinesque), channel catfish; *I. melas* (Rafinesque), black bullhead; *Lepomis cyanellus* Rafinesque, green sunfish; *Micropterus* sp., bass; and *Perca flavescens* (Mitchill), yellow perch.

Frog and toad remains and snake vertebrae and ribs have been found but have not been studied. Turtle remains were reported by Schultz (1965) and include *Chelydra serpentina* (Linnaeus), snapping turtle; *Chrysemys picta* (Schneider), painted turtle; *Pseudemys* cf. *P. scripta* (Schoepff), elegant slider; and *Trionyx* sp., softshell turtle. Bird remains have been found but are unpublished.

The mammals reported by Schultz (1965) include *Sorex cinereus* Kerr, masked shrew; †*Paramylodon harlani* (Owen), Harlan ground sloth; *Cynomys ludovicianus* (Ord), blacktail prairie dog; *Citellus richardsoni* (Sabine), Richardson ground squirrel; *Citellus* cf. *C. tridecemlineatus* (Mitchill), thirteen-lined ground squirrel; *Geomys* cf. *G. bursarius* (Shaw), eastern pocket gopher; *Perognathus hispidus* Baird, hispid pocket mouse; *Perognathus* sp., a small pocket mouse; *Dipodomys* cf. *D. ordi* Woodhouse, Ord kangaroo rat; *Onychomys* cf. *O. leucogaster* (Wied-Neuwied), northern grasshopper mouse; *Reithrodontomys* cf. *R. megalotis* Baird, western harvest mouse; *Microtus pennsylvanicus* (Ord), meadow vole; *Microtus* (*Pedomys*) *ochrogaster* (Wagner), prairie vole; *Sylvilagus* sp., cottontail; †*Camelops kansanus* Leidy, Kansas camel; *Equus* cf. *E.* (*Asinus*) *conversidens* Owen, Mexican ass; *Equus* cf. *E. niobrarensis* Hay, Niobrara horse; and *Equus* sp., horse. Some of these mammals are shown in Figure 3.

Pollen studies of the Butler Spring horizon by Kapp (1965) showed no *Picea*, less *Pinus*, and much more *Juniperus* sim. and deciduous tree pollen than in the underlying Adams horizon. Pollen of Ambrosineae and other Compositae is scarce, whereas grass is well represented. Pollen of aquatic plants such as *Typha* and *Sparganium* were locally abundant and fungal remains were abundant in all of the samples studied.

Paleoecology and correlation. On the basis of pollen and faunal evidence as well as stratigraphic position, a late-Illinoian age is assigned to the Butler Spring biota. The presence of less pine and more juniper and deciduous tree pollen suggests a warmer climate than that in which the Adams biota lived.

The mammalian fauna supports a late-Illinoian age and includes a mixture of forms from both glacial and interglacial assemblages. The masked shrew, meadow vole, and Richardson ground squirrel are characteristic of other Illinoian faunas in the area (Schultz, 1965, p. 257–58), whereas the ground sloth, prairie dog, and pocket mouse are found in the subsequent Sangamon interglacial Cragin Quarry local fauna (Hibbard and Taylor, 1960).

Inferences about the environment in which the Butler Spring local fauna lived can be made from a study of the present-day habitats of the

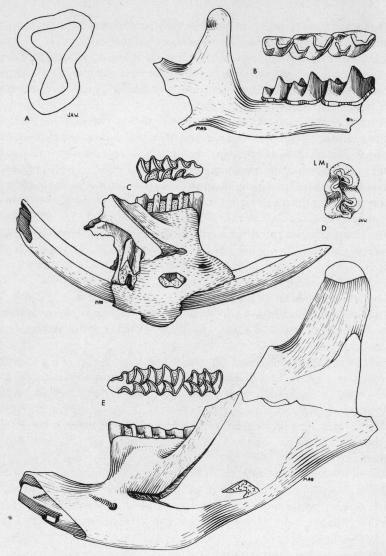

FIG. 3. Pleistocene mammals. A, *Paramylodon harlani*, left fifth upper tooth, occlusal view, ×1. B, *Sorex cinereus*, right jaw with M_1-M_3, lateral and occlusal views, ×10. C, *Microtus (Pedomys) ochrogaster*, right jaw with M_1, lateral and occlusal views, ×6. D, *Onychomys* cf. *O. leucogaster*, left M_1, occlusal view, ×8. E, *Microtus pennsylvanicus*, left jaw with M_1-M_2, lateral and occlusal views, ×6. (All specimens from University of Michigan Museum of Paleontology.)

species represented as fossils or of their closest living relatives. A medium-sized stream larger than the present Cimarron River is indicated by the gar and the channel catfish. The variety and abundance of aquatic mollusks and the presence of pond and river turtles indicate a quiet, shallow-water environment with dense beds of submergent aquatic plants in a protected area away from the main stream. At a nearby locality, the presence of a thin, freshwater marl, the abundance of large bird eggs, numerous remains of the meadow vole *Microtus pennsylvanicus*, and the presence of the masked shrew *Sorex cinereus* indicate an immediate marshy habitat surrounded by a moist, low meadow with a good grass cover. The western harvest mouse *Reithrodontomys megalotis* indicates areas of tall grass and weeds; the prairie vole *Microtus (Pedomys) ochrogaster* probably lived on the drier grassy slopes and nearby uplands.

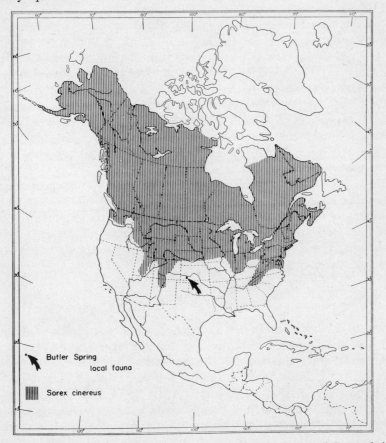

Butler Spring
local fauna

Sorex cinereus

Fig. 4. Present distribution of *Sorex cinereus*, masked shrew (Burt and Grossenheider, 1964, p. 4). Black dot marks location of Butler Spring and Robert local faunas, Meade County, Kansas.

A dry, upland prairie habitat not far from the environment of deposition is indicated by the presence of ground squirrel, prairie dog, pocket mouse, kangaroo rat, grasshopper mouse, and cottontail. Open grasslands are also suggested by horse, camel, and ground sloth.

Conclusions about the climate in which the Butler Spring local fauna lived are derived partly from the nature of the local habitat and partly from the present distribution of living species represented in the fauna. The larger size of the ancient stream implies a much greater or more effective rainfall than that of today. Markedly cooler summers than those of today are indicated by a number of species of mollusks that are at present more northern in distribution. Such distributions suggest that the winters were no more severe than those of Nebraska today.

Evidence derived from the mammals supports the conclusions based on the mollusks and fishes. Three of the species are more northern forms and do not occur in Meade County, Kansas, at present. *Sorex cinereus*, the masked shrew, now extends as far south as central Nebraska (Fig. 4 and Burt and Grossenheider, 1964, p. 4). The ground squirrel *Citellus richardsoni* is found to the west in Colorado and to the north in the eastern Dakotas (Burt and Grossenheider, 1964, p. 106). The meadow vole *Microtus pennsylvanicus* occurs as far south as Jewell County, Kansas (Fleharty and Andersen, 1964; Burt and Grossenheider, 1964, p. 192). The remainder of the extant mammals inhabit Meade County at the present time.

There is no area of sympatry in which all of the extant species occur today. The largest number of molluscan species now live in the Central Lowlands of the eastern Dakotas. In addition, all the extant mammals may be found somewhere within North Dakota. However, a larger mammalian fauna is needed before the position of the present equivalent community can be located more accurately.

CRAGIN QUARRY LOCAL FAUNA

Description. Above the Butler Spring horizon is a sequence of reddish buff silts and lacustrine deposits containing a large and diverse vertebrate fauna which is assigned to the Sangamon interglacial. This fauna contains many forms found in the Cragin Quarry local fauna reported by Hibbard and Taylor (1960) from the Big Springs Ranch 15 miles to the north and is considered to be equivalent in age. The fauna (Schultz, unpublished) includes numerous mollusks (unstudied) and freshwater ostracodes. The latter were identified by Edwin D. Gutentag and include *Candona nyensis* Gutentag and Benson, *C. truncata* Furtos, *Cyprideis littoralis* Brady, *Ilyocypris bradyi* Sars, and *Cyprinotus* sp.?

The amphibians include *Ambystoma* cf. *A. tigrinum* (Green), tiger salamander; *Bufo woodhousei* Girard, Woodhouse toad; *Rana* sp., frog; and *Scaphiopus* sp., spadefoot toad. Reptiles are represented by extinct species of the tortoises *Gopherus* and *Geochelone*. Numerous snake, lizard, fish, and bird remains are either unstudied or await publication.

The mammals are by far the largest and most varied group represented and include:

Cynomys ludovicianus (Ord), blacktail prairie dog

Citellus cf. *C. tridecemlineatus* (Mitchill), thirteen-lined ground squirrel

Geomys sp., pocket gopher

Perognathus hispidus Baird, hispid pocket mouse

Perognathus sp., a small pocket mouse

Castor canadensis Kuhl, Canada beaver

Onychomys cf. *O. leucogaster* (Wied-Neuwied), northern grasshopper mouse

Reithrodontomys megalotis Baird, western harvest mouse

Peromyscus progressus Hibbard, Plains deer mouse

Microtus pennsylvanicus (Ord), meadow vole

M. (Pedomys) ochrogaster (Wagner), prairie vole

Canis (Aenocyon) dirus (Leidy), dire wolf

Vulpes velox (Say), swift fox

Mustela cf. *M. frenata* Lichtenstein, longtail weasel

Mephitis sp., skunk

Panthera atrox (Leidy), fierce jaguar

†*Mammuthus* sp., mammoth

Lepus sp., jackrabbit

Sylvilagus sp., cottontail

†*Platygonus* cf. *P. compressus* Le Conte, Le Conte peccary

†*Camelops kansanus* Leidy, Kansas camel

†*Camelops* sp., camel

†*Tanupolama macrocephala* (Cope), extinct llama

Odocoileus sp., deer

†*Capromeryx furcifer* Matthew, forked pronghorn

Equus cf. *E. (Hemionus) calobatus* Troxell, stilt-legged ass

Equus cf. *E. scotti* Gidley, Scott horse

Equus cf. *E. (Hemionus)*, Asiatic ass

Paleoecology and correlation. The large fauna from deposits overlying the Butler Spring horizon is considered to be equivalent to the Cragin Quarry local fauna of early Sangamon age partly on the basis of stratigraphic position and partly on the basis of the nature of the fauna and its similarity to that on the Big Springs Ranch (Hibbard and Taylor, 1960). A thick, massive caliche that overlies this faunal horizon at both localities probably formed during a prolonged period of aridity in middle Sangamon time. Similar but older caliches elsewhere in the region can be related to older faunas and appear to have formed during Aftonian and Yarmouth times.

No pollen has been reported from these deposits in the Butler Spring area, and the molluscan fauna has not been published. Remains of the large land tortoises, *Gopherus* and *Geochelone*, suggest a warm interglacial climate (Hibbard, 1960, p. 22). All the extant mammals are found in Meade County, Kansas, at present and nearly all of them are characteristic of upland habitats, suggesting that the local habitat was primarily a grassy upland with occasional trees and shrubs. Mammals of lowland or marshy habitats are rare in the fauna. Two teeth of the beaver *Castor canadensis* were recovered, and the presence of the western harvest mouse *Reithrodontomys megalotis* at one locality suggests a moist habitat with tall grass and weeds. However, no remains of muskrats, water rats, cotton rats, or rice rats were recovered.

ROBERT LOCAL FAUNA

Description. The fourth fauna, called the Robert local fauna, lies at the top of the stratigraphic section near Butler Spring and 80 ft above the Cragin Quarry faunal horizon. It occurs in a dark gray zone, 1 ft thick, that apparently represents an old marsh deposit (Schultz, unpublished). Land snails from this bed yielded a radiocarbon date of 11,100 ± 390 years B.P. (SM-762). The Robert local fauna is therefore of late-Wisconsin age.

The fauna includes large numbers of *Succinea ovalis* Say as well as other mollusks (unstudied). Freshwater ostracodes identified by Gutentag (personal communication) include *Eucypris meadensis* Gutentag and Benson, *Cypridopsis vidua* (O. F. Müller), *Ilyocypris bradyi* Sars, and *Candona* sp.

The remains of fish, frogs, toads, turtles, snakes, lizards, and birds have been found but have not been studied. The mammalian fauna consists entirely of microvertebrates and includes

Sorex cinereus Kerr, masked shrew
S. palustris Richardson, water shrew
Blarina b. brevicauda (Say), shorttail shrew
Cynomys ludovicianus (Ord), blacktail prairie dog

Citellus cf. *C. richardsoni* (Sabine), Richardson ground squirrel
C. tridecemlineatus (Mitchill), thirteen-lined ground squirrel
Geomys sp., pocket gopher
Thomomys cf. *T. talpoides* (Richardson), northern pocket gopher

Reithrodontomys sp., harvest
  mouse
Peromyscus cf. *P. maniculatus*
  Wagner, deer mouse
Microtus pennsylvanicus (Ord),
  meadow vole
M. (*Pedomys*) *ochrogaster*
  (Wagner), prairie vole

Synaptomys cooperi Baird,
  southern bog lemming
Zapus cf. *Z. hudsonius*
  (Zimmermann), meadow
  jumping mouse
Sylvilagus sp., cottontail

No pollen has been reported from this horizon.

Paleoecology and correlation. The Robert local fauna implies a cool climate because of the presence of more northerly species of mammals, such as *Sorex cinereus*, masked shrew; *S. palustris*, water shrew; *Blarina b. brevicauda*, shorttail shrew; *Thomomys* cf. *T. talpoides*, northern pocket gopher; *Microtus pennsylvanicus*, meadow vole; and *Zapus* cf. *Z. hudsonius*, meadow jumping mouse, whose present ranges do not extend as far south as Meade County, Kansas. The fauna indicates a marshy habitat surrounded by a moist low meadow with a good grass cover. A dry upland prairie grassland not far from the immediate environment of deposition is indicated by fragmentary remains of prairie dog, ground squirrel, and cottontail. Of special interest is the first appearance of the ostracod *Eucypris meadensis*, which lives today in an artesian spring-fed stream in Meade County State Park (Gutentag and Benson, 1962, p. 25). Mollusks have been collected but have not been studied. The radiocarbon date of 11,100 ± 390 years B.P. was obtained by dating shells of *Succinea ovalis* Say. Hence, a late-Wisconsin age is assigned to the fauna.

The climate in Meade County during this time may be compared with that of the area in which the greatest number of extant forms are living today, namely a small section in the extreme northeastern corner of South Dakota and an adjacent portion of Minnesota (Fig. 5). The mean annual temperature in this region is 42°F, whereas in southwestern Kansas it is 55°F (Visher, 1954, map 3). In South Dakota the average difference between January and July mean temperatures is about 62°, compared with 45° in Kansas for the same period (Visher, 1954, map 328). The precipitation in both areas (Visher, 1954, map 492) is about 22 inches per year, with 75–80% occurring in the period of April through September. Because of the lower temperature in South Dakota, however, evaporation there is far less than in southwestern Kansas.

In summary, the Robert local fauna lived in Meade County, Kansas, during late-Wisconsin time in a climate with summers similar to those of extreme northeastern South Dakota today. The average annual

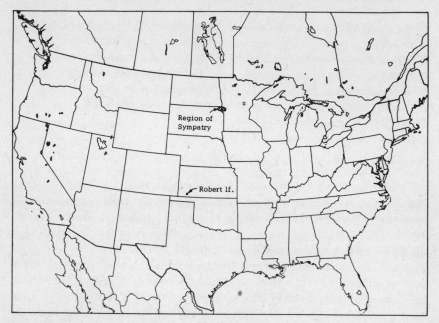

Fig. 5. Present region of sympatry (shaded zone) of eleven extant mammalian species represented in the Robert local fauna (late Wisconsin) in Meade County, Kansas.

temperature was probably about 42°F. The precipitation averaged about 25 inches per year, with at least 75% falling during the summer, and there was less evaporation in the area than there is today.

On the basis of stratigraphic, faunal, and pollen evidence, the four superimposed faunas described here are assigned to the middle or maximum Illinoian, late Illinoian, Sangamon, and late Wisconsin respectively. These faunas differ greatly in the number and diversity of species represented, thus reflecting differences in the conditions of preservation, the size of the deposits, and the degree to which the deposits were sampled and the specimens studied. Pronounced differences in faunal composition result from northward and southward migrations or population shifts and thus reflect changes in climate and local habitats in the region under study. All these factors make it difficult to establish any sort of extinction chronology for the late Pleistocene on the basis of this study alone.

With respect to the percentage or number of extinct species in each fauna, several points are worth noting. Of fifty-three species and twenty-nine genera of mollusks in the Butler Spring local fauna, only two species are extinct and all genera are living today. Among the fishes, amphibians, and reptiles that have been identified from the faunas thus far, only three species of turtle, belonging to the genera †*Emydoidea*, *Geochelone*, and *Gopherus*, are now extinct.

Of all the vertebrates, the mammals show the greatest number of extinct forms. In the Adams and Cragin Quarry local faunas, in which a large percentage of the known vertebrate fauna consists of the larger mammals, there is a high percentage of extinct forms. In the Adams local fauna these include *Megalonyx*, *Castoroides* cf. *C. ohioensis*, *Mammuthus*, *Camelops*, and *Equus scotti*. In the Cragin Quarry local fauna, six of twenty-nine genera and thirteen of thirty-four species of vertebrates are now extinct. These are primarily such large mammals as *Canis dirus*, *Panthera atrox*, *Mammuthus*, *Platygonus*, *Camelops*, *Tanupolama*, *Capromeryx*, and several species of *Equus*. The only small mammal included is *Peromyscus progressus*. In contrast, the Robert local fauna contains fifteen identified species of vertebrates; all of these are small mammals. No extinct forms have been noted.

References

Burt, W. H., and Grossenheider, R. P., 1964, *A field guide to the mammals*: Boston, Houghton Mifflin, 283 p.

Fleharty, E. D., and Andersen, K. W., 1964, The meadow vole *Microtus pennsylvanicus* (Ord) in Kansas: *Kansas Acad. Sci. Trans.*, v. *67*, p. 129–30

Frye, J. C., and Schoff, S. L., 1942, Deep-seated solution in the Meade basin and vicinity, Kansas and Oklahoma: *Amer. Geophys. Union Trans.*, p. 35–39

Gutentag, E. D., and Benson, R. H., 1962, Neogene (Plio–Pleistocene) freshwater ostracodes from the Central High Plains: *Kansas Geol. Survey Bull. 157*, p. 1–60

Herrington, H. B., and Taylor, D. W., 1958, Pliocene and Pleistocene Sphaeriidae (Pelecypoda) from the central United States: *Univ. Michigan Mus. Zool. Occasional Paper 596*, p. 1–29

Hibbard, C. W., 1939, Notes on some mammals from the Pleistocene of Kansas: *Kansas Acad. Sci. Trans.*, v. *42*, p. 463–79

————— 1943, *Etadonomys*, A new Pleistocene Heteromyid rodent, and notes on other Kansas mammals: *Kansas Acad. Sci. Trans.*, v. *46*, p. 185–91

————— 1949a, Pleistocene stratigraphy and paleontology of Meade County, Kansas: *Univ. Michigan Mus. Paleont. Contrib.*, v. *7*, p. 63–90

————— 1949b, Techniques of collecting microvertebrate fossils: *Univ. Michigan Mus. Paleont. Contrib.*, v. *8*, p. 7–19

————— 1960, An interpretation of Pliocene and Pleistocene climates in North America: *Michigan Acad. Sci. Ann. Rept.*, *62*, p. 5–30

Hibbard, C. W., and Taylor, D. W., 1960, Two late Pleistocene faunas from southwestern Kansas: *Univ. Michigan Mus. Paleont. Contrib.*, v. *16*, p. 1–223

Kapp, R. O., 1965, Illinoian and Sangamon vegetation in southwestern

Kansas and adjacent Oklahoma: *Univ. Michigan Mus. Paleont. Contrib.*, v. *19*, p. 167–255

Mudge, B. F., 1879, The new sink-hole in Meade County, Kansas: *Kansas City Rev. Sci.*, v. *3*, p. 152–53

Schultz, G. E., 1965, Pleistocene vertebrates from the Butler Spring local fauna, Meade County, Kansas: *Michigan Acad. Sci., Arts. Lett., Papers*, v. *50*, p. 235–65

———— Unpubl. MS., The geology and paleontology of a late Pleistocene basin in southwest Kansas

Smith, C. L., 1958, Additional Pleistocene fishes from Kansas and Oklahoma: *Copeia*, no. *3*, p. 176–80

Smith, H. T. U., 1940, Geological studies in southwestern Kansas: *Kansas Geol. Survey Bull.*, v. *34*, 212 p.

Taylor, E. H., 1943, An extinct turtle of the genus *Emys* from the Pleistocene of Kansas: *Univ. Kansas Sci. Bull.*, v. *29*, p. 249–54

Tihen, J. A., 1955, A new Pliocene species of *Ambystoma*, with remarks on other fossil Ambystomids: *Univ. Michigan Mus. Paleont. Contrib.*, v. *12*, p. 229–44

———— 1958, Comments on the osteology and phylogeny of Ambystomatid salamanders: *Florida State Mus. Biol. Sci. Bull.*, v. *3*, p. 1–50

———— 1962, A review of New World fossil Bufonids: *Amer. Midland Naturalist* v. *68*, p. 1–50

Visher, S. S., 1954, *Climatic atlas of the United States:* Cambridge, Harvard Univ. Press, 403 p.

CYNTHIA IRWIN-WILLIAMS

Paleo-Indian Institute
Eastern New Mexico University
Portales, New Mexico

ASSOCIATIONS OF EARLY MAN WITH HORSE, CAMEL, AND MASTODON AT HUEYATLACO, VALSEQUILLO (PUEBLA, MEXICO)

Abstract

The fluviatile deposits laid down at Hueyatlaco indicate the existence of a stream or streams in the locality over a period of some time. The favorable conditions repeatedly attracted to the site a large variety of now extinct fauna including mammoth, mastodon, horse, camel, four-horned antelope, smilodon, dire wolf, tapir and glyptodon.

Early hunters were likewise attracted to the spot, where several times during the prehistoric period represented they killed and butchered the available game. There is accordingly no longer reason to question the co-existence of man with horse, camel, mastodon, and four-horned antelope, nor the suitability of these animals as game. The lack of such associations therefore no longer constitutes a valid argument against human agency in Pleistocene extinction. Each of the hunts represented was an isolated instance in time, providing a brief but informative glimpse of the relevant technology and typology.

Taken as a whole, the evidence yields a skeletal framework for a very early prehistoric sequence. The exact chronology of this sequence must await the results of geologic, radiocarbon, and other studies now under way.

The question of the origin, age, and character of the earliest inhabitants of the Western Hemisphere is an intriguing one, and it presents one of the major unsolved problems of New World prehistory. Scientific opinion has shifted dramatically in the past several decades from the belief widely held prior to 1926 that man in this hemisphere dated back no more than a few millennia; modern estimates range from 11,000 to more than 30,000 years ago. Whatever the ultimate solution, it is apparent by now that the earliest Americans were few in number and left relatively little imperishable material culture. It is also evident that man in this remote period was only one of a large and varied fauna in a complex environment and that he should be studied in the context of the available knowledge of contemporary geology and paleontology. The interaction of these early human and animal populations, and the part played by man in faunal extinction, are basic problems in the study of the late-Pleistocene. Lack of direct association of man with certain now extinct fauna has frequently been employed in the argument against human agency in the extinction. It is evident in the light of the current research that arguments on the subject, whatever their conclusion, should be based on positive rather than negative evidence.

The Valsequillo Reservoir near Puebla (Puebla, Mexico) had long been known as an area that offered excellent opportunities for Pleistocene (and earlier) research. Professor Juan Armenta Camacho had carried out surficial reconnaissance in the region for many years and had amassed a large collection of archaeological and paleontological materials. His studies led him to conclude that certain bone and stone objects possibly of human manufacture had originated in the Valsequillo Formation, which had also produced an extensive extinct faunal assemblage featuring camel, horse, glyptodon, mastodon, mammoth, four-horned antelope, and dire wolf. If those objects definitely of human manufacture were in fact derived from this formation, and if it proved to be of the antiquity suggested by the fauna, research here might provide vital clues concerning the earliest inhabitants of the New World.

In 1962 Professor Armenta and I conducted an intensive search for artifacts *in situ* in the Valsequillo Formation. After a wide survey, four localities at which there was definite evidence of this association were identified and tested. The results of this research have been summarized in an earlier report (Irwin-Williams, 1963). It was evident on the basis of these results that further research could profitably be done on the archaeology of the Valsequillo Formation. In addition, the character of the data indicated that an interdisciplinary approach, involving geology and paleontology as well as archaeology, would be of considerable value. Accordingly, the expanded 1964 Valsequillo project included extensive research in geology by Harold Malde (U.S. Geological Survey) and paleontology by Clayton Ray (U.S. National Museum) as well as continued archaeological studies. The project was made possible by a grant from the National Science Foundation.

Archaeological attention in 1964 focused on the Hueyatlaco locality, which had already produced numerous artifacts in association with mammoth, mastodon, horse, camel, four-horned antelope, etc. The site comprised two resistant outcrops, Stations 1 and 2, separated by a small recent arroyo. The outcrops evidently represented a fine-grained fluviatile deposit, a local facies of the Valsequillo Formation. Microstratigraphic subdivisions of this deposit were recognized in 1962, but detailed study awaited the 1964 work. Of importance, however, was the immediate observation that the character of the faunal remains and artifacts indicated that they had been moved little or no distance from their point of origin. The fossils were unworn, in some cases unbroken and occasionally still articulated, and the artifacts were likewise fresh, with no sign of water wear. The materials employed in artifact manufacture were of nonlocal origin.

The 1964 excavations at Hueyatlaco produced, as had been hoped, clarification of the stratigraphic relations of the two stations and evidence of multiple occupation of the locality through time, and they yielded a much more comprehensive picture of the internal stratigraphy. The composite picture is one of a long series of alluvial deposits, many of which are highly fossiliferous, and a few of which contain evidence of the hunting, butchering, and camping activities of Early Man. Because of its controversial character and because of the importance of documenting unquestionable associations of this kind, the field evidence is here presented in some detail.

STRATIGRAPHY

Unit A, the uppermost recognized, was a recent dark loam, covering much of the earlier deposit (Fig. 1). Culturally it yielded a few obsidian flakes, recent ceramics, and glass. It was obviously separated from the earlier deposits by a considerable hiatus marked by an erosional contact.

The uppermost early deposit, Unit B, was a fine-grained sandy silt within which two minor subdivisions were recognized. It contained a few fossils (principally horse and camel) but no cultural remains.

Deposit C at the south end of the site (Station 1) represented the effects of a small, shallow stream that cut a distinct southwest-trending channel in earlier strata. The numerous recognizable divisions reflect the deposition of complexly bedded sediments within the channel: the uppermost of these produced abundant fossils, principally horse, and camel. Culturally it yielded the largest assemblage of artifacts available from any single deposit at the site. All the artifacts recovered from Station 1 in 1962 lay in this stratum (a large bifacial cutting tool, a scraping edge-and-perforator, a cutting edge on a blade or flake, a concave end scraper-and-perforator, and a gouge or scraper on an oblong flake). One of these, a concave scraper-and-perforator, was

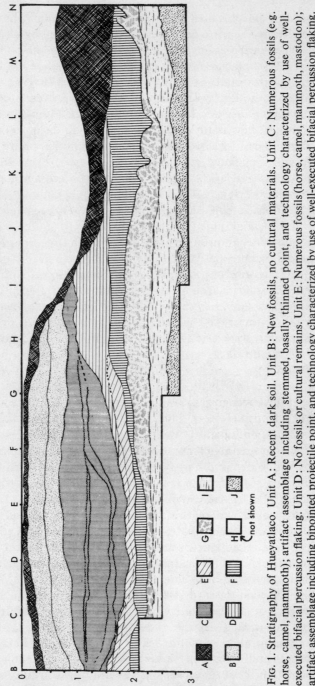

FIG. 1. Stratigraphy of Hueyatlaco. Unit A: Recent dark soil. Unit B: New fossils, no cultural materials. Unit C: Numerous fossils (e.g. horse, camel, mammoth); artifact assemblage including stemmed, basally thinned point, and technology characterized by use of well-executed bifacial percussion flaking. Unit D: No fossils or cultural remains. Unit E: Numerous fossils (horse, camel, mammoth, mastodon); artifact assemblage including bipointed projectile point, and technology characterized by use of well-executed bifacial percussion flaking. Unit F: Rare fossils; no cultural remains. Unit G: Rare fossils; no cultural remains. Unit H: Local feature; fossils, and single slightly retouched flake. Unit I: Numerous fossils (including horse, camel, mammoth, mastodon); artifact assemblage including edge-trimmed projectile points on blades or flakes, and technology characterized by percussion and pressure edge trimming, and lack of bifacial or total-coverage unifacial flaking. Unit J: Rare fossils; no cultural remains.

uncovered about 1 cm from a horse mandible. A tooth belonging to the mandible occurred about 2 cm from the parent fossil, illustrating the small magnitude of displacement characteristic of the deposit. The major part of this subdivision of Unit C had been removed in 1962, but from the remaining portion excavated in 1964 an unretouched flake and a percussion-flaked stemmed projectile point were recovered. The correlative stratum at the north end of the site (Station 2) also yielded a crude percussion-flaked lanceolate object, probably a projectile point. Several of the other subdivisions of Unit C at both stations produced fossils but no cultural remains.

Unit D consisted of a very fine clayey silt with a massive structure and light tan color. It occurred principally as a thick deposit in the southwest end of the trench between the two stations. It produced no artifacts or fossils.

Deposit E at Station 1 comprised a series of alluvial sediments in a shallow channel cut in earlier strata. The channel represented was apparently considerably broader than that cut by the stream responsible for Unit C, and its orientation was distinct. That the current was occasionally somewhat stronger is shown by lenses and layers of coarse sand and small gravel. In mid-channel it was possible to recognize a maximum of eight subdivisions (1E1–1E8). Elsewhere near the channel edges only two (1E1 and 1E8) were distinguishable. The uppermost subdivision (1E1) was composed of a fine-bedded gray sand with a layer of pale yellow coarse sand, with fine gravel at its upper boundary. It contained fossil remains and one heavily calcinated chert flake, which will be described below. The next subdivision yielded very abundant fossil remains, including prominently horse, camel, four-horned antelope, and mammoth. In one section of the site was uncovered a part of a semiarticulated horse skeleton definitely associated with two artifacts, and probably with two more: among the group of semiarticulated horse ribs and vertebrae was uncovered a well-made bifacial bipointed projectile point. Directly under an associated vertebra occurred the tip of a well-made bifacial knife or point. Nearby were a fragment of a thick bifacial tool and a core or utilized core. The remaining subdivisions of Unit E at the southern end of the site (Station 1) were fossiliferous but contained no cultural materials.

Unit 2E at the northern end of the site (Station 2) may correlate as a whole with the deposits (1E) just discussed, or it may predate them. It was completely separated from the latter by arroyo cutting, and it presented a distinct group of three subdivisions. The uppermost of these (2E1 and 2E2) comprised layers of variable yellowish-gray sand and grit with few fossils. The lowermost subdivision (2E3) was a thin layer

at the base of 2E2, composed of yellowish-gray coarse sand and grit.
It contained very numerous fossils, principally mastodon, horse, camel,
and mammoth. Many of the mastodon bones were concentrated in a
small area, may be from the same animal, and were associated with
evidence of human activities. The mandible and maxilla of the mastodon
were apparently purposely split and fragmented. Two artifacts were
discovered in direct association with these fragments. In 1962 a small
chopping tool made on a pebble was discovered deeply imbedded in a
fragment of mandible next to the tooth row. Nearby a thin flake, with a
burin-like spall removed from one edge, was recovered in 1964 *in situ*
between the cusps of one of the molar teeth.

Unit F consisted of a pale yellowish-gray sandy clay and silt conforming to what may have been a very broad shallow channel partly
cut into and partly simply overlying older sediments. It contained
relatively few fossils and no cultural remains.

Unit G comprised a pale yellow gravel and very coarse sand or grit,
faintly bedded and locally displaying flow rolls at the base. Fossils were
moderately abundant, including mastodon, mammoth, camel, and horse.
No objects of cultural interest were recovered.

Deposit H was a very localized phenomenon, occurring only at
Station 2: it comprised a massive pale yellowish-gray fine sand, well
sorted and peppered with dark grains. It evidently represented the fill of
a narrow channel at the base of Deposit 2G, running in a general
southwest–northeast direction. Many fossil remains of camel and horse
concentrated in the channel were accompanied by a single lightly
retouched flake scraper.

Like Unit G, Unit I could be traced throughout the length of the site.
The deposit consisted of yellowish very fine sand and silt, gritty clay,
and grit. It contained inclusions of a distinctive volcanic ash. Relatively
little work has been done on the unit. Abundant fossil remains were
recovered (horse, camel, mastodon, four-horned antelope, etc). Of
considerable interest were six man-made objects: first, a unifacial
pointed object, probably a projectile point, *in situ* under a large ungulate
rib at Station 1; second, a unifacial pointed piece, very probably a
projectile point, found adjacent to and nearly touching a fragmentary
proboscidean acetabulum at Station 2; a third probable projectile
point found under a large ungulate rib a short distance from the second;
an unretouched flake found nearby; a second flake recovered in the
trench between Stations 1 and 2; and a third at Station 1.

Unit J has been little tested but appears to comprise a yellowish coarse
sand and very fine gray clay. It produced a few fossils and no cultural
remains.

It is hoped that more work at the site will enable us to subdivide further the stratigraphic units and furnish a still more detailed description. It is evident, however, that the stratigraphy represents a series of fluviatile deposits that provide a local relative chronology for the associated cultural materials. It is now possible to describe these complexes in detail and to comment on the character of the remains.

CULTURAL REMAINS

Unit C, the uppermost early artifact-bearing stratum at Hueyatlaco, produced the largest assemblage recovered from the site. The diagnostic single item is a well-made projectile point, asymmetrical, broadly stemmed, with a concave base and abrupt shoulders (Plate I; 1). It was fashioned by well-controlled percussion with a soft hammer followed by occasional terminal pressure retouch. The point is much thicker than the base, which was apparently deliberately thinned: the scars of wide, shallow, thinning flakes struck from the basal end are visible on both faces; they were evidently removed before the final point was formed, as all other visible flake removals post-date them. A burin-like, possibly fortuitous removal at the tip of the piece completes the picture. Also recovered in the same group of artifacts and fossils at Station 1 was a thick bifacial piece with cutting edge likewise produced by well-controlled soft-hammer percussion. It displays long, lateral edge removals of a distinctly burin-like character. The concave scraper-perforator (Pl. I, 2) recovered adjacent to the horse mandible was made on a flake with a faceted striking platform and worked with soft-hammer percussion. A straight end scraper made by the same technique was recovered nearby. Also recovered in the same group were several retouched flakes and blades.

The correlative stratum at Station 2 yielded a single bifacially worked leaf-shaped point. Of interest are the extreme carination of the piece and the rather crude percussion technique used in its production. In conclusion, then, the small but varied assemblage recovered from Unit C is evidence of a rather well-controlled bifacial percussion technique, employed to produce projectile points, cutting edges, scrapers, perforators, and possibly burins. Although most of the raw material is rather poor and full of irregularities, a few pieces suggest the possibilities of a blade industry and the preparation of striking platforms. The character of the assemblage as a whole indicates the importance of hunting and processing game, specifically horse, camel, four-horned antelope, and mammoth. A brief hunting camp at or near a kill-site best fits these conditions.

PLATE I. 1, Unit 1C: Bifacial, basally thinned projectile point. 2, Unit 1C: Concave end scraper. 3, Unit 1E: Bifacial knife point. 4, Unit 1E: Bifacial bipointed projectile point. 5, Unit 2I: Edge-trimmed unifacial projectile point, made on blade. 6, Unit 1I: Edge-trimmed projectile point on flake. 7, El Horno: Edge-trimmed pointed flake, recovered among semiarticulated mastodon ribs.

The single cultural item from Stratum E1 is a small, heavily calcinated random flake. Notable internal fractures were caused by excessive heating.

The next major cultural group proceeds from Stratum 1E2 at Station 1. Outstanding is the large bifacial bipointed piece (Pl. I; 4) recovered in direct association with the partially articulated portion of a horse. The artifact was made by soft-hammer percussion with terminal pressure retouch. It is strongly carinated on one face. In the same group the point of a projectile point or knife was recovered under a vertebra of the same horse (Pl. I; 3). The specimen appears to have been made by the same technique and shows some secondary resharpening. The bifacial fragment recovered nearby reflects a similar technology. The core fragment seems to be from a simple discoidal nucleus. Undoubtedly this assemblage represents a single horse "kill" and possibly a brief camp.

At Station 2 the stratum that may relate to E at Station 1 or may predate it produced several artifacts in direct association with the split and butchered mastodon bones and teeth. One is a crude percussion-worked, sinuous-edged pebble tool that had been driven into a fragment of mandible below the tooth row. The butt end shows battering. Another is a simple percussion flake with a lateral edge removal or burin blow along one side (the piece was found between the cusps of a mastodon molar). Brief hunting and butchering activities are evidently represented.

Deposit I produced six cultural objects: the first is a projectile point associated with a large ungulate rib at Station 1. The piece was made on a flake (Pl. I; 6), and has a roughly triangular form, with a short asymmetrical basal projection on one side that may constitute a stem. The edges and base of the piece have been lightly trimmed by shearing and crude pressure retouch.

The second probable point recovered adjacent to the proboscidean acetabulum was made on a blade and is asymmetrically diamond-shaped in overall form (Pl. I; 5). The contracting stem has been formed by abrupt, coarse, probably stone-percussion flaking; the point, generally conforming to the outline of the original flake, has been lightly trimmed by shearing. All work is confined to the edges—none reaches far across either face. Of interest is the lower third of the stem, which has been heavily ground and polished both at the lateral edges and across the ventral face. The attitude of the polishing stroke is generally at right angles to the long axis of the piece.

The third point from Unit 2I was found under a large ungulate rib a short distance from the second and has several similarities. The piece was made on a thin blade, and the point has been lightly trimmed by pressure shearing. The existing base is formed by an abrupt hinge fracture. In addition, some simple edge-trimmed flake scrapers were recovered from this stratum. The artifacts indicate the existence of a technology and typology radically different from those of the succeeding levels. There is evidence of a rather well-controlled blade-producing technique. At the same time the advanced percussion and pressure-flaking extending across the surfaces of the artifact and characteristic of the materials in Units C and E are totally absent. Both points and scraping edges were prepared with little secondary work, but polishing was occasionally employed. The character of the assemblage indicates hunting and butchering activities involving mastodon, mammoth, horse, camel, four-horned antelope, etc.

In view of the distinctive typologic group comprised by the points from Unit I, the single projectile point from the nearby site of El Mirador should be noted. The piece was recovered in association with horse, mammoth, camel, mastodon, etc., and shows detailed similarities to the first of the specimens from Unit G (e.g. manufactured on a blade, stem formed by heavy abrupt percussion, point trimmed by shearing, etc.). Also of possible relevance is the broad, rather blunt-pointed flake (Pl. I; 7) found between the semiarticulated mastodon ribs at the El Horno mastodon kill in 1962 (Irwin-Williams, 1963).

From comparative typology I would make a few tentative suggestions on cultural affinities and the position of the material in the larger context of New World prehistory. (1) The material evidently antedates the cultural sequence recovered by MacNeish (1963) at nearby Tehuacan, whose earliest radiocarbon date falls near the beginning of the seventh millennium B.C. (2) The pseudo-fluted point from Unit C is unrepresented at Tehuacan but bears some resemblances to specimens from the El Inga Site in Ecuador (W. Mayer-Oakes, personal communication, 1965). (3) The bipointed projectile point from Unit 1E is similar in some respects to a variety of types from widely scattered localities (e.g. Lerma, Cascade, Ayampitin), but the great formal, spatial, and temporal range makes valid correlation impossible. (4) The early group of crudely edge-worked points on blades or flakes is essentially unparalleled in relevant New World literature. Comparisons to certain material from various Old World Upper Paleolithic assemblages must remain at present without historic connotation. The significance of the pieces lies in the substantiation of the existence, at least in this region, of a unifacial flake or blade-point industry preceding the large bipointed forms often believed to be prototypical of much of subsequent projectile-point development in the New World.

References

Irwin-Williams, Cynthia, 1963, Preliminary report on investigations in the
    Valsequillo region, Puebla, Mexico: Departmento de Prehistoria, Instituto
    Nacional de Antropologia e Historia
MacNeish, R. S., 1963, Second annual report of the Tehuacan Archaeological–
    Botanical Project: Andover, Mass., Phillips Acad.

KAZIMIERZ KOWALSKI

Institute of Systematic Zoology
Polish Academy of Sciences
Kraków, Poland

THE PLEISTOCENE EXTINCTION
OF MAMMALS IN EUROPE

Abstract

*In Europe—which, because of its geographic situation had only limited possi-
bilities of faunal migrations between north and south—the Ice Age caused the
impoverishment of fauna. This depletion affected all groups of mammals (except,
perhaps, bats), but it especially influenced the larger mammals, which were not
able to survive in the small refuges of southern Europe. Their limited populations
approached more easily the point at which the lack of genetic plasticity, accidental
change of sex ratio, epidemics, and other factors could bring about their extinction.*

*The hunting activity of primitive man, even over a very long period, does not
necessarily bring the extinction of his prey. Extensive literature concerning the
hunting habits of European Paleolithic man indicates that all large phytophagous
mammals were hunted, but it is very difficult to judge the size of the human
population and the degree to which hunting influenced the population of particular
species. Reindeer, red deer, aurochs, wild horse—all extensively hunted—were
able to survive, whereas some mammals, which were probably more difficult for
primitive man to kill (mammoth, woolly rhinoceros, cave bear) or were of no
interest to him (lion, hyena), became extinct. When the density of the prey
dropped, man with his primitive weapons could not get enough game and began to
hunt other animals, migrated, or finally died of hunger. Not only in the forests of
Europe but also in the steppes of Asia, numerous herds of large herbivorous
mammals were able to persist in spite of centuries of hunting, until in modern
times they were decimated with firearms. The extinction (local, at least) of many
mammals was caused only by the development of agriculture and/or the raising
of domestic animals, which make the human population independent of game
animals.*

*In the forest zone of Europe, only extensive deforestation in the late Middle
Ages brought the danger of extinction to such species as the aurochs and the
bison. No doubt, the preservation of the European bison, as well as of other*

mammals (such as beaver, elk, red deer, and brown bear) is now possible only through active protection.

The extinction of large mammals at the end of the last glaciation is less dramatic in Europe than in North America. Its main cause, in Europe, was the disappearance of the specific habitat represented by the "steppe–tundra" of the Pleistocene. The species connected with this biotope became extinct, while others were able to survive in the zone of warm steppes (horse and saiga) or of tundra (reindeer).

The extinction of large mammals in Europe needs further research, especially in radiocarbon dating and the stratigraphy of Paleolithic sites, chiefly in caves.

A comparison of Europe and North America can be of interest in the study of the Pleistocene extinction of animals. Although the climatic and geologic conditions were similar, man coexisted much longer with the fauna in Europe than in North America, where the primitive human population appeared dramatically at the decline of the Ice Age. Knowledge of the composition of the European Quaternary mammalian fauna is extensive but still far from complete.

Using Guilday's conclusions (this volume) I shall try to distinguish between extinction through evolution—the disappearance of a species through its transformation into another—and true extinction—the disappearance of an evolutionary line without descendants. The first type of extinction proves the success of a phylogenetic branch, and the second, its failure; the types are therefore antithetical. Sometimes, however, this distinction is not simple. In the fauna of European small mammals, the rodent genus *Mimomys* disappeared in the early-Pleistocene and was replaced in northwestern Europe by a similar genus, *Arvicola*, representing a higher stage in the development of molar teeth. It was recently discovered, however, that in southeastern Europe *Arvicola* appeared earlier and coexisted during a long period with the latest species of *Mimomys*. *Arvicola* apparently developed from one of the kinds of *Mimomys*, probably outside Europe. In some regions, then, *Mimomys* experienced "extinction through evolution." In regions characterized by the "true extinction" of the genus, *Arvicola* appeared when some species of *Mimomys* still existed and perhaps caused their extinction through active competition. The same systematic unit can therefore disappear partly through evolutionary and partly through true extinction.

When discussing the Quaternary fauna of European mammals and its changes, we must keep in mind the peculiar geographic situation of that continent. Europe, which could be considered a peninsula of Asia, is wide open to migrations from the east. At the same time, the Mediterranean Sea and the European mountain ranges that run mostly from east to west limit to a great degree the possibility of migration from north to south. In the cold periods typical of the Pleistocene, elements of the thermophilous fauna could find shelter in the forest regions of Southern Europe, but the Pyrenean and Apennine regions and the Balkan peninsulas were limited in space and were isolated.

In spite of many studies, we still do not know enough about the change in the European Quaternary faunas to be able to correlate them with the pattern of four glaciations described by geologists. Many paleozoologists, primarily in unglaciated regions of Europe, prefer therefore to use special names for the periods of the faunal development, independent of the names of glaciations (Kretzoi, 1965). Farther north the connection of the faunal changes with the course of glaciations, especially later ones, is evident.

The Ice Age, as a whole, caused the extinction of the rich Pliocene fauna of Europe, with its numerous subtropical elements. In the oldest Pleistocene (Villafranchian) a successive extinction of many evolutionary lines occurred (e.g. of mastodonts, tapirs, gazelles, *Parailurus*, and many groups of deer). The same is true for small mammals and also for many plants (*Azolla, Tsuga, Pterocarya, Magnolia*). Even in the Mediterranean regions, where the modern climate is similar to that of Central Europe in the late Pliocene, the fauna is much poorer as a result of Pleistocene extinction. The family Soricidae, for example, is represented in the Pliocene of central Europe by the genera *Sorex, Neomys, Crocidura, Petenyia, Beremendia, Blarinoides, Petenyiella, Soriculus*, and *Allosorex*, with numerous species. In similar climatic conditions in southern Europe, the same family recently contained only four genera: *Sorex, Neomys, Crocidura*, and *Suncus*, with few species. This depletion of the fauna of Europe has not yet been compensated, because migrations in the Holocene were somewhat restricted by geographical barriers. In eastern Asia, where the migrations from north to the south were not impeded, the fauna of Insectivora is much more diversified—comparable to that of preglacial Europe. Many ecological niches are therefore unoccupied in the recent fauna of Europe, particularly in glaciated islands (e.g. Ireland, where moles, voles, and many other mammalian species are lacking, and their niches apparently empty).

The spatial limitations of the refuges of European thermophilous fauna during the Ice Age had different effects on different species. In many cases large mammals became extinct, whereas small mammals were able to survive in limited territories but were sometimes no longer able to expand when conditions improved. For example, water-moles (Desmaninae), widely distributed over Europe in the Pliocene and early Pleistocene (Fig. 1), now live only in two refuges, the Pyrenees (*Desmana*

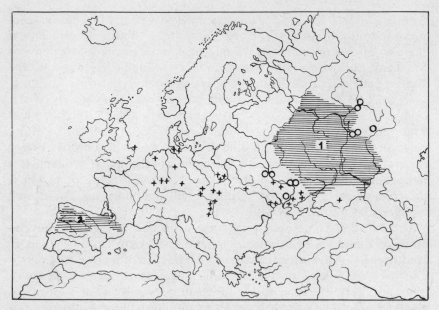

FIG. 1. Distribution of *Desmana* in Europe. 1, *D. moschata;* 2, *D. pyrenaica;* +, fossil localities; O, localities of recent artificial introduction. Once continuous areas are now limited to refuges because of the Ice Age.

pyrenaica) and the Ukraine (*D. moschata*). Other species, when the ice retreated, managed to spread again into central and northern Europe. The moles (*Talpa europaea*) of Europe are clearly differentiated into three forms—western, central, and eastern—each of which expanded in postglacial time from other refuges (Fig. 2) (Stein, 1963).

Thus the Pleistocene brought the gradual extinction of many mammalian elements of the "warm" Tertiary fauna. Particular species died out during different glaciations. The extinction affected large as well as small mammals, although the latter had a better chance of surviving in refuges; sometimes their isolation during cold periods brought the evolution of new subspecies or even species.

The influence of the last glaciation (Würm) on the fauna of Europe is better known than that of earlier glaciations. The small-mammal fauna of the last interglacial (Eemian) in Europe is very similar to the recent one. Not a single species of micromammalia of the European Eemian forest fauna is completely extinct today, but the fauna of the large mammals contains extinct as well as extant species, the former including the forest elephant *Palaeoloxodon antiquus*, the rhinoceros *Dicerorhinus kirchbergensis*, the hippopotamus *Hippopotamus amphibius*, and fallow deer *Dama dama*. The hippopotamus and fallow deer

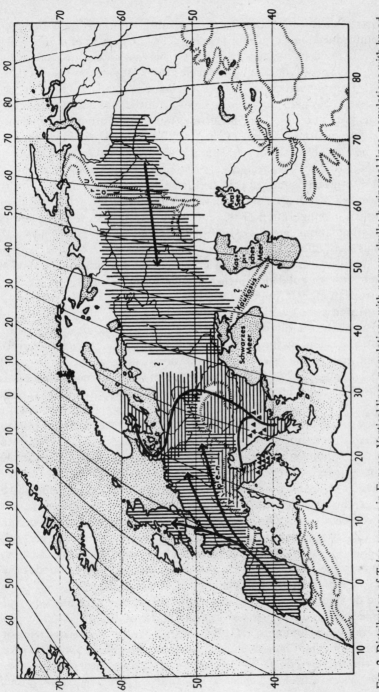

Fɪɢ. 2. Distribution of *Talpa europea* in Europe. Vertical lines, populations with narrow skulls; horizontal lines, populations with broad skulls; solid triangles, *T. e. romana*. Repopulation of central Europe after glaciations starting from different refuges (after Stein, 1963).

disappeared in Europe but still exist outside its limits (fallow deer was later reintroduced in Europe for hunting). The extinction of these species can be explained by changes in climate. The last glaciation limited the suitable biotope of enumerated species to a few areas in southern Europe. The population of elephant and rhinoceros in each of the refuges was so small that even a local factor could bring extinction. Man was present at that time in Europe, but there is almost no evidence for his role in extermination of animals; at such a primitive stage of culture he probably was not a real danger for animals as large as elephant or rhinoceros. The disappearance of these large mammals at the end of the last interglacial is the last episode in the great extinction of European fauna caused by Quaternary climatic changes in conjunction with the geographical configuration of the continent. According to the quantity of their populations and the extent of individual areas, large and small mammals were affected differently by the climate.

The Würm glaciation brought a great extension of the Scandinavian ice sheet and the total destruction of the fauna of northern Europe. The European climate became increasingly continental as a permanent center of high atmospheric pressure developed above the ice sheet and the extent of the sea diminished (because of the glaciation of the Baltic and North seas and by the isostatic lowering of the world oceanic level). The great area of the "steppe–tundra" was formed, connecting central and western Europe with the steppe regions of western and central Asia, with the rich fauna of open country. The composition of the rodent fauna, as well as pollen analysis, shows that the character of Pleistocene vegetation at the time was different from that of the recent Arctic tundra in northern Europe and from that of the recent dry steppes in southeastern Europe and southwestern Asia. In the Pleistocene of central Europe the lemmings *Lemmus* and *Dicrostonyx*, which are now limited to the tundra, lived side by side with the small hamsters (*Cricetulus*), now confined to the steppes far south of the lemming area. *Microtus gregalis*, a characteristic rodent of the Würm glaciation in Europe, lives today in the tundra and steppe regions of central and eastern Asia. In general, Pleistocene vegetation from the time of the Würm glaciation is most analogous to the vegetation of eastern Asia, which developed under the influence of a cold continental climate in middle latitudes.

At the decline of the Würm glaciation in Europe, small mammals of the steppe and tundra retired toward the north or southeast, unable to live in rapidly developing forests. During the same period six species of the larger mammals became extinct: mammoth (*Mammuthus primigenius*), woolly rhinoceros (*Coelodonta antiquitatis*), Irish deer (*Megaceros giganteus*), cave bear (*Ursus spelaeus*), cave lion (*Panthera spelaea*), and cave hyena (*Crocuta spelaea*). The extant European bison (*Bison bonasus*) may be the direct offspring of the extinct primitive bison (*Bison priscus*). The musk-ox (*Ovibos maschatus*) became extinct in the Old World but still lives in America. The wild horse (*Equus przewalskii*) lives in Central Asia; the saiga antelope (*Saiga tatarica*) persisted in Asia as well as on the outskirts of Europe; and the reindeer (*Rangifer tarandus*) was preserved partly as a wild, partly as a domestic animal, in extreme northern Eurasia.

Mammoth (*Mammuthus primigenius*) developed from the steppe elephant (*Elephas trogontherii*) in the penultimate (Riss) glaciation. Its makeup is well known, because, besides numerous skeletal remains, many specimens with soft parts are preserved in the permafrost of Siberia, and one in a layer of silt, impregnated with petroleum and salt in Starunia on the northern slope of the Carpathians (Garutt, 1964). The mammoth's adaptation to the cold climate is provided, for example, by its long hair, small ears, and tusks shaped to drive away snow in the search for food in winter. The geological situation of mammoth fossil remains, and the accompanying fauna and flora indicate that it lived in an open, woodless landscape.

The mammoth, whose Old World fossil remains are distributed from the Atlantic to the Pacific and from the Arctic to the Mediterranean and to Central Asia, probably had a more limited range at any one time. Carbon-14 dates of mammoth corpses discovered in Siberia range between 32,500 and 39,000 years B.P. (Heinz and Garutt, 1964) and thus suggest correlation with the relatively warm oscillation (Göttweig) within the last glaciation. The mammoth from Taimyr was dated by the above authors as 11,450 ± 250 years B.P. (an indication of the Allerød warmer oscillation at the end of the last glaciation). The stratigraphy of the European sites with mammoth also suggests its disappearance at the end of the Würm.

There is no doubt that Upper Paleolithic man hunted the mammoth. Probably it was done mostly by stampede, but pitfalls were also in use. The important role of mammoth in the engravings and sculptures of primitive man in Europe also suggests that it was hunted (Fig. 3).

The problem of the extinction of mammoth has been widely discussed. Earlier hypothetical explanations—e.g. that its extinction was a consequence of its poor adaptation to the conditions of life (Neuville, 1921)— now have only historical significance. The suppositions about a larger number of specimens with pathological characteristics immediately before extinction find no confirmation in an analysis of broader materials (Kubiak, 1965). The role of Paleolithic hunters in the extinction of mammoth is difficult to estimate, but mammoth was probably never

Fig. 3. Paleolithic drawings of mammoth (*Mammuthus primigenius*) (from different sources, after Garutt, 1964).

the principal game of human groups, and the traces of human colonization of Siberia at that time are very scarce.

The most logical explanation of the extinction of mammoth seems to be that it was caused by the disappearance of its habitat at the end of the Würm glaciation. The Pleistocene vegetation during this glaciation was different from that of the northern tundra as well as from that of the present steppes in Europe and western Asia. The adaptation of mammoth to the cold climate precluded life in the southern steppes, with their warm and dry summers. The present tundra, north of the Polar Circle, is very different from the steppe–tundra of the late Pleistocene; long polar nights and abundant snowfall in winter create quite peculiar conditions of life. Central and western Europe, independent of temperature changes in the Pleistocene, had the same geographical latitude as today and, of course, no polar night; the snowfall was probably slight. The development of forests in the postglacial of Europe and Siberia (Vangengeim, 1961) restricted the habitat of mammoth and finally caused its extinction.

The history of the woolly rhinoceros (*Coelodonta antiquitatis*) is similar. It is also known from the corpses with soft parts preserved in the permafrost of Siberia and particularly in the salt- and petroleum impregnated silt in Starunia on the northern slope of the Carpathians (Fig. 4). Hypsodont molars, reduced front teeth, and the anatomy of the upper lips prove that it was a grass-eating animal; the long hair and the paleontologic and geologic data suggest its adaptation to cold climate. The woolly rhinoceros did not develop until the late Pleistocene. The remains found on the Indigirka River in Siberia are radiocarbon-dated at 38,000 years B.P. (Heinz and Garutt, 1964); the species became extinct in Europe at the end of the Würm glaciation. Although *Coelodonta antiquitatis* was sometimes represented in Paleolithic art, few data suggest that it was hunted in the older Stone Age. In all probability its extinction may be explained by the disappearance of its proper habitat, as in the case of mammoth.

The Irish deer (*Megaceros giganteus*) also developed in the late-Pleistocene. Its older forms (*M. g. antecedens*), provided with smaller antlers, could have been forest animals, but later populations (*M. g. giganteus*) lived in an open landscape, as is suggested at least by the enormous antlers, which probably would have made movements in forests impossible. The Irish deer, probably confined to milder climate than was the contemporaneous reindeer, was still living in the Allerød, when the warming climate caused the expansion of forests, and died out in the beginning of the Dryas period about 10,000 years B.P. (Guenther, 1960).

The European primitive bison (*Bison priscus*) undoubtedly inhabited open, treeless country of the late Pleistocene. Its relation to the extant forest bison (*B. bonasus*) needs further study (Degerbøl and Iversen, 1945). *B. priscus* might be the ancestor, or both forms might have lived side by side in different habitats, and *B. priscus* might have disappeared with the decline of grassland in Europe at the end of the last glaciation, while the other persisted. The steppe bison was important game for primitive man and was often represented in Paleolithic art.

Musk-ox (*Ovibos moschatus*) developed from a species adapted to mild climate and was not confined to an arctic environment until the late-Pleistocene. Its extinction in the Old World is difficult to explain,

Fɪɢ. 4. Woolly rhinoceros (*Coelodonta antiquitatis*), specimen preserved with soft parts in Starunia, now in the museum of the Institute of Systematic Zoology, Kraków (photo by L. Sych).

because it seems to have been well adapted to life in the Eurasian tundra. Few data indicate that it was hunted by Paleolithic man; its distribution was probably always limited to the sparsely inhabited periglacial regions.

In spite of voluminous literature devoted to the wild horse, a clear picture of its extinction in Europe is difficult to obtain. In the open areas of the late Pleistocene, the wild horse was very common and was a principal prey of Paleolithic hunters. In the postglacial, the range of the wild horse contracted, beginning with western Europe, and *Equus przewalskii* now lives only in the semideserts of Central Asia (Mohr, 1959), on the verge of extinction. Historical data prove the existence of wild horses in the Ukrainian steppes as late as the middle of the nineteenth century. These horses were described as a separate species, *Equus gmelini*, but they were more probably feral. Vetulani (1933) supposed that the peculiar forest form of *E. gmelini* was extant in western Europe in the Middle Ages and in Poland as late as the eighteenth century. Primitive domestic horses in Poland were, according to him, the direct descendants of these wild horses, and he started to reconstruct this form by breeding the Polish primitive strain in wild conditions (Vetulani, 1948). In these continuing experiments, the animals have developed some apparently primitive characters (not necessarily related to ancestral genotypes). The existence of a sylvan form of the horse seems ecologically improbable, and the supposed "forest tarpan" was probably a feral domestic horse. The postglacial development of forests made the existence of the wild horse in western and central Europe impossible, and the final limitation of its area to the semideserts of central Asia was the result of predation by man.

The extinction of the cave bear (*Ursus spelaeus*), the remains of which are very abundant in European caves, has been widely discussed (Kurtén, 1958). The cave bear appeared during the Mindel Glaciation, survived two later glaciations and two interglacials, and became extinct at the end of the Würm glaciation; its latest remains are from Würm III. It coexisted in Europe with the brown bear (*U. arctos*), but these two species were seldom found together. The cave bear can be distinguished from the brown bear by its larger dimensions, by many details of the skeleton and dentition, and by its probably strictly phytophagous habits.

Abel's suggestion (1929) that the extinction of the cave bear was anticipated by the appearance of degenerated, pathological specimens and by the excess of male births was rejected by Kurtén (1958) as being based on fragmentary or erroneously interpreted material. Another hypothesis holds Paleolithic man responsible for the extermination of the cave bear; according to some archaeologists, tribes of Mousterian man living mostly in mountains specialized in hunting bears. These suppositions also lack an adequate factual basis, and they concern the time of the Eemian Interglacial, long before the process of extinction of cave bear began. Kurtén (1958) stated that the extinction of cave bear was gradual,

and that in many regions the species disappeared long before its final
extinction elsewhere. Before extinction, isolated populations with a
limited number of specimens revealed the effects of inbreeding in the
diminished variation of morphological characters.

There is no defense for Kurtén's explanation that cave bears disappeared because of the occupation by Paleolithic man of caves used as
bear dens. Most of the remains of cave bear were found in caves only
because of the much better conditions for the preservation of bones in
cave sediments than outside, and not because of a narrow specialization
of cave bear to wintering in caves. In the Tatra Mountains of Poland,
where brown bears still exist, their subfossil remains are found exclusively
in caves. The animals hibernate sometimes in caves, but even in the
calcareous Tatras, where caves are abundant, the hibernation places are
generally located below uprooted trees in the forests. The preservation of
bones outside caves, especially in mountains where erosion prevails over
accumulation, is very unlikely. Because of the highly irregular distribution of caves in Europe, it is hard to imagine a species of large mammals
that depended on the caves for its existence. In the same period, Paleolithic settlement of caves, except for some regions in France and Spain,
was irregular and sporadic. In Poland the caves with bear remains have
only scarce traces of human habitation, and the high mountains (e.g.
Tatra), where cave bear remains are numerous, had no Paleolithic
habitation at all. The competition for shelter between Paleolithic man
and cave bear, therefore, cannot be the principal cause of the reduction
of the once numerous population of this animal to scattered small
groups as a prelude to extinction. Here again, the disappearance of
the proper biotope seems to be the most probable cause of extinction.

Cave lion (*Panthera spelaea*) and cave hyena (*Crocuta spelaea*) became
extinct in the final stages of the Würm glaciation. They were probably
never regularly hunted by Paleolithic man.

Most large Pleistocene forest mammals able to survive the last glaciation still live in Europe: the elk (*Alces alces*), red deer (*Cervus elaphus*),
roe deer (*Capreolus capreolus*), wild boar (*Sus scrofa*), and others. Two of
them have faced extinction in historical times: the aurochs (*Bos primigenius*), which disappeared in the wild and survived only as domestic
cattle, and the European bison (*Bison bonasus*), which was saved through
active protection when it was on the verge of extinction.

Aurochs appeared in Europe during the Holsteinian or Great Interglacial (Mindel–Riss) and survived two glaciations. It was principally a
member of the forest community but was also able to live in open areas.
Aurochs was distributed through Europe except for the northernmost
regions, northern and temperate Asia and northern Africa. Important

game of Paleolithic man, it was hunted also in later prehistoric times, in antiquity, and in the Middle Ages. The rapid process of its extinction began in the late Middle Ages, probably as a result of extensive deforestation. In the thirteenth century aurochs were still present in France, Germany, Sweden, Poland, and Russia. In the fourteenth and fifteenth centuries they virtually disappeared as wild animals (Fig. 5). In Poland the hunting of aurochs was early reserved for monarchs and sovereigns, who prohibited general hunting of the animal and commanded their protection (e.g. Prince Boleslaus of Masovia in 1298, Prince Ziemovit of

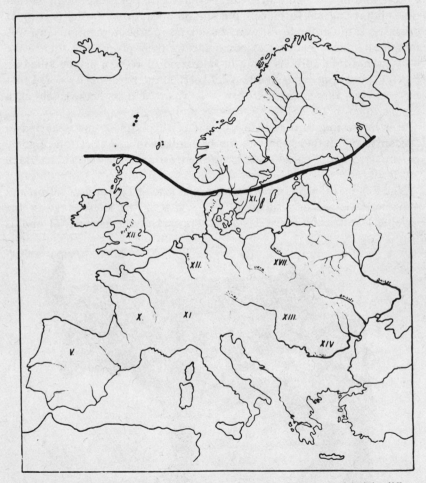

FIG. 5. The dates (by century) of extinction of aurochs (*Bos primigenius*) in different countries in Europe. Solid line, northern limit of distribution (after Lukaszewicz, 1952).

Masovia in 1359, and others). In the sixteenth century, aurochs existed only in the Jaktorov primeval forest near Warsaw, where, under official protection, they were guarded by the inhabitants of surrounding villages, who gave them hay in winter. Reports about the aurochs reservation in the Jaktorov Forest—prepared by royal officers between 1510 and 1630 and published recently by Lukaszewicz (1952)—offer interesting data not only on the final extinction of *Bos primigenius* in spite of its protection but also on their biology and on the ecological conditions in which the last of them lived. In 1557 there were still about fifty in Jaktorov Forest, and in 1562 there were thirty-eight (eleven of them bulls). At that time all hunting was forbidden, but the deforestation and the grazing of domestic cattle in the forest were disastrous for the population. In 1599, in spite of numerous decrees commanding their protection, there were only twenty-four aurochs living in Jaktorov. In 1601, a plague killed all but four, only one of them a female. In 1630 the royal officer noted that he found no aurochs in Jaktorov, but was told that the last animal, a female, died in 1627 (Fig. 6).

It is interesting to note that in Germany experiments were started to "reconstruct" aurochs through the hybridization and selection of some primitive forms of cattle resembling their extinct ancestors (Lengerken, 1953).

The European bison (*Bison bonasus*) was also on the verge of extinction. Whether it is the descendant of *B. priscus* or the latest stage of the independent line of forest bison, it appears in the postglacial and is

FIG. 6. A drawing of aurochs (*Bos primigenius*) from A.D. 1557 (after Lengerken, 1953).

limited in its distribution to Europe. In antiquity it was common on the whole continent. The process of its extinction began in the Middle Ages, and in the eighteenth century it disappeared everywhere but in the Bialowieza primeval forest and in the Caucasus. The lowland population from Bialowieza and the mountain population in the Caucasus were in contact via the forested valley of the Don as late as the Middle Ages. The Caucasian bison is smaller and of darker color.

In 1803 Bialowieza, then in the Russian empire, was declared the private hunting ground of the czars. The herd of bison numbered 300 to 500. In 1860 there were as many as 1,500 bison there, but at the beginning of World War I the population once more declined to 750. The main reason for the weak birth rate and the later decline of the bison population was the destruction of forest vegetation in Bialowieza by an excessive number of red deer, maintained there for hunting purposes. During the nineteenth century some of the bison from Bialowieza were sent to different places as gifts from the czars. Four specimens presented to Pszczyna in Silesia propagated in 1920 to a herd of seventy head.

During World War I, when the Russian army retreated, German soldiers killed about 500 bison; the remainder were exterminated by poachers when Bialowieza passed successively into the hands of the Lithuanian, Russian, and Polish armies. At the same time, the Pszczyna herd, the largest outside Bialowieza, was decimated, and only three specimens survived; other bison lived in zoological gardens. There were at that time no more than fifty purebred European bison in the world. Their breeding in Poland, Germany, Sweden, and the Netherlands increased the number in 1939 to ninety-four. During World War II some centers of bison breeding were destroyed, but the herd in Bialowieza survived. Now there are several hundred bison in Europe, 213 of them in 1966 in Poland, where some live in reservations, and three herds live freely in large forest regions such as Bialowieza.

The Caucasian herd of bison numbered about a thousand specimens in the second half of the nineteenth century. It diminished gradually, and the last known specimen was killed in 1926. Only one Caucasian bison, a bull, was transported from Russia before World War I; it mated with the lowland bison from Bialowieza and left numerous hybrid offspring. The issue of this line is kept separately and now forms quite a large population in Russian and Polish reservations.

We can hope that the European bison has been saved as a species, although it is almost completely dependent on continuous protection from man.

References

Abel, O., 1929, *Paläobiologie und Stammesgeschichte:* Jena, Fischer, 423 p.

Degerbøl, M., and Iversen, J., 1945, The bison in Denmark: København, *Danmarks geol. unders.* R. *II*, no. *73*, p. 1–62

Garutt, W. E., 1964, Das Mammut: Wittenberg, *Neue Brehm-Bücherei*, no. *331*, p. 1–140

Guenther, E. W., 1960, Funde des Riesenhirsches in Schleswig-Holstein und ihre zeitliche Einordnung, p. 201–06, in *Steinzeitfragen d. Alten u. Neuen Welt;* Zotz-Festschrift: Bonn, L. Röhrscheid Verlag

Guilday, J. E., Differential extinction during late-Pleistocene and Recent times (this volume)

Heinz, A. E., and Garutt, W. E., 1964, Determination with radioactive carbon of the absolute age of fossil remains of mammoth and woolly rhinoceros found in permanently frozen ground of Siberia (in Russian): Moscow, *Doklady Akad. Nauk SSSR*, v. *154*, p. 1367–70

Kretzoi, M., 1965, Die Nager und Lagomorphen von Voigtstedt in Thüringen und ihre chronologische Aussage: Berlin, *Paläont. Abh.*, *A*, v. *II*, no. *2/3*, p. 585–660

Kubiak, H., 1965, Examples of abnormalities in the dentition of fossil elephants (in Polish): Krakow, *Folia. Quatern.*, v. *19*, p. 45–61

Kurtén, B., 1958, Life and death of the Pleistocene cave bear: *Acta. zool. fennica*, v. *95*, p. 1–59

Lengerken, H., 1953, Der Ur und seine Beziehungen zum Menschen: Wittenberg, *Neue Brehm-Bücherei*, no. *105*, p. 1–80

Lukaszewicz, K., 1952, The Ure-ox (in Polish): Krakow, *Ochrona Przyrody*, v. *20*, p. 1–33

Mohr, E., 1959, Das Urwildpferd: Wittenberg, *Neue Brehm-Bücherei*, no. *249*, p. 1–144

Neuville, H., 1921, On the extinction of the Mammoth: *Smithson. Inst. Ann. Report for 1919*, p. 327–38

Stein, G., 1963, Unterartengliederung und Nacheiszeitliche Ausbreitung des Maulwurfs, Talpa europaea L.: Berlin, *Mitteil. zool. Mus.*, v. *39*, p. 379–402

Vangengeim, E. A., 1961, Paleontologic materials for the stratigraphy of the Quaternary sediments of northeastern Siberia (in Russian): Moscow, *Trudy geol. inst. Akad. Nauk SSSR*, v. *48*, p. 1–182

Vetulani, T., 1933, Zwei weitere Quellen zur Frage des europäischen Waldtarpans: Berlin, *Zeit. Säugetierkunde*, v. *8*, p. 281–82

Vetulani, T., 1948, Premières observations sur la régénération du Tarpan sylvestre européen dans la forêt vierge de Bialowieza: *Bull. int. Acad. Pol. Sci. Lettres, Cl. Sci. Math. Nat.*, B, 1947, p. 1–22

N. K. VERESHCHAGIN

Zoological Institute
Academy of Sciences of the U.S.S.R.
Leningrad, U.S.S.R.

PRIMITIVE HUNTERS AND
PLEISTOCENE EXTINCTION
IN THE SOVIET UNION

Abstract

Archaeological investigations undertaken in the ranges of the U.S.S.R. have yielded very important documentary material for studying the mammalian fauna. In different Pleistocene epochs nearly all the U.S.S.R. was inhabited by the mammoth fauna, characterized by mammoth, cave hyena, cave lion, horse, reindeer, giant and red deer, bison, saiga, etc. In the Russian plain, along the rivers in Siberia, in karst regions of the Crimea, Caucasus, the Urals, Middle Asia, East Siberia, and the Far East, game animals were used by Paleolithic and later tribes. Numerous remains found in the Crimea and in the Ukraine testify to the hunting for large animals.

The largest sites of the Upper Paleolithic are estimated by radiocarbon to be 9,000–14,000 years old.

At the boundary between the Paleolithic and Neolithic, complex changes in the fauna and ecological assemblages of different areas took place. About ten species of large Pleistocene mammals became extinct in the U.S.S.R. Other species underwent a reduced distributional range in the Holocene. Some species, having expanded their range into the taiga, increased their populations (moose, brown bear, beaver). In the Neolithic, man began to exploit fish and marine mammals.

The main reason for the absolute extinction of animals of the mammoth complex and for the reduction of range in some species is the change in climate and terrain, especially the change in the regime of winter weather. The destructive effect of man supplemented and intensified the influence of climatic factors.

As many as two thousand years ago ancient Greek and Roman philosophers wondered how our distant ancestors had mastered the animal world. Titus Lucretius thought primitive man had been a brutal plunderer:

Consectabantur silvestria saecla ferarum
Missilibus saxis et magno pondere clavae;
Multaque vincebant, vitabant pauca latebris . . .
De Rerum Natura

This view has been confirmed by abundant archaeological evidence and observations from all continents.

Primitive man penetrated Europe and northern Asia (thoughout the U.S.S.R.) from the south with fully formed meat-eating habits and adaptations. The earliest traces of artificial splitting of bones of antelope and deer by pre-Chellean anthropomorphous creatures were found in the upper Pliocene (Villafranchian) deposits of the Black Sea coast and Taman peninsula (Vereshchagin, 1957). In a study of more than 200 Paleolithic and Neolithic nomadic camps in the U.S.S.R. containing Quaternary animal remains, as many as 73 species of animals were found in Paleolithic sites and 60 species in Neolithic (Table 2).

Primitive Paleolithic peoples hunted in general for the mammals listed below. Possibly they also hunted Pisces, Amphibia, Reptilia, Aves, and Insecta, although proof is lacking. As a rule, Insectivora, numerous Rodentia, and birds (especially Passeriformes)—the bones of which were found in the camps of nomads—served as food for owls rather than for primitive man. In the Paleolithic the composition of animals and methods of hunting for them differed to a certain extent from those of western Europe (Lindner, 1937). These differences were caused, in general, by another paleogeographical situation and by its history. The factors of animal extinction there were quite different.

PALEOLITHIC CULTURES

Paleolithic tribes as well as modern ones depended on hunting for their supply of animal food and generally took the most abundant and available animals. The specific assemblage of animals usually reveals clearly the characteristic features of paleo-landscapes of a given region and broadens or defines more precisely our view of the faunistic complexes obtained from studying non-Paleolithic burials. The composition of animals over geographical regions and paleo-landscapes is as follows:

1. *The Caucasian isthmus.* Predominant in the Lower Paleolithic in karst regions were cave bear, red deer, and goat, with lesser numbers of European bison, sheep, boar, horse, rhinoceros, cave lion, leopard, cave hyena, wolf, Siberian red dog, glutton, marmot, porcupine, ape, and salmon. In the Upper Paleolithic, cave bear was hunted only in the mountains. In the foothills of the Caucasus, European bison, goat,

horse, red and giant deer, and boar made up most of the game. Saiga, chamois, and moose occurred rather seldom. Cave hyena, large and small cats, and glutton were very rare. Brown bear began to be hunted (Vereshchagin, 1959).

2. *The Crimea.* In the Lower Paleolithic of the northern Crimean hills game was rather varied, including primarily giant deer, ass, horse, mammoth, saiga, bison, and cave bear; more rarely it included red deer, reindeer, woolly rhinoceros, cave lion, cave hyena, wolf, fox, arctic fox, corsac fox, glutton, European hare, boar, and ibex.

In the Upper Paleolithic, people generally hunted horse and ass. In cave deposits, remains were found of mammoth, woolly rhinoceros, cave hyena, and cave bear, which had nearly vanished by the end of this period. As on the Caucasian isthmus, brown bear began to be hunted (Gromov, 1948).

3. *Russian Plain.* During the Middle Pleistocene there was in the southern part of the plain a huge reserve of large animals nearly untapped by man—"steppe" mammoth, Merck's and woolly rhinoceroses, elasmotherium, horse, ass, and camel (Vereshchagin, 1953). The assemblage of the animals is traced from Mousterian times, i.e. from the beginning of the Upper Pleistocene. At that time ancient inhabitants of the Prut, Dniester, Volga, and Ural river basins hunted in general for horse, bison, mammoth, saiga, and wolf (Pidoplichko, 1954; Panichkina, 1953; Vereshchagin and Kolbutov, 1957; Tchernysh, 1959; David, 1961).

In the Upper Paleolithic, primitive people killed for meat and skins as many as 23 species of animals. In most archaeological excavations the bones and remains of horse, arctic fox, Don hare, mammoth, red deer, reindeer, brown bear, and wolf are the most abundant. The remains of glutton, cave lion, woolly rhinoceros, saiga, tur, bison, and musk-ox, are, as a rule, very rare. Kitchen refuse and other remains of animals are usually mixed with layers of loess-like colluvium in gently sloping ravines opening into river valleys. The Mesin site on the Desna River (Pidoplichko, 1959) and the Kostionki XII site on the Don (Vereshchagin, 1961) indicate the species, which, in varying proportions, are found throughout the Upper Paleolithic. Different indices of the number of any particular species in various nomadic camps are generally explained by the character of accumulation of bone-containing areas, not by differences in the megafaunal complexes or in seasonal or specialized hunting by different primitive tribes. Excavation in the Kostionki XIV site (Fig. 1) revealed dissimilar accumulation of remains in different stratigraphic layers in the Upper Paleolithic. However, some families and whole tribes undoubtedly hunted particular species, especially when conditions

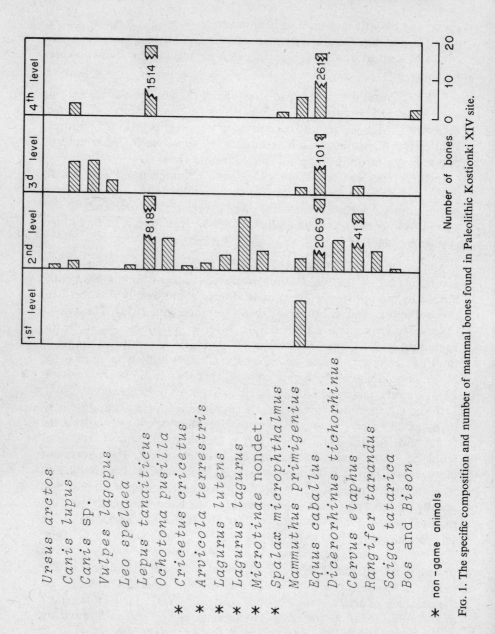

FIG. 1. The specific composition and number of mammal bones found in Paleolithic Kostionki XIV site.

favored such hunting. For example, at the Elsejevichi site (north of Briansk), the hunting of arctic fox, mammoth, wolf, and brown bear prevailed, and ungulates were rare (Fig. 2).

Tundra, forest–steppe, and steppe animals were very abundant, whereas forest and forest–steppe species (brown bear, moose, roe, beaver, and especially lynx, which later became the inhabitant of taiga) were nearly lacking. Very few of these animals existed in the middle belt of the Russian plain in that epoch. True desert inhabitants, such as camel, cheetah, and goitered gazelle, were absent, although in the Mindel-Riss Interglacial large camels were distributed in the Russian Plain up to the latitude of Kazan. Want of skill or desire could not be the reason for specialized hunting. Ancient people of the Russian plain hunted rather successfully for such dangerous and nimble animals as wolf, brown bear, and cave lion. Where there were no cave refuges, they apparently built huts, covered them with skins (Fig. 3), and surrounded them with skulls and bones of mammoths. Excavations revealed about ten species of marine, freshwater, and terrestrial mollusks, which apparently served as decoration (Shovkoplias, 1965).

In the middle belt of the plain, tribes specialized exclusively in the hunting of reindeer or bison. In the valley of the middle course of the Bug, Dniester, Dnieper, Don, Volga, and Ural rivers, various animals were hunted. However, on the Black Sea coast, tribes generally hunted bison. The fate of these bison hunters is unknown.

In the Upper Paleolithic, primitive tribes pursued mammoth, horse, and reindeer far north to the low reaches of the Pechora, latitude 65–80°N.

4. *The Ural Mountains.* The people lived here in foothill caves and hunted cave bear, reindeer, moose, wolf, glutton, sable, mammoth, woolly rhinoceros, and willow grouse. Roe and saiga, though very rare, penetrated north to the upper reaches of the Pechora during the development of steppe terrains (Vereshchagin and Kuzmina, 1962; Guslitzer and Kanivetz, 1965).

5. *Middle Asia.* In foothill Paleolithic camps near Samarkand, the remains of horses, ass, red deer, camel, and tur were found. In the spurs of the Gissar ridge (the Mousterian sites Aman-Kutan and Teshik Tash) the usual game were mouflon, Siberian ibex and roe, Buchara deer, brown bear, cave hyena, leopard, fox and wolf, marmot, porcupine, and Horsfield's terrapin (Bibikova, 1958; Gromova, 1949).

6. *Eastern Kazakhstan and the Altai.* In the Upper Paleolithic, primitive man hunted horse, bison, Pamir argali, mongolian ass, Baikal yak,

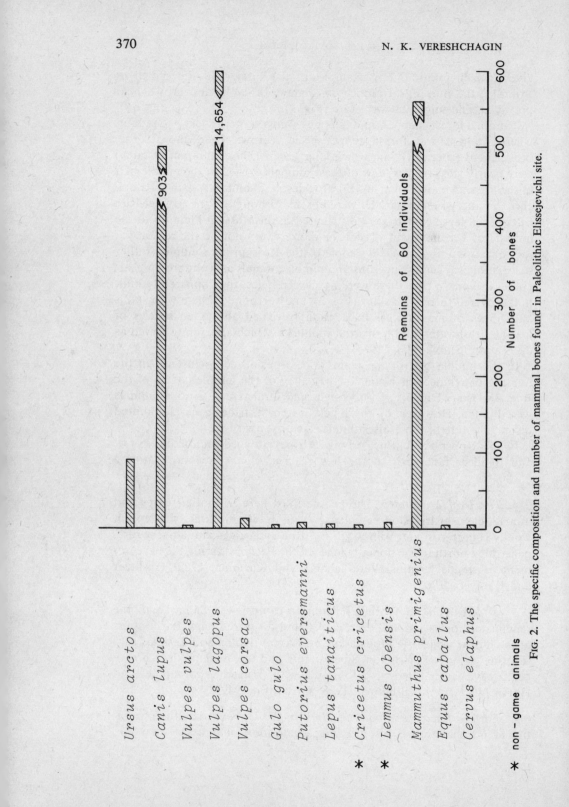

Fig. 2. The specific composition and number of mammal bones found in Paleolithic Elissejevichi site.

camel, antelope (*Spirocerus*), brown bear, roe, rhinoceros, mammoth, heath cock, and snow cock (Vereshchagin, 1956).

7. *Eastern Siberia and southern Yakutia.* In the Upper Paleolithic along the Yenisei River and its tributaries and along the Angara, primitive

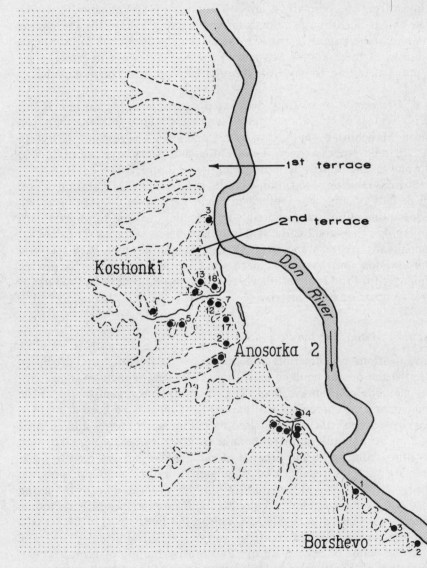

FIG. 3. The scheme of location of Upper Paleolithic camps on the Don south of Voronezh (according to Rogachev, 1957).

man hunted mostly reindeer, horse, primitive bison, Arctic fox, blue hare, woolly rhinoceros, and mammoth. In the valley of the Angara River fowling was quite common (Gromov, 1948). The composition of animals killed by ancient inhabitants of Yakutia has not been investigated well, but it is thought to be close to the complex of large animals known from the permafrost levels (Chersky, 1891; Vangengeim, 1961). The abundance of large animals in northeastern Asia in the Pleistocene enabled some primitive tribes to settle widely over cold forest–steppe plateaus of the Okhotsk Sea and in the Korjak area and later, in the Upper Paleolithic, to penetrate through Beringia into North America.

8. *Ussuriland*. In recently discovered caves along the Suchan River, broken bones of typical mammoth fauna were found—mammoth, horse, bison, Manchurian wapiti, cave hyena, and wolf. The remains of "south-Asiatic relic" species (tiger, raccoon-dog, axis deer) have been found so far only in the uppermost postglacial layers (unpublished data).

So mass hunting of Paleolithic hordes of large animals is beyond doubt. In periods of hunger, cannibalism might very possibly have occurred. The hunting of birds was undoubtedly secondary to hunting of mammals. Of reptiles, they apparently used only Horsfield's terrapin. The data for amphibians are very uncertain. Of fish, Paleolithic man used for food only salmon and barbel (Kudaro Cave in the Caucasus); the use of mollusks for food is doubtful. There are no available data on food use of grasshoppers or other insects.

Hunting Methods in the Paleolithic

Investigations of numerous bone remains of Paleolithic and Neolithic hearths and of stone and bone weapons deter hasty conclusions on hunting methods of primitive man. The ingenuity and inventiveness of the hunters and an appreciation of animal food must have been aroused in part by seasonal starvation, caused by falling temperatures and changes of flora. Apparently, primates became carnivorous because of starvation resulting from fights and because of cannibalism, as observed in other animals, such as rodents and ungulates. Primitive man probably ate the remains of lion and leopard kills until he elaborated his own hunting methods for large animals. Undoubtedly, the abundance of herbivorous animals as well as a terrain convenient for hunting were of decisive significance for primitive hunters.

In the plains of eastern Europe and Siberia the life of primitive man was connected with river valleys. Large herds of mammoth, rhinoceros, roe, giant deer and reindeer, and boar roamed from south to north and back along the valleys and floodplains of the rivers. The inhabitants of steppe watersheds preferred meadows and forests of floodplains, especially in dry periods or when the ground was covered with ice crust, because then elk, bison, tur, horse, and even saiga and camel fed upon branches of bushes and trees. It was possible to find animals trapped in bogs. It was easy to drive frightened animals into ravines or narrow pools, and pitfalls could be conveniently arranged. In winter animals could be driven onto the slippery ice of floodland lakes.

In the lower and early-Upper Paleolithic, hunting for large animals was collective. All active members of a family, a group of families, or the whole tribe took part in battues. The success of a battue depended on the number of beaters (which were not many in the primitive community) and on the terrain. In the Lower Paleolithic of the Crimea, the success of a battue of Pleistocene asses was determined by the availability of narrow canyons and isolated plateaus surrounded by steep limestone ravines. In the Bakhchisarai area near the Mousterian camp Staroselje, where the remains of many hundreds of asses have been found, rushing herds of ungulates were ambushed in narrow canyons by the hunters and killed with stones, cudgels, boar spears, and spears. The pasture plateau Dzhugut-Kala, surrounded by vertical ravines 30–40 m high, was an ideal place for regular hunting of herbivorous animals. The frightened animals fell to their death into the ravine (Fig. 4).

It is evident that Chellean and Acheulian man was capable of killing the largest herbivorous animals including bear, elephant, rhinoceros, and European bison. But the nature of their hunting weapons is not yet known. Stone instruments of the Lower Paleolithic, large hand points (ovates) and rough narrow-pointed stones, were unsuitable for killing and cutting up large animals. A sleepy cave bear could be killed with a huge wooden cudgel or boar spear, but certainly not a mammoth or European bison. Only in the Acheulian-Mousterian of the Caucasus, Crimea, or the Russian plain did primitive people begin to use flattened flint points, sometimes with barbs on the back edge and suitable for attachment in a spear shaft (Fig. 5). Darts also began to be used here only at the end of the Lower Paleolithic.

At the beginning of the Upper Paleolithic the spear was improved considerably, but collective hunting continued to develop. Remains found in Kasennaja balka near Amvrosievka indicate that during a hunt nearly 1,000 bison were run through with at least 270 spears tipped with flint and 35 tipped with bone (Pidoplichko, 1953). Upper Paleolithic people undoubtedly hunted mammoth by driving them into deep ravines with sheer walls, which are common in deep loess areas of the first steppes around the Black Sea coast.

FIG. 4. The ravines of Dzhugut (Tchufut)—Kala, a probable hunting site for Pleistocene asses (photo by N. K. Vereshchagin, 1959).

In Siberia, primitive hunters obtained reindeer meat, skins, tendons, and horns during the migrations of thousands of herds, like those at the Angara River and near the Malta site on the Birjusa River.

Deer, arctic fox, hare, and willow grouse were undoubtedly known to man during the Upper Paleolithic, but whether they were trapped is not clear. Ungulates and even large predatory animals could be caught by means of strap loops with a hanging device. Data are lacking in both the Lower and Upper Paleolithic on the use of fire and torches as well as fences and nets in hunting.

Fig. 5. Flint head of a dart from Upper Paleolithic Kostionki site (according to Rogachev, 1957).

Hunting for large animals contributed to the growth of collective habits and to a feeling of interdependence among primitive peoples. The spear, the main weapon in the Upper Paleolithic, was improved with better flint and bone heads and a shaft. Small straight birch, pine, juniper, and fir trees, found nearly full grown along the banks of rivers and lakes, were apparently used for spear and dart shafts.

The flint head was improved through flattening, the reduction of the cross-sectional area, and the strengthening of the working edges by retouching. The main technical qualities of the flint head, i.e. the thickness and angle of the cutting edge, were elaborated by primitive masters as early as the Upper Paleolithic and remained unchanged until the Bronze Age. Evidently nothing better could be attained with this material (Fig. 6). The angle of the working edge along the first 10–12 mm of the blade was about 36° to 38° in both early and late heads. In the most thoroughly chipped heads the angle was reduced to 27° (Fig. 6).

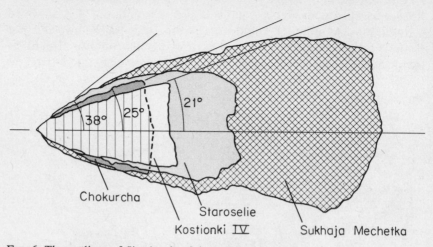

FIG. 6. The outlines of flint heads of darts from Mousterian and Upper Paleolithic sites of the Russian plain.

Narrower bone heads were prepared from reindeer horns and bison metapodials and were sometimes supplemented with cutting flint plates fastened to the heads. Besides increasing cutting ability, these plates kept the spear embedded in the wound and thus intensified bleeding (Fig. 7).

FIG. 7. Dart head made of horn wall of reindeer with longitudinal slots; Upper Paleolithic camp on the Yenisei River.

The invention of a light missile dart created unlimited possibilities for hunting the largest and most dangerous animals by one or several hunters. Collective driving of animals over a precipice or into an ice crevasse or a bog was not necessary if such darts were available. A spear thrown 20 to 30 meters by a skillful hand could pierce the pectoral wall between ribs and reach the vital organs. By piercing and tearing the abdominal wall near the groin a running animal might be disemboweled. The supposition is borne out by the well-known drawing of a disemboweled European bison on one of the walls of the Lascaux Cave in the Dordogne (Fig. 8).

In the U.S.S.R. and elsewhere, there are many descriptions of large animal bones that were pierced with flint, bone spear heads, or darts (Fig. 9). Injuries caused by such weapons are evident in the condition of certain fossil bones (Tasnadi-Kubacska, 1962).

Specialized hunting of mammoth in the Upper Paleolithic of the Russian plain probably followed closely the invention of a narrow flint tip capable of piercing the abdominal wall of elephant and rhinoceros.

Fig. 8. A dart pierced the abdominal wall of the primitive bison; Lascaux Cave, Dordogne, France.

In the Ukrainian Upper Paleolithic, hunters prepared large cutting instruments made of chalky flint, as is indicated by findings in Novgorod-Seversk on the Desna River. These gigantic flint weapons (Pidoplichko, 1941), 12×45 cm and 4–8 kg, could be used as axes for cutting meat. Their marks can be seen on mammoth bones. These flint axes, fastened to a long shaft, could be used for finishing off large wounded animals. Many hundreds of broken, and less often unbroken, bones and dozens of mammoth skulls as well as peculiar blocks of these bones are found in the Upper Paleolithic camps of the Russian Plain. They have caused bewilderment and misunderstanding among zoologists and archaeologists. The blocks were usually described as remains of huts (Fig. 10).

FIG. 9. A dart head made of reindeer horn went through the shoulder blade of the bison; camp on the Yenisei River.

Without spades for digging large pits or metal spear heads for cutting the Achilles tendon or piercing the pectoral or abdominal wall, it is not definitely known how hunters could have killed mammoth and rhinoceros. Russian paleontologists and artists have unsuccessfully attempted depictions of the armament of primitive man and hunting for mammoth and other thick-skinned animals by means of pitfalls and stone darts (see frieze, "Stone Age" by Vasnetsov, 1956, the pictures preserved in museums, and illustrations in certain papers).

FIG. 10. Blocks of bones and skulls of mammoths in the Mesin site, the Desna River (photo by N. K. Vereshchagin in 1956).

Some paleontologists and archaeologists (see Gromov's review, 1948) have made the valid suggestion that Upper Paleolithic hunters generally used the carcasses of mammoth that had died in floods or from some other natural cause.

Artificial hills of mammoth skulls and bones found along the Don, Desna, and Dnieper rivers (sites of Kostionki, Mesin, and Kirillovskaja, among others) might originally have been mounds around Paleolithic huts, or ritual hills. Maceration of carcasses must have required several years, however, and a decaying mammoth carcass certainly would not have been suitable for mounds near wigwams. Alternatively, such accumulations could have been formed after spring floods in ravines near the western high banks of the rivers.

The cleaned and dried skull of an adult mammoth with small tusks weighs at least 100–110 kg, so that dragging it even over ice for several kilometers would have been quite difficult. A lower jaw of an adult Indian elephant (and mammoth) weighs 30–35 kg. Thus the construction of such mounds could have taken place only if there was an abundance of bone material, which would suggest animal epizooty caused by natural calamities.

Kornietz (1961), who examined bones on the Mesin site, claims, without going into details, that primitive hunters actively hunted mammoth and destroyed whole herds on occasion. Usually, communities of hunters armed with spears hunted for solitary mammoth or small groups of females with young, as African Negroes now hunt elephants. The hunters pierced the abdominal wall of the mammoth and then pursued them, sometimes for many kilometers; but hunters could obtain only single specimens at a time. They hunted woolly rhinoceros in the same manner.

Along the river valleys of the Russian plain, hunting for Proboscidea and ungulates was especially successful when the ground was firmly covered with frozen snow crust. Animals, weakened by lack of food, gathered in the forests or brush of floodplains, where they were killed with spears and cudgels.

The best account of the evolution of hunting and weapons in the Paleolithic is given by the archaeologist Zamiatnin (1960). Data on hunting methods in the Paleolithic for ungulates, Proboscidea, and predators in various regions of the U.S.S.R. are still fragmentary and speculative. It is difficult to calculate the number of animals killed, even if they were butchered and eaten in the same place; leavings were usually thrown away, pilfered by predatory animals, decomposed, or washed away. Therefore, the numbers of animals given in Table 1 might be at

TABLE 1. *The Number of Killed Animals, as Indicated by Remains Found in Some Camps and Slaughter Areas*

Sites	Number of Animals
Amvrosievka ravine, South Ukraine	1,000 bison
Staroselje site in the Crimea	435 asses (at least 1,200 over the whole territory)
Mesin site on the Desna River	44 mammoths
Elissejevichi site of the Briansk region	60 mammoths
	285 arctic foxes
Anosovka II site on the Don River	32 mammoths
Kostionki XIV site on the Don River	500 hares
	120 horses

least tenfold greater, if the one- to two-year existence of each camp is kept in mind. According to our observations in modern open camps of fishermen and hunters, with dogs, cats, and magpies, no more than 1 per cent of the cleaned bones remain or are trampled into the soil layer. In Paleolithic camps, leavings were pilfered by arctic foxes, rodents, and birds. Within the U.S.S.R., however, no camps have been found with as great a number of bone remains as in those of western Europe.

In the Paleolithic, as in the present, many more animals were killed than were needed. In Amvrosievka, for instance, if a stampede was successful, the hunter used only the uppermost few bison out of several dozens of animals that fell into the ravine. The same was true of the slaughtered herds of horses, asses, reindeer, and mammoth.

The human population of the Crimea and Russian Plain in the Upper Paleolithic hardly exceeded ten to fifteen thousand. For their frugal feeding (2 kg per day = Eskimo ration; see Mowat, 1963) they needed about 120,000 reindeer, 80,000 horses, 30,000 bison, or 10,000 mammoths (mainly semi-adult beasts) per year. The absolute number is not relatively high, as the herds of different ungulates in the Pleistocene probably consisted of hundreds of thousands of head, but the effect of primitive hordes upon the animal world went far beyond the destruction of large animals for basic needs. Dwelling near the habitual watering places or near migration routes of large animals, primitive people often disturbed the behavior pattern of large herds and consequently might have unintentionally caused death among the animals.

Bone layers in ancient alluvium of the Volga, Ural, Indigirka, Viluj, and Kolyma rivers indicate that natural mortality (often caused by disaster) of large animals in the Pleistocene was sometimes enormous. According to my calculations in 1949–51 on bone-bearing beaches of the lower Kama, middle Volga, and lower Ural (Mysy, Undory, Tungus, Janvartzevo, and others), only 0.1 to 0.5% of Pleistocene bones bear the traces of artificial marks or splits made by flint spears. Presumably, most animals died independently of man. Recent examples (e.g. Pannonik lowland) seem to indicate that natural animal deaths were caused not only by snowfalls or by the lack of food but also by vast autumn and spring floods.

The Use of the Carcass

After a large animal was killed, the main operations were to flay the animal, take out choice pieces, especially the liver, and cut flesh from the skeleton. For this purpose, primitive men used flat flint or obsidian tools with thin, sharp blades. Less suitable were knife-like plates, leaf-like points, and chisels blunted by retouching. The traces of cuts on bones show that, for flaying and cutting, primitive people generally used large fragments of cores and knife-like plates rather than finished instruments. Cuts made by them are pure and straight, without the lateral parallel lines that are inevitable when a retouched blade is used for cutting. Used and blunt blades were thrown away.

There are no trustworthy data available on preservation of meat by sun-drying or by smoking. However, in frosty periods, meat reserves could remain in caves and pits for several months, and in permafrost regions for hundreds of years. Primitive hearths contain animal bones intentionally smashed with stones for the extraction of the marrow (Fig. 11). Animal carcasses were also used in different ways; e.g. large animal

FIG. 11. Phalange of red deer, smashed for the extraction of the marrow; Fatma-Koba Cave in the Crimea (drawing by N. K. Vereshchagin).

skins were apparently used for lining huts and covering roofs. Strips were apparently cut from the skins of horses and deer for nooses; tendons were used for sewing clothes and blankets made of skins of deer, hare, and arctic fox. Bones and skulls of mammoths were used for strengthening mud huts (Rogachev, 1955) and for the construction of funeral chambers (Boriskovsky, 1956) and sacrifice hills (Fig. 12).

Bones of mammoths were good fuel, as suggested by charcoal found in Chokurcha and Kostionki sites. Tusks were used for making knives, awls, and needles, and molars were used for smashing bones. Needles and awls were made also of splint bones of horses and of the femur, tibia, and

FIG. 12. Block of bones found in camp Anosovka II. Lower jaws of mammoths (photo by N. K. Vereshchagin).

capral bones of arctic fox. Thin needles were cut from radial bones of arctic fox and tusks of mammoth. Mammoth tusks and cylindrical bison bones, as well as reindeer horns, served as spear tips and darts (Fig. 7). Necklaces, bracelets, pendants, and decorations on leather cloaks were made of drilled fangs of arctic fox, of incisors of wolf, beaver, and saiga, of mammoth tusks and sometimes of seashells (*Buccinum superabile, Cerithium vulgatum,* and *Nassa reticulata*). Bones and skulls of some animals were used as fetishes and luck charms for ceremonial rites. In the Kostionki IV site, skulls of cave lions were found on the roofs of some huts (Rogachev, 1955) or in ritual bone hills.

Animals and Art of Paleolithic Man

Within the U.S.S.R., animals were the main subject of primitive imitative paintings. Their images are represented in sculpture and in drawings of the Upper Paleolithic of the Russian plain, Urals, and Siberia.

In all, there were recorded thirteen species of realistically portrayed animals or of their heads. The most common are drawings of bear, cave lion, bison, musk-ox, camel, mammoth, and horse. Figurines closely resemble birds (Mesin site), including flying geese (Malta, Buret sites) (Avramova, 1962; Shovkoplias, 1965). Bone, marl, limestone, clay, and probably wood were used for pieces of sculpture, but such artifacts have disappeared. The most expressive art works are sculptural portraits of cave lion and cave bear. Animals were drawn on stones, tusks of mammoths, and walls of caves (Kapova Cave, southern Urals, Fig. 13).

The walls of Kapova Cave have very interesting drawings representing mammoth and rhinoceros in ocher (Bader, 1963). Sometimes such drawings on rocks and figurines had a ritualistic meaning, especially in the Neolithic (Bader, 1941, 1954).

NEOLITHIC CULTURES

The transition from the Paleolithic to the Mesolithic and then to the Neolithic and metal cultures throughout the U.S.S.R. is accompanied by a further improvement of hunting weapons. The pressure of human tribes upon the animal world continued to increase in spite of the appearance of domestic animals (dogs, horse, pig, goat, sheep, and cow). The invention of the bow, boat, net, snare, trap, hook, and harpoon and the use of dogs and horses for searching out, impounding, and driving animals created new possibilities for hunting fur-bearing animals and wild ungulates and for exploiting untapped resources of fish, seals, whales, and other marine animals (Fig. 14).

Individual weapons, especially the spear, continued to improve in the Neolithic. Archaeologists, however, have not been able to explain the displacement of large flint weapons of the Paleolithic by microliths in the Mesolithic. Among numerous microlithic weapons found in steppes on the northern border of Caspij, only small arrow points were suitable for hunting. The possibility that flint tips of spears and darts replaced bone and horn tips during that period is increased by the fact that polished axes were already made of slate and nephrite, which yielded easily to treatment.

FIG. 13. Paleolithic drawing made with ocher on the wall of Kapova Cave (according to Bader, 1963).

In the Neolithic and post-Neolithic the number of drawings on rocks in taiga, steppe, and desert zones increased. There are published and partially studied Neolithic drawings on the rocks of Lake Onega and the White Sea (Ravdonikas, 1936–38; Kühn, 1956), in the Ukraine (Bader, 1941), in Transcaucasia (Vereshchagin and Burchak-Abramovich, 1948), in Ciscaucasia (Markovin, 1958), in Middle Asia (Grach, 1957; Marikovsky, 1953; Sokolov, 1964); and in eastern Siberia (Grach, 1957; Rygdylon, 1955; Skalon, 1956; Okladnikov, 1959). As in the

Paleolithic, the cave paintings are presumed to represent animals and hunters. The specific composition of animals in these drawings resembles, as a rule, the zoogeographical pictures of today.

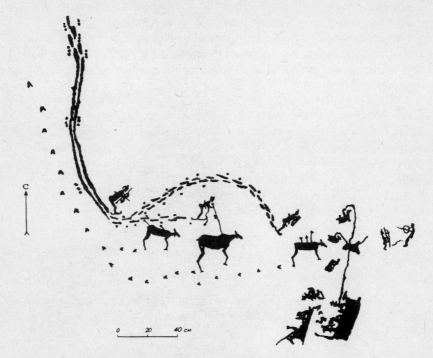

FIG. 14. Neolithic drawings on coastal granite of the White Sea; hunting for moose and white whale (according to J. A. Savvateev).

Changes in the Faunal Composition

Throughout northern Asia, the composition of animals changed considerably at the boundary of the Paleolithic and the Neolithic (Table 2, below). The main faunal changes are listed below. Throughout the U.S.S.R. a large group of animals of a typical upper-Pleistocene mammoth complex became extinct:

Ursus spelaeus   Equus hidruntinus
Crocuta spelaea   Capra sp. (prisca?)
Leo spelaea   Ovibos moschatus
Mammuthus primigenius   Spirocerus kjachtensis
Lepus tanaiticus   Poephagus baicalensis
Dicerorhinus tichorhinus

Nevertheless, many Pleistocene species withstood the increasing influence of man and passed into the Holocene with ecological and morphological changes. They include, for instance, cosmopolitan carnivora and some ungulates:

Canis lupus	*Capreolus*
Vulpes vulpes	*Cervus elaphus*
Ursus arctos	

Some game animals, having survived the Pleistocene, changed their locale completely in the Holocene. These changes are rather complicated, because some of them occurred passively under the influence of terrain changes, whereas others were carried out under the influence of active physiological and morphological adaptations of species (Vereshchagin, 1963).

Such animals as arctic fox, glutton, beaver, and reindeer became extinct in the south but were quickly acclimatized in the north, in the zone of taiga and partly of tundra:

Vulpes lagopus	*Castor fiber*
Gulo gulo	*Rangifer tarandus*

European hare, boar, and lynx from the zone of southern mountains and forest–steppes ranged farther to the north:

Lepus europaeus
Felis lynx
Sus scropha

The distribution of steppe inhabitants that were widespread in the Pleistocene (Paleolithic)—horse, bison, primitive tur, and red deer—was sharply reduced. These animals were represented by new forest and steppe forms:

Equus caballus	*Bison priscus, B. bonasus*
Bos primigenius	*Cervus elaphus*

Corsac fox, camel, saiga, and sheep receded into steppes, plateaus, and mountains of central Asia:

Vulpes corsac	*Poephagus baikalensis,*
Camelus knoblochi, C.	*P. gruniens*
bactrianus	*Ovis gmelini, O. cicloceros*
Saiga borealis, S. tatarica	*Ovis ammon*

In the Holocene, the southern limits of the U.S.S.R., the Caucasus, Central Asia, and Ussuriland were inhabited by new species of south-Asiatic origin: jackal, raccoon dog, striped hyena, lion, tiger, wild ass, axis deer, and goitered gazelle:

Canis aureus	*Panthera tigris*
Nyctereutes procyonoides	*Equus hemionus*
Hyaena hyaena	*Cervus nippon*
Leo leo	*Gazella subgutturosa*

Not only the distribution but also the number of large animals changed greatly. Populations of brown bear, beaver, boar, lynx, blue hare, roe, and especially of moose increased in the Holocene to such an extent that

Ursus arctos	*Sus scrofa*
Castor fiber	*Capreolus capreolus, C.*
Felis lynx	*pygargus*
Lepus timidus	*Alces alces*

together with Anseriformes and Tetraonidae (*Tetrao urogallus* L. and *Tetrao tetrix* L.), Pisces, Pinnipedia, and Cetacea, could provide the subsistence for Neolithic and later tribes over vast territories of northern Asia.

CAUSES OF EXTINCTION

In reviewing changes within the megafauna, we found that the nature and cause of its disappearance were considerably more complicated than we formerly believed. A major point is that different reasons apply to different territories, and extinction cannot be explained by anthropogenic influence alone.

Primitive man played, as a rule, an auxiliary or complementary role in the destruction and change of the complex of large animals in the Pleistocene. Man's influence upon the mortality of mammoth, horse, bison, and saiga in the Russian Plain was far greater than it was over vast territories of Siberia. In the Pleistocene, in the north of Yakuti an influence of primitive man upon animals has yet to be proved. The accumulations of mammoth bones and carcasses of mammoth, rhinoceros, and bison found in frozen ground in Indigirka, Kolyma, and Novosibirsk islands bear no trace of hunting or activity of primitive man. Here large herbivorous animals perished and became extinct because of climatic and geomorphic changes, especially changes in the regime of winter snow and increase in depth of snow cover.

TABLE 2. The Animals hunted by man in the Paleolithic, Neolithic, and Bronze Age in the Soviet Union

P = Paleolithic, N = Neolithic and the beginning of the Bronze Age. For the Neolithic and later cultures the author has used his own unpublished data and those of Zalkin (1956, 1960, 1962) and Bibikova (1953). †, Species entirely extinct; †ₚ, species which became extinct in a given area, or species partially surviving or domesticated.

Species	Russian plain	Crimea	Caucasus	Urals and western Siberia	Kazakhstan and Altai	Middle Asia	East Siberia	Ussuriland
Primates								
†ₚ *Macaca* sp.; macaque			P					
Carnivora								
Canis lupus Linnaeus; gray wolf	PN	PN	PN	PN	PN	PN	PN	PN
Canis sp.	N	N	N					
Cuon sp.; red dog			P				N	
Vulpes vulpes L.; red fox	PN	PN	PN	PN	PN	PN	PN	PN
Vulpes corsac L.; corsac fox	PN	PN	P		N		N	
Vulpes lagopus L.; arctic fox	PN	P		PN			P	
Nyctereutes procyonoides Schrenck; raccoon-dog								N
Ursus arctos L.; big brown bear	PN	PN	PN	PN	PN	PN		PN
† *Ursus spelaeus* Rosenmüller; cave bear	PN	P	PN	P				
Selenarctos tibetanus G. Cuvier; black bear								N
† *Crocuta spelaea* Goldfuss; cave hyena	P	P	P	P	P	P	P	P
† *Leo spelaea* Goldfuss; cave lion	P	P	P	P		P	P	P
†ₚ *Leo leo* L.; lion	N		N?					
Uncia uncia Schreber; snow leopard						P		
Panthera pardus L.; leopard	N	N	PN			PN		PN
Panthera tigris L.; tiger			N?			N?	N?	N
Felis silvestris Schreber; wild cat	N	PN	PN					
Felis libyca Forster; spotted cat	PN	P				PN		
Felix lynx L.; lynx	PN	P	PN	N			N	
Meles meles L.; badger	PN	PN	PN	N				PN
Gulo gulo L.; wolverine	PN	P	P	PN			P	P
Martes martes L.; pine marten	N		P	N				
Martes foina Erxleben; stone marten		PN	PN			P		
Martes zibellina L.; sable				PN			PN	N
Mustela eversmanni Lesson; polecat	PN	PN	P	P				

TABLE 2.—continued

Species	Russian plain	Crimea	Caucasus	Urals and western Siberia	Kazakhstan and Altai	Middle Asia	East Siberia	Ussuriland
Mustela putorius L.; European polecat	P							
Lutra lutra L.; common otter	P		PN	PN				
Pinnipedia								
†ₚ *Phoca groenlandica* Erxleben: harp seal (on Baltic Sea)	N							
Phoca vitulina L., common seal	N							
Phoca caspica Gmelin; Caspian seal			N			N		
Cetacea								
Tursiops truncatus Montague; bottlenosed dolphin	N							
Delphinus delphis L.; common dolphin	N		N					
Proboscidae								
† *Mammuthus primigenius* Blümenbach; mammoth	P	P	P	P	P		P	P
Lagomorpha								
Lepus europaeus Pallas; European hare	P	PN	PN					
Lepus timidus L.; blue hare	N			N			PN	
† *Lepus tanaiticus* Gureev; Tanais hare	P	P		P				
Lepus tolai Pallas; Tolai hare						PN		
Lepus sp.							P	N
Ochotona pusilla Pallas; steppe pika	PN	PN		PN				
Rodentia								
Hystrix leucura Syxes; Indian porcupine			PN			PN		
Castor fiber L.; beaver	PN	PN	PN	PN				
Marmota bobac Müller; Bobak marmot	PN	PN	PN	PN	PN			
Marmota sp.							PN	
Citellus rufescens Keyserling et Blasius; red-cheeked souslik	P	P						
Citellus pygmaeus Pallas; little souslik	P	P	N					
Perissodactyla								
† *Dicerorhinus tichorhinus* Fischer; woolly rhinoceros	P	P	P	P	P		P	

TABLE 2.—continued

Species	Russian plain	Crimea	Caucasus	Urals and western Siberia	Kazakhstan and Altai	Middle Asia	East Siberia	Ussuriland
† *Rhinoceros* sp.			P					P
†ₚ *Equus caballus* (*fossilis*) L.; tarpan	PN	P	P	P	PN	PN	P	P
† *Equus hidruntinus* Regalia; Pleistocene ass	P	P	P			P		
Equus hemionus Pallas; wild ass	N	N	N		PN	N	PN	
Artiodactyla								
Sus scrofa L.; wild boar	PN	PN	PN			P		PN
†ₚ *Camelus knoblochi* Poljakov; wild camel					N	P	P	
Moschus moschiferus L.; musk deer							N	PN
† *Megaceros euryceros* Aldrovandi; great-horn deer	P	P	P					
Cervus elaphus L. (s. lato);	PN	PN	PN	P	P	PN	PN	PN
Rangifer tarandus L.; reindeer	P	P		PN			PN	
Cervus nippon Temminck; Sika deer								N
Capreolus capreolus L.; roe	PN	PN	PN					PN
Capreolus pygargus Pallas; Siberian roe	PN			PN	PN	PN	PN	N
Alces alces L.; moose	PN		PN	PN			PN	PN
Saiga tatarica L.; saiga	P	P	P	P	N		P	
Rupicapra rupicapra Gray; chamois			PN					
Capra caucasica Güldenschtaedt; Caucasian goat			PN					
Capra sibirica Pallas; Siberian goat						PN	PN	
† *Capra* sp.	P	P						
Ovis orientalis Gmelin; Asiatic mouflon	P	P	PN			P		
Ovis ammon L.; Argali			P		PN		P	
†ₚ *Ovis* cf. *gmelini* Blyth; Gmelini mouflon			PN					
† *Ovis* sp.	P	P	PN					
†ₚ *Ovibos moschatus* Zimmerman; musk-ox	P				P		N	
†ₚ *Bison priscus* Bojanus (s. lato); bison	PN	P	PN	P	P	P	PN	P
†ₚ *Bos primigenius* Bojanus; European tur	PN		PN			PN		
† *Spirocerus kjachtensis* Pavlova; Kijachta antelope					P		P	
†ₚ *Poephagus baikalensis* Vereshchagin; wild yak					P		P	

Radiocarbon analysis of bone remains and charcoal in the Upper Paleolithic camps of the Russian Plain has shown that the latest Paleolithic camps with bones of mammoth are 9,000–14,000 years old.

Meanwhile, the latest remains of mammoth from the Arctic zone (Taimir peninsula) are also dated as 11,500 years old. Most carcasses and soft tissues of mammoth found in the frozen ground (Sanga-juriakh, Berezovka, Mokhovaja, the delta of the Lena River, and the Verchnaja Gyda) are dated now as 29,000 to 44,000 years old (see Garutt, 1965). Hence it follows that, in the Arctic, mammoth died and were preserved without the intervention of man, not only on the boundary of the Pleistocene and the Holocene but also within the whole late-Pleistocene (Würm).

Large, active, carrion-feeding carnivores (cave lions and hyenas) disappeared after the extinction of herbivores or somewhat earlier. It is more difficult to explain the extinction of cave bear. According to our observations, in the northern Urals the populations of these animals became extinct owing to changes in the flood regime: in early spring many animals were drowned in caves by rivers swollen with thawed snow. The predatory and competitive roles of primitive hunters in the extinction of this animal were also very significant.

The descriptions of layers containing bones, skeletons, and carcasses of mammoth, rhinoceros, bison, and horse in the basins of the Indigirka, Vilui, Jana, and Kolyma rivers suggest that the animals died in winter, generally in great numbers and thus catastrophically.

The corpses of herbivores were swept away with the floods into depressions. In summer these carcasses formed in boggy areas the so-called "mammoth horizon," a thick layer consisting of bones, skulls, tusks, peat, and tree trunks interlocked by permafrost. Such a horizon is sometimes exposed by river erosion for a distance of many kilometers and is characterized by a putrid smell. The detailed study of such bone accumulations could confirm or reject the conclusions on the nature of mortality of Pleistocene animals (mammoth and bison in particular) in northeastern Siberia.

The great importance of snow cover in the life of mammals and birds was proved by investigations undertaken by Formozov (1962) and Nasimovich (1955). Not low temperatures alone but cold weather accompanied by snow and ice-encrusted ground cause mass hunger and epizooty of wild and domestic ungulates in steppes of Kazakhstan and deserts of Central Asia today (Sludsky, 1963).

Decisive in the life of small and large mammals, according to Formozov (1962), is the presence or absence of ice crust. Compact crusts are formed when changeable winter weather prevails, with alternating snowfalls, thaws, and freezes. The ice excludes air from small rodents, cuts the feet of ungulates, and deprives horses, musk-oxen and reindeer of their food. Thus climatic changes, in particular winter thaws and winds without a considerable increase of snowfall, could have a fatal significance for many Pleistocene animals.

According to investigations of Gussev (1956), who described the Quaternary deposits of northwestern Siberia, the accidental mortality and preservation of carcasses of mammoth and rhinoceros might take place only in "baidzherakh" terrain, which has layers of fossil ice eroded deeply at places by summer streams. The beds of such glacial streams, covered from above with a crust of ice and with a thin layer of silt, were on occasion insidious traps for heavy animals. Moreover, beyond all question these rare ideal instances can be related to the "mammoth level." General geologic sections (Gussev, 1956) show that the beds of silty streams eroded into the continental ice were formed at the same level as layers of "bone-bearing or lake bog deposits" (Fig. 15).

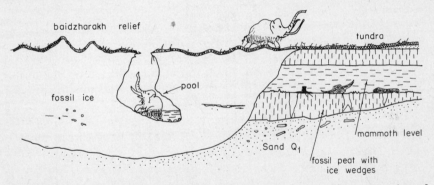

FIG. 15. Death and preservation of the mammoth in the secondary excavation of fossil ice.

Apart from the snow factor and the development of vast spring floods, the development of dense, dark coniferous forests in post-glacial time deprived ungulate animals of winter and spring food. In north-eastern Siberia the marine transgression could have exerted a fatal influence.

Instead of east-European and Siberian plains, vast areas of which were suitable in the Pleistocene for supporting steppe populations of horse, bison, and saiga, in the Holocene there remained only belts of forest steppe, steppe, and semidesert stretching to the south from the taiga zone. However, it was just here (from the Carpathians to Man-churia) that herd ungulates began to feel the pressure of mounted tribes

394

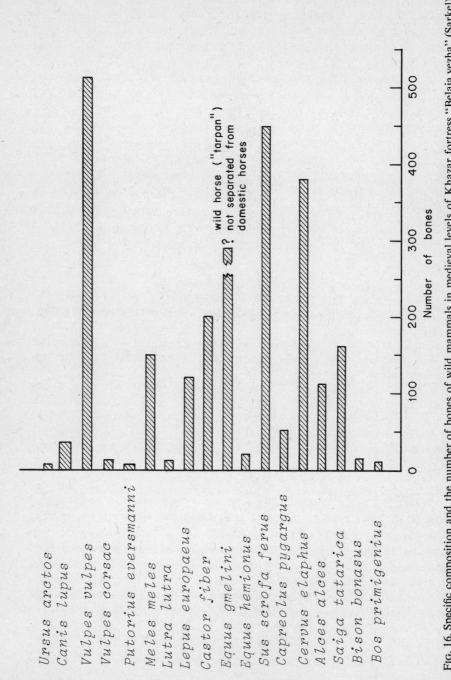

Fig. 16. Specific composition and the number of bones of wild mammals in medieval levels of Khazar fortress "Belaja vezha" (Sarkel) on the Don River; in the steppe zone.

of steppe nomads. Within the Holocene this pressure gradually increased, and at the time of the great Hun and Tartar–Mongolian campaigns, it apparently reached its culmination. Some historical data are now available (Kirikov, 1959) on the hunting of nomadic tribes in the Middle Ages (Fig. 16).

It is to be noted that, after great but short-term migration of nomads, the populations of large animals in steppe and forest–steppe did not disappear but relatively quickly returned to their former numbers. In the southern part of eastern Europe the abundance of steppe and forest animals was maintained throughout the Middle Ages.

Only the extraordinary increase of the human population, the development of cattle-breeding, and cultivation of the steppes in the nineteenth century brought the destruction of the Holocene assemblage of large animals in the Russian steppes.

References

* Not in Russian

Abramova, A. Z., 1962, Paleolithic art on the territory of the U.S.S.R., p. 1–84, *in Archaeology of the U.S.S.R.*, A4–3: Moscow–Leningrad, U.S.S.R., Inst. Archaeology

Bader, O. N., 1941, Petroglyphs of the stone grave, p. 297–322, *in Paleolithic and Neolithic of the Ukraine*, vol. *I:* Kiev, Acad. Sci. Ukr. S.S.R., Inst. Archaeology

—— 1954, Sacrifice place under rock pictures on the Vishera river: *Soviet Archaeol.*, v. *21*, p. 241–58

—— 1963, Paleolithic drawings of Kapova Cave (Shulgan-Tash) in the Urals: *Soviet Archaeol.*, no. *1*, p. 125–34

Bibikova, V. I., 1953, The fauna of the early tripol settlement Luka-Vrublevetzkaja: Moscow, Acad. Sci. U.S.S.R., *Materials and Investigations of Archaeology of the U.S.S.R.*, no. *38*, p. 411–58

—— 1958, Some notes on the fauna from the Mousterian cave Aman-Kutan I: *Soviet Archaeol.*, no. *3*, p. 229–33

Boriskovsky, P. I., 1956, Excavations of Paleolithic dwellings and burial place in Kostionki II in 1953: *Soviet Archaeol.*, v. *25*, p. 173–88

Chersky, I., 1891, A description of the collections of post-Tertiary mammals collected by Novosibirsk expedition in 1885–1886: Appendix to v. *65*, *Notes of the Acad. Sci.*, v. *I*, p. 1–706

David, A. I., 1961, The remains of mammals from excavations of Paleolithic site "Starye Druitory": Acad. Sci. U.S.S.R., Moldavian Branch, *Izvest.*, no. *3* (81), p. 53–60

Formozov, A. N., 1962, The study of ecology of mammals under the conditions of snowy and frosty winters of the north of Eurasia, p. 102–11, *in*

Symposium Theriologicum: Praha, Acad. Sci., Int. Symp. Methods Mammallogical Invest. (Brno, 1960), Proc.

Garutt, V. E., 1965, Fossil elephants of Siberia: *Sci. Res. Inst. Geol. Arctic, Proc.,* v. *143,* p. 106–30

Grach, A. D., 1957, Petroglyphs of Tuva. I: Acad. Sci. U.S.S.R., *Mus. Anthropol. Ethnog., Proc.,* v. *17,* p. 385–429

Gromov, V. I., 1948, Paleontological and archaeological explanations of the stratigraphy of Quaternary deposits in the U.S.S.R. (Mammals, Paleolithic): Moscow, *Inst. Geol. Sci., Proc.,* v. *64,* Geol. ser., v. *17,* p. 1–520

Gromova, Vera, 1949, Pleistocene fauna of mammals from the Teshic-Tash Cave: *Moscow State Univ. Inst. Anthropology, Proc.,* p. 87–99

Gusev, A. I., 1956, The mammoth level. Materials on the Quaternary geology and geomorphology of the U.S.S.R., n.s., *I:* Leningrad, *All-Union Geol. Inst.,* p. 169–77

Guslitzer, B. I., and Kanivetz, V. I., 1965, *Caves of Pechora Ural:* Moscow, Nauka, 133 p.

Kirikov, S. V., 1959–1960, Changes in the animal world in natural zones of the U.S.S.R. (XIII–XIX). *Steppe zone and forest steppe* (1959), 174 p.; *Forest zone and forest tundra* (1960), 155 p.: Moscow, Acad. Sci. U.S.S.R., Inst. Geog.

Kornietz, N. L., 1961, On the reasons for mortality of the mammoth on the territory of the Ukraine: Kiev, Acad. Sci. Ukr. S.S.R., Inst. Zoology, Ph.D. Thesis, p. 1–17

Kühn, H., 1956, *The rock pictures of Europe:* London, Sidgwick and Jackson, 226 p.*

Lindner, K., 1937, *Die Jagd der Vorzeit. I:* Berlin and Leipzig, Walter de Gruyter, 435 p.*

Marikovsky, P. I., 1953, Methods and objects of hunting after the cave drawings of Chulak Mountains (Kazakhstan): *Zool. Jour.* v. *32,* p. 1064–74

Markovin, V. I., 1958, Cave pictures in foothills of northeastern Dagestan: *Soviet Archaeol.,* no. *1,* p. 147–62

Mowat, R., 1963, *Folk of the caribou land:* Moscow, Ed. Foreign Literature, 313 p.

Nasimovich, A. A., 1955, *The role of regime of the snow cover in the life of ungulate animals in the U.S.S.R.:* Moscow, Acad. Sci. U.S.S.R., Inst. Geog., 402 p.

Okladnikov, A. P., 1959, *Shishkinskaja rock pictures:* Irkutsk, Ed. Irkutskoe kniznoe izdatelstvo, 209 p.

Panitschkina, M. Z., 1953, Exploration of the Paleolithic on the Middle Volga: *Soviet Archaeol.,* v. *18,* p. 233–64

Pidoplichko, I. G., 1941, Late Paleolithic site Novgorod-Severskij, p. 65–106, *in Paleolithic and Neolithic of the Ukraine,* vol. I: Kiev. Acad. Sci. Ukr. S.S.R., Inst. Archaeology

———— 1953, Amvrosievskaja Paleolithic site and its peculiarities: Kiev, Acad. Sci. Ukr. S.S.R., Inst. Archaeology, Brief Repts., v. *2*

—— 1954, On the Ice Age: Kiev. Acad. Sci. Ukr. *S.S.R.*, *Inst. Zoology Publ.*, v. *3*, 217 p.

—— 1959, New data on the fauna of the Mesin site: Kiev, Acad. Sci. Ukr. S.S.R., Inst. Archaeology, Brief Repts., v. *8*, p. 104–08

Ravdonikas, V. I., 1936–1938, Rock pictures of Onega Lake and the White Sea: *Acad. Sci. Ukr. S.S.R., Inst. Anthropology, Archaeology, and Ethnography, Archaeol. ser.*, v. *9*, no. *1*, 1936, 205 p.; v. *10*, no. *2*, 1938, 168 p.

Rogachev, A. N., 1955, Kostionki IV—The site of Old Stone Age near the village of Kostionki on the Don River: Moscow, Acad. Sci. U.S.S.R., *Materials and Investigations on Archaeology of the U.S.S.R.*, no. *45*, 163 p.

—— 1957, Multilevel sites of Kostionki-Borshevsk region on the Don River and the problem of the development of culture in the Upper Paleolithic on the Russian Plain: Acad. Sci. U.S.S.R., *Materials and Investigations on Archaeology of the U.S.S.R.*, no. *59*, p. 9–134

Rygdylon, E. R., 1955, Notes on rock pictures near the village of Matkechika on Abakan: All-Union Geog. Society, *Izvest.*, v. *87*, p. 558–60

Shovkopljas, I. G., 1965, *The Mesin site:* Kiev, Ed. Naukova Dumka, 325 p.

Skalon, V. N., 1956, Reindeer stones of Mongolia and the problem of the origin of reindeer breeding: *Soviet Archaeol.*, v. *25*, p. 87–105

Sludsky, A. A., 1963, "Dzhuty" (starvation of animals) in European–Asiatic steppes and deserts: Acad. Sci. Kazakh. S.S.R., *Inst. Zoology, Proc.*, v. *20*, p. 1–88

Sokolov, V. E., 1964, On the former areas of ungulates of Kazakhstan according to rock pictures: *Moscow Natur. Society, Bull.*, v. *69*, no. *1*, p. 113–17

Tasnadi-Kubacska, A., 1962, *Paläopathologie. Pathologie der vorzeitlilichen Tiere:* Jena, Fischer, 269 p.*

Tchernysh, A. P., 1959, Late Paleolithic of Middle Pridnestrovje: Moscow, Acad. Sci. U.S.S.R., *Comm. on the Study of the Quaternary Period*, v. *15*, 214 p.

Vangengeym, E. M., 1961, Paleontological explanation of the stratigraphy of the anthropogenic deposits in northeastern Siberia: Moscow, Acad. Sci. U.S.S.R., *Inst. Geology, Proc.*, v. *48*, p. 1–182

Vasnetzov, V. M., 1956, *Stone age:* Moscow, Goskultprosvetizdat, p. 1–13

Vereshchagin, N. K., 1953, Great *"burial-grounds" of animals in the valleys of the rivers of the Russian Plain: Priroda*, no. *12*, p. 60–65

—— 1956, On the former distribution of some ungulates in the region of junction of European Kazakhstan and central Asiatic steppes: *Zool. Jour.*, v. *35*, no. *10*, p. 1541–53

—— 1957, The remains of mammals from lower Quaternary deposits of the Taman peninsula: *Acad. Sci. U.S.S.R., Inst. Zoology, Proc.*, v. *22*, p. 9–74

—— 1959, *Mammals of the Caucasus. The history of the fauna:* Moscow–Leningrad, Acad. Sci. U.S.S.R., Inst. Zoology, 703 p.

—— 1961, On the typology of burial places of the remains of terrestrial vertebrates in Quaternary deposits: *Moscow, Acad. Sci. U.S.S.R., Dept. Geol.Geog. and Comm. Quaternary Research, Materials of the Symposium on the Quaternary Period*, v. *I*, p. 377–87

———— 1963, The main features of the formation of the Holarctic theriofauna in the Anthropogene: *Zool. Jour.*, v. *42*, no. *11*, p. 1686–98

Vereshchagin, N. K., and Burchak-Abramovich, N. O., 1948, Drawings on the rocks of southeastern Kabristan: *Izvestija of the State Geog. Soc.*, v. *80*, p. 507–18

Vereshchagin, N. K., and Kolbutov, A. D., 1957, The remains of animals in a Mousterian site near Stalingrad, and the stratigraphic position of the Paleolithic level: Acad. Sci. U.S.S.R., *Inst. Zoology, Proc.*, v. *22*, p. 75–89

Vereshchagin, N. K., and Kuzmina, I. E., 1962, *Excavations in the caves of northern Ural:* Moscow, *Priroda*, no. *3:* p. 76–78

Zalkin, V. I., 1956, *Materials on the history of cattle breeding and hunting in Ancient Russia:* Moscow, Acad. Sci. U.S.S.R., Materials and Investigations of Archaeology of the U.S.S.R., no. *51*, 183 p.

———— 1960, *The history of cattle breeding in northern Prichernomorje:* Moscow, Acad. Sci. U.S.S.R., Materials and Investigations of Archaeology of the U.S.S.R., no. *53*, 109 p.

———— 1962, *On the history of cattle breeding and hunting in eastern Europe:* Moscow, Acad. Sci. U.S.S.R., Materials and Investigations on Archaeology of the U.S.S.R., no. *104*, 130 p.

Zamjatnin, S. N., 1960, Some problems of the study of economy in Paleolithic: Leningrad, Acad. Sci. U.S.S.R.. *Inst. Ethnography (Miklukho-Maklai), Proc.*, n.s., v. *54*, 80–108

D. A. HOOIJER

Rijksmuseum van Natuurlijke Historie
Leiden, Netherlands

PLEISTOCENE VERTEBRATES
OF THE NETHERLANDS ANTILLES

Abstract

Recent discoveries of vertebrate remains in cave and fissure deposits in late-Quaternary limestones in Curaçao and Aruba include the extinct muskrat Megalomys curazensis Hooijer; *this Antillean genus of cricetines became extinct (on Martinique) at the beginning of the present century. Fossil teeth of* Sigmodon (*cotton rat*) *and* Zygodontomys (*cane rat*) *have been found in Aruba. The capybara,* Hydrochoerus hydrochaeris (*Linnaeus*), *found fossil in Curaçao, must be considered endemic on the island, pointing to a climate moister than that prevailing at present. A small ground sloth,* Paulocnus petrifactus Hooijer, *found on the Tafelberg Santa Barbara, Curaçao, belongs to the group of late-Pleistocene and early-historic megalonychids of Cuba, Hispaniola, and Puerto Rico but differs more from these northern Antillean forms than they differ among themselves. The skeletal elements were found associated, and a snout even has teeth in occlusion, indicating mummification prior to fossilization, and hence a dry climate at the time of its deposition. Remains of gigantic land tortoises* (Geochelone) *in Curaçao are the first found in the southern Antilles.*

Until about ten years ago, the Netherlands Antilles Aruba, Curaçao, and Bonaire were a blank as far as fossil vertebrates are concerned. Then P. H. de Buisonjé of the Geological Institute of Amsterdam University found large numbers of cricetine teeth embedded in phosphatic cave fillings in a marine limestone at Devil's Bluff, eastern Curaçao, 50 m above the present sea level. The material proved to be very fragmentary, consisting mainly of isolated teeth and small portions of bone; this was described in 1959 as belonging to a large extinct muskrat, *Megalomys curazensis* Hooijer (1959). The genus *Megalomys*, largest of the New World mice (Cricetinae), is exclusively insular, occurring in the fossil state in the Lesser Antilles Barbuda and Curaçao, and found living only in Martinique and Santa Lucia, whence some specimens reached the museums of London, Paris, and Leiden. The extermination of *Megalomys* was brought about at the beginning of the present century. The Curaçao form is very close dentally to the Martinique and Santa Lucia forms, indicating that the genus extended along the island arc for a distance of at least some 750 miles in the past. The origin of *Megalomys* is obscure, and, although it is clearly an oryzomyine rodent, no such large forms have yet been found on the South or North American continent. Recently the genus has been reported from the Galápagos (Niethammer, 1964).

Subsequent explorations by Mr. de Buisonjé in the Netherlands Antilles have shown that *Megalomys* occurs at an altitude of 40–65 m above sea level, in at least seven localities on Curaçao, and in Aruba, to the west of Curaçao. Unlike Curaçao Aruba is separated from the coast of South America by a shallow sea that probably withdrew during Pleistocene low sea-level stages (Koopman, 1958). It is of considerable interest to note that the Aruba *Megalomys* (a tooth from Seru Canashito is shown in Plate I, Fig. 4) is rather small, just at the lower limit of variation in size of the Curaçao type.

The island of Aruba proved to have a richer rodent fauna than either Curaçao or Bonaire, for jaws and teeth of two further cricetines were found in deposits at Seru Canashito and Isla: *Zygodontomys* cf. *brevicauda* (Allen and Chapman), the cane rat (Plate I, Fig. 1), and *Sigmodon* cf. *hispidus* (Say and Ord), the cotton rat (Figs. 2, 3). The remains differ in no way from recent specimens which are still living in Venezuela, Colombia, and Panama, *Sigmodon* in the southern United States, and *Zygodontomys* in Trinidad. If the species to which the fossil teeth have been provisionally assigned made their entry into the island some time during a glacial stage of low sea level, as seems most likely, there has not been sufficient time for the island populations to differentiate from the parent stock to a degree that shows up in the dentition. However, the external characters, which are forever lost to us, may have been different from the mainland forms.

Neither *Sigmodon* nor *Zygodontomys* has thus far been found in the fossil collections from Curaçao. A sub-Recent *Oryzomys*, representative of the *concolor* group (subgenus *Oecomys*: Hershkovitz, 1960), has been found in the cave of Hato and in one at plantation Noordkant, Curaçao, in association with the post-Columbian *Rattus* that led to its extermination on the island.

The island of Bonaire has thus far provided fossil material of a cricetine very similar to *Thomasomys*, to which genus it may or may not belong. There are no other species of cricetines represented in the samples from Bonaire, and nothing like "*Thomasomys*" has yet been found in

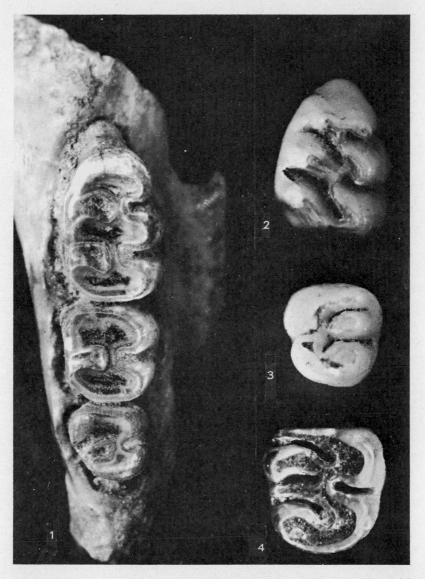

FIG. 1. *Zygodontomys* cf. *brevicauda* (Allen and Chapman), right maxillary with M¹⁻³ from Pleistocene cave deposit at northeast side of Seru Canashito, Aruba; elevation 50 m; × 18.
FIGS. 2, 3. *Sigmodon* cf. *hispidus* (Say and Ord). Fig. 2: right M¹; Fig. 3: left M³. From Pleistocene fissure-filling at east point of Isla, Aruba; elevation 50 m; × 18.
FIG. 4. *Megalomys curazensis* Hooijer, right M² from Pleistocene cave deposit at northwest side of Seru Canashito, Aruba; elevation 45 m; × 18.

Aruba or Curaçao. Although we have more material than in 1959, the generic assignment of the Bonaire form is still uncertain. It is, at any rate, a species that has the basic dental pattern of present-day *Thomasomys*, and in this connection it is noteworthy that in its complex glans penis and single-chambered stomach this form would resemble the postulated ancestral cricetine type (Hershkovitz, 1962, p. 21). The Bonaire form may be an offshoot from the original stock that formed a relict on the island. The genus is badly in need of revision.

Before leaving the subject of the cricetines a few words may be said about the mode of origin of the microfauna in these cave deposits. The fauna usually is quite monotonous; more than one species of cricetine is hardly ever present. The deposits are very rich; more than 150 molars of *Megalomys* have been extracted from one cubic decimeter at Seru Mainshi (de Buisonjé, 1964, p. 75). It would seem clear that the material has been derived from owl pellets, as is the case in the australo-pithecine caves of South Africa (de Graaff, 1961, p. 61), which locally abound in remains of small animals. Barn owls that roost on ledges in a cave cause a considerable accumulation of regurgitated pellets below the roost; the disintegrating pellets mix with erosional debris from the cave and are bound by calcium carbonate. The owl responsible for the *Megalomys* deposits has not yet been found. The fact that worn last molars are far from rare in the *Megalomys* samples indicates that adults were eaten. Their body length was probably as much as 30 cm. In the pellets of the living race of barn owl of Curaçao, *Tyto alba bargei* (Hartert), we find remains only of much smaller animals, sub-adults of *Rattus rattus* (Linnaeus) with an adult body length averaging 20 cm (Husson, 1960, p. 96). The presence of an owl much larger than the living island race at the time of formation of the rodent deposits in Curaçao may thus be postulated. Such owls, considerably larger than *Tyto alba bargei*, now inhabit Venezuela and Columbia (Voous, 1957, p. 180), and have been recorded, e.g., from cave deposits in Haiti (*Tyto ostologa* Wetmore, 1922) and the Bahamas (*Tyto pollens* Wetmore, 1937). The largest owl of the world, *Ornimegalonyx* Arredondo (1961, p. 20), has been found in Cuba, but it is still undescribed.

A rather surprising find in Curaçao was a set of teeth of a juvenile individual of the capybara, *Hydrochoerus hydrochaeris* (Linnaeus), indistinguishable from the recent form, never before recorded from the Antilles. The specimens were found by H. J. MacGillavry and P. H. de Buisonjé in a cave deposit at an altitude of 160 m above sea level on Tafelberg Santa Barbara. The capybara is an animal rather partial to water, living in the Guiana lowlands wherever there is a river or lake with sufficient succulent vegetation. It is much too dry in Curaçao for a capybara to live there today, and, in the absence of geological evidence for a wetter climate in Curaçao at the time, I postulated (Hooijer, 1959, p. 24) that the juvenile capybara teeth might have belonged to a stranded individual that somehow, by floating or swimming, or in native boats, had made the crossing from Venezuela. Further explorations by Mr. de Buisonjé have, however, left no doubt that the *Hydrochoerus* deposit was formed during a wet period; several teeth are embedded in cave pearls (formed in not quite stagnant pools of water supersaturated with carbonate of lime: Hess, 1929), and there are flowstones (de Buisonjé, 1964, p. 76) indicating wet climatic conditions. It is therefore evident that the capybara should be considered endemic on the island, and that it had found an agreeable habitat there at some time.

In 1961, during blasting operations in the top levels on Tafelberg Santa Barbara, a great many fossilized bones of a small ground sloth were found at a level 150 m above sea level. Several tons of rock were saved by Mr. Paul Stuiver of the Mijnmaatschappij Curaçao at Newport and shipped to Holland. When treated with acetic acid by Paul de Buisonjé, the most promising blocks proved to contain a skull and many limb bones still articulated, which were described as *Paulocnus petrifactus* Hooijer (1962). Ground sloths of a peculiar type have been known from Cuba (Matthew and Paula Couto, 1959), Hispaniola (Miller, 1929; Hoffstetter, 1955; Hooijer and Ray, 1964), and Puerto Rico (Anthony, 1918, 1926), and comprise a number of genera, *Megalocnus*, *Mesocnus*, *Microcnus*, and *Acratocnus*. These sloths range in size from that of a sheep to that of a woodchuck. In skeletal characters they are very similar to the Miocene Santacruzean ground sloths like *Hapalops* or *Eucholoeops*, and they are supposed to have been derived from such forms (Simpson, 1956), spreading in the Antilles from South or Central America in the Miocene or early Pliocene at the latest. Although belonging to the group of northern Antillean sloths, *Paulocnus* differs more from these forms than they differ among themselves, as the study of a snout with the teeth in occlusion has shown (Hooijer, 1964). The most notable characteristics of the Curaçao ground sloth are: upper caniniform teeth curved, trigonal in section as in *Acratocnus* but with the internal face wider than the external; lower caniniform teeth straight, trigonal in section, with internal surface convex, not concave as in *Megalocnus*, *Acratocnus*, and *Mesocnus*; symphyseal tongue moderately elongated, shorter than in *Mesocnus* but longer than in *Acratocnus*; first cheek teeth subquadrate, but longer anteroposteriorly on the internal than on the external face, instead of the reverse as in *Megalocnus* and *Acratocnus*. Unlike *Acratocnus* the skull has no sagittal crest and the carpus and tarsus are unspecialized as in all the Antillean genera.

The condition of the *Paulocnus* deposit is quite different from that of the *Hydrochoerus* deposit, with shrinkage cracks instead of flowstones and cave pearls. The preservation of the remains in their natural position suggests mummification before fossilization set in, and thus points to a dry climate at the time the deposit was formed. Because the *Hydrochoerus* deposit is definitely later than the *Paulocnus* deposit, we have evidence for at least one climatic oscillation (dry–wet–dry), but it has not been possible to tie this in with the sequence of climatic changes deduced from other evidence (de Buisonjé, 1964, p. 76).

At various sites in Curaçao, bones have been found that belong to land tortoises of the genus *Geochelone* (Hooijer, 1963). These were gigantic forms, with carapace lengths of 60 to 80 cm, and specifically distinct from the Cuban *Geochelone cubensis* (Williams, 1950). Geographically, however, land tortoises are not very interesting, as they swim so well. One has been found by Simpson (1942) in deposits in Patagonia formed before the late-Tertiary land bridge existed; it must have reached South America overseas and not by any land connection.

Finally, mention should be made of a fossil fish tooth that came from a block of matrix of the Tafelberg Santa Barbara similar to those containing *Paulocnus* teeth and bones. The specimen is a central upper tooth of a file fish, *Balistes* spec., closely resembling recent specimens of that genus. File fishes of the genus *Balistes* occur in tropical and warm seas, browsing on the corals on which they subsist.

No means have yet been found for dating the various bone-bearing deposits in the Netherlands Antilles reviewed above. The faunal remains would seem to suggest a late-Pleistocene date; though now extinct on the islands, several of the rodents are little or no different from species still living in adjacent areas. The ground sloths of the Greater Antilles became extinct only after the advent of man, several millennia B.C. (Rouse, 1960). It is hoped that with the progress of our knowledge we shall learn when the various elements did spread to the islands, and whence they came.

References

Anthony, H. E., 1918, The indigenous land mammals of Porto Rico, living and extinct: *Amer. Mus. Nat. Hist. Mem.*, n.s. v. *2*, p. 329–435
———— 1926, Mammals of Porto Rico, living and extinct—Rodentia and Edentata: *N. Y. Acad. Sci., Scient. Survey Porto Rico and the Virgin Islands*, v. *9*, P. 2, p. 97–241

Arredondo, O., 1961, Descripciones preliminares de dos nuevos generos y especies de edentados del Pleistoceno Cubano: *Bol. Grupo Explor. Cient.*, v. *1*, p. 19–40

Buisonjé, P. H. de, 1964, Marine terraces and sub-aeric sediments on the Netherlands Leeward Islands, Curaçao, Aruba and Bonaire, as indications of Quaternary changes in sea level and climate: *Proc. Kon. Ned. Akad. Wet. Amsterdam*, ser. B, v. *67*, p. 60–79

Graaff, G. de, 1961, A preliminary investigation of the mammalian microfauna in Pleistocene deposits of caves in the Transvaal system: *Pal. africana*, v. 7, p. 59–118

Hershkovitz, P., 1960, Mammals of Northern Colombia, preliminary report no. 8: Arboreal rice rats, a systematic revision of the subgenus Oecomys, genus Oryzomys: *U.S. Nat. Mus. Proc.*, v. *110*, p. 513–68

—— 1962, Evolution of Neotropical cricetine rodents (Muridae) with special reference to the phyllotine group: Fieldiana: *Zoology*, v. *46*, 524 p.

Hess, F. L., 1929, Oölites or cave pearls in the Carlsbad Caverns: *U.S. Nat. Mus. Proc.*, v. *76*, art. 16, 5 p.

Hoffstetter, R., 1955, Un mégalonychidé (Edenté Gravigrade) fossile de Saint-Domingue (Ile d'Haiti): *Mus. Hist. Nat. Paris Bull.*, ser. 2, v. *27*, p. 100–04

Hooijer, D. A., 1959, Fossil rodents from Curaçao and Bonaire: *Studies fauna Curaçao*, v. 9, p. 1–27

—— 1962, A fossil ground sloth from Curaçao, Netherlands Antilles: *Proc. Kon. Ned Akad. v. Wet. Amsterdam*, ser. B, v. 65, p. 46–60

—— 1963, Geochelone from the Pleistocene of Curaçao, Netherlands Antilles: *Copeia*, 1963, no. *3*, p. 579–80

—— 1964, The snout of Paulocnus petrifactus (Mammalia, Edentata): *Zool. Med. Mus. Leiden*, v. *39*, p. 79–84

Hooijer, D. A., and Ray, C. E., 1964, A metapodial of Acratocnus (Edentata: Megalonychidae) from a cave in Hispaniola: *Proc. Biol. Soc. Washington*, v. *77*, p. 253–58

Husson, A. M., 1960, *De zoogdieren van de Nederlandse Antillen:* Fauna Nederlandse Antillen no. 2, The Hague (Nijhoff), 170 p.

Koopman, K. F., 1958, Land bridges and ecology in bat distribution on islands off the northern coast of South America: *Evolution*, v. *12*, p. 429–39

Matthew, W. D., and de Paula Couto, C., 1959, The Cuban Edentates: *Amer. Mus. Nat. Hist. Bull.*, v. *117*, p. 1–56

Miller, G. S., 1929, A second collection of mammals from caves near St. Michel, Haiti: *Smithson Inst. Misc. Coll.*, v. *81*, 30 p.

Neithammer, J., 1964, Contribution à la connaissance des mammifères terrestres de l'île Indéfatigable (=Santa Cruz), Galapagos: *Mammalia*, v. *28*, p. 593–606

Rouse, I., 1960, *The entry of Man into the West Indies:* Yale Univ. Publ. in Anthrop., no. *61*, 26 p.

Simpson, G. G., 1942, A Miocene tortoise from Patagonia: *Amer. Mus. Novit.*, no. *1209*, 6 p.

————— 1956, Zoogeography of West Indian Land Mammals: *Amer. Mus. Novit.*, no. *1759*, 28 p.

Voous, K. H., 1957, *The birds of Aruba, Curaçao, and Bonaire:* Studies on the fauna of Curaçao and other Caribbean islands, v. *7*, 1957 (Natuurwetenschappelijke Studiekring voor Suriname en de. Nederlandse Antillen. Uitgaven no. 14, 1957), The Hague, Nijhoff.

Wetmore, A., 1922, Remains of birds from caves in the Republic of Haiti: *Smithson. Inst. Misc. Coll.*, v. *74*, 4 p.

————— 1937, Bird remains from cave deposits on Great Exuma Island in the Bahamas: *Harvard Univ. Mus. Comp. Zool., Bull.* v. *80*, p. 427–41

Williams, E., 1950, Testudo cubensis and the evolution of Western Hemisphere tortoises: *Amer. Mus. Nat. Hist. Bull.*, v. *95*, p. 1–36

R. BATTISTINI and P. VÉRIN

Laboratoire de Géographie et Centre d'Archéologie
Université de Tananarive
Madagascar

ECOLOGIC CHANGES
IN PROTOHISTORIC MADAGASCAR

Abstract

A vast and underpopulated island, Madagascar is nevertheless a place where one is tempted to invoke the action of man in order to explain the otherwise enigmatic faunal change. Not only faunal extinction but also reduction of forests and the acceleration of erosion may be attributed to prehistoric man. The arrival of the first Malagasy and their itinerary are still unknown. However, it is possible on the basis of radiocarbon evidence to confirm the existence of a human population at least along the coasts by the end of the first millennium.

THE ROLE OF MAN IN THE TRANSFORMATION OF THE FLORA
AND THE SYSTEM OF EROSION

At present the forest occupies only the eastern slope of the island, a region characterized by an accentuated relief and a heavy rainfall (Fig. 1). Much reduced by burning, the tall humid forest persists in its primitive state only along the very sharp dropoff in the high valleys. The hills closer to the coasts bear a secondary forest (*savoka*). Its western limit also seems artificially determined, with only a few wooded patches remaining in the highlands. The western highlands are dominated by a prairie of grasses, with only a few more or less extensive wooded patches that have generally been preserved in privileged topographic locations (such as the eastern slopes of the Ankaratra), in the holy places (such as Ambohimanga), or in those places least suitable for agricultural or pastoral life (such as the other side of the calcareous or sandstone cuestas, or the karst of Antsingy). It is no longer a question of a large humid forest but indeed of different forest formations, passage having been made progressively toward the west and south to a drier and drier tropophile forest. The residual character of such wooded patches is in numerous cases indisputable.

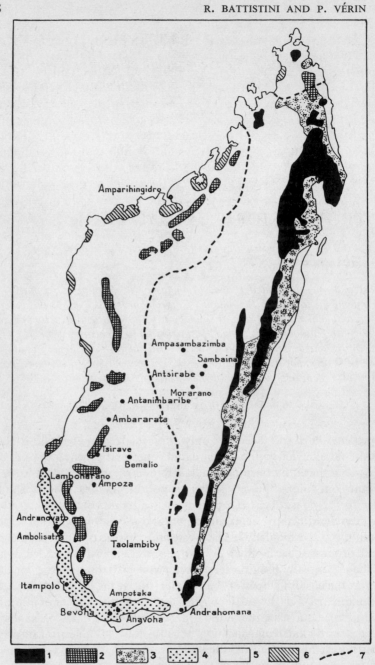

FIG. 1. Map of vegetation and locations of subfossil sites: 1, rain forests; 2, tropophile forest; 3, savoka; 4, bush; 5, herb savanna; 6, mangrove; 7, limit of windward and leeward flora.

Some of the most characteristic remains are those of the forest of Ambohitantely (literally, the mountain where there is honey) on the Ankazobe tampoketsa about 100 km north of Ambohitantely, at an altitude of 1,500–1,600 m. This forest covers about 2,000 hectares and is completed by about another 1,000 hectares of small, scattered forest patches that have been preserved especially in the heads of more humid small valleys on the edge of the plateau. This latter, a transitional area between the rainy forest of the eastern cliff region and the tropophile forest of the west and southwest, contains rather tall trees on the slopes and in the bottoms of the humid valleys; it passes to a far lower and clearer formation on the plateau, where the ground is hardened by a ferruginous shell. This transition, visible over a few hundred yards or even less, suggests the variety of primitive wooded formations that must have covered the highlands. All the intermediary passages would have existed between the true forest and simple, sparse coppices, in terms of the pedological conditions of humidity and relief. The character of the remains of these wooded areas of Ambohitantely has been explained well by Bastian (1964, p. 34), who describes

> the signs of the process of regressive evolution that translate the fragility, the weakness, the instability, and the rupture of the climatic equilibrium. Then, when the morphological conditions appear favorable to the development of a plant cover (the breaking apart of the old shield because of a resumption of erosion), the domain of the forest thins out rapidly and the specific characteristics of the formation are blurred before the invasion of the secondary species that have been substituted. The author recognizes the importance of the man-made erosion in this evolution. The hypothesis of a drier climatic cycle cannot be proved. The pedologists cannot find any traces of a recent shielding, even of recent formation of concretion on the newly deforested soils (Riquier, 1954). The value of the evidence and current observations must make us accept the decisive role of the intervention of man, especially by the practice of setting fire to wooded areas.

Another quite characteristic group of wooded relics is that of the Zomitsa wood and its annexes, east and north of Sakahara, in the southwestern part of the island. This is a tropophile formation of a tendency markedly drier than that of Ambohitantely, and including baobabs in particular and large Euphorbiae. The countryside contains rounded ridges, rather soft and quite extensive, in very thick sandy soils weathered from Isalo 3. Nothing in topography, pedology, or even hydrology explains why the forest has been preserved there. Every year grass fires come and lick its rim and thus contribute to a progressive reduction of the forest patches, whose residual character is certain. Here again a notable difference may be seen between the aspect of the vegetation in the bottom of the more humid talwegs, on the slopes, and on top of the hills where the tall trees vanish as the coppice invades.

Forest remains such as these from Ambohitantely or Zomitsa type are destined to disappear in the very near future, unless effective measures of conservations are taken immediately; otherwise, they will become grassland dotted with ronier, satrana (*Hyphaene coriacea*), and sakoa (*Spondia dulcis*), and other trees. The residual forest patches preserved in privileged topographic situations or on the reverse sides of karst that are unsuitable to cattle raising and to agriculture (such as the forests of Antsingy, Maintirano, and Namoroka) will on the other hand be able to persist much longer.

Did the forests formerly cover all or a part of these expanses now devoted to prairie? One has serious reasons to think they did, along with Perrier de la Bathie (1921, notably p. 3) and Humbert (1927). The existence of residual wooded patches is already convincing. But there are stronger pedological arguments. Most pedologists estimate indeed that the lateritic soils covering the highlands and most of the west could have been formed only under a wooded type of vegetation. One must then admit that over large expanses the forest has disappeared, in a relatively recent epoch.

Perrier de la Bathie, in his study of the Marotampona site between Betafo and Antsirabe, has observed that here on the sides of a vast swampy depression traversing the Andrantsay River, caused by a recent blow from the volcano Antsifotra, there exist rich deposits of vegetal debris of all kinds containing a few fragments of the bones of *Aepyornis* and *Hippopotamus*. The layer consists of a mass of trunks, branches, and fruit covered by 60–100 cm of silt or alluvial material. In its upper part are a few pieces of charred wood. Perrier de la Bathie has managed to identify among these remains twenty-eight vegetal species, including:

> *Cephalostachyos Chapelieri*, bamboo creeper
> *Dracaena angustifolia*, 4–8 m tall
> *Chrysalidocarpus*, from Menankazo, palm tree, 12–15 m tall
> *Ravensara* sp., tree, 10–12 m tall
> *Cupania* sp., tree, 10–15 m tall
> *Cipadessa* sp., small tree
> 2 *Dombeya* spp., trees, 8–10 m tall
> 2 Leguminosae, a tree and a liana
> 1 Menispermaceae, small tree

All these essentially forest species except *Cupania* are characteristic of the forests of the eastern slope of the central area, and the *Dracaena* is rather ubiquitous. The region is at present completely barren, and one must go at least 70 km east of Antsirabe to find a similar forest. The author then concludes:

1. at the moment of deposition of this vegetal alluvium the upper basin of the Andrantsay was still wooded;
2. the forests that then covered this basin were very much similar to those seen at present 70 km east of Antsirabe;
3. at Marotampona, just as at Antsirabe, the bones of the sub-fossils disappear at a level where one begins to find burnt wood, i.e. human traces;
4. the three periods of destruction of the forest, of the coming of man, and of the disappearance of the subfossils are contemporaneous;
5. the forest has disappeared from the region only since the time necessary for the deposit of a layer of clay less than 1 m thick, that is to say, at a date apparently going back no further than about three centuries.

(Perrier de la Bathie, 1914, p. 137–38)

The reopening of this site is now being considered in order to make the stratigraphy precise and to verify by the radiocarbon method the age attributed to the vegetal debris by Perrier de la Bathie. Important debris of burned vegetation has been observed in somewhat comparable conditions by J. T. Hardyman in the region of Lake Alaotra; it also deserves to be analyzed.

In the cross section of the Vohitrandriana on the eastern bank of Lake Alaotra, a paleosol, containing charcoal debris, perhaps contemporaneous with the first deforestation, is now covered by 2–4 m of sandy clay corresponding to the alluvial cones of ancient *lavaka*[1] that are now fixed. The samples, which were quantitatively insufficient, would no doubt have given an idea of the time of deforestation in this region (Battistini and Vérin, 1966).

It is interesting to note that the traditions in the Sihanaka country, as in Imerina and at Betsileo, preserve the memory of a "great fire" (*afitroa* or *afotroa*), of which the conditions and the time are totally imprecise. On this subject, see Dandouau and Chapus (1952, p. 45) for the Sihanaka country, and for the Betsileo see Rainihifina (1958, p. 19), who writes: "When the Vazimba arrived here the forests that covered

1. A form of accelerated erosion of lateritic slopes, extremely common in Madagascar.

the countryside were set afire. The fire burst out suddenly during a period of dryness, but its cause is not known. The people call this fire *afotroa*." J. Dez has pointed out to us that the word *troa* was picked up by Deschamps in Antaisake, where it means the clearing of vegetation in order to make a rice field.

In this study of the vegetational history of Madagascar, knowledge of traditions, a stratigraphic study of the sections, identification of the vegetal debris, and absolute dating must concur to shed light on the problem. Paleopalynology, unfortunately still in its infancy on Madagascar, will soon give important additional support to these different research techniques, thanks to the studies of H. Straka; indeed, to be able to identify the fossil pollen picked up in the sections and to compare them to that of the present flora of the region, it is necessary to possess reference collections, now in the course of publication (see de Waard, 1961, and Palynology in Africa, 1962).

In the regions of the island possessing a thick mantle of lateritic weathering formed under forest cover, deforestation has brought about a break in the equilibrium and the appearance of accelerated forms of erosion of the lavaka type in the soft lateritic soils. Certain parts of the highlands—Alaotra–Mangoro and the Andriba region—present a relief of lateritic hills literally gnawed away by the lavaka. Abrupt, branching gashes are cut into the thick mantle of soft lateritic soils on gneissic slopes. In certain regions around Ambatondrazaka, Andriba, and Maevatanana, the density of the lavaka is such that there is no longer anything between them but very narrow traces of the slopes. The study of the forms of accelerated erosion on the highlands proves also that the erosion is more complex than it had first appeared to be. There exist several generations of lavaka, the most ancient ones appearing in the slopes as old scars so well filled that only a very keen eye can distinguish them. Although figures concerning the rapidity of the evolution of present lavaka can give only a very imprecise idea of their rate of formation and scarring, these features might be more than 1,000 years old.

THE DISAPPEARANCE OF SUBFOSSILS

In Madagascar, subfossils are understood as types, sometimes of great size, that apparently disappeared only a few centuries ago. Among the extinct types one can cite

Testudo grandidieri
Aepyornis: (*maximus, medius, hildebranti*)
Mullerornis
Hippopotamus lemerlei

Cryptoprocta ferox spelea (larger than the present species)
Several large Lemurids, such as:
Megaladapis
Palaeopropithecus
Archeolemur
Hadropithecus
Daubentonia (*Cheiromys* larger than the present species)

Remains of these subfossils have been found in a large number of sites, distributed as much on the highlands as in the east and in the extreme south (for the location, see Fig. 1, according to J. Piveteau's treatise on paleontology). Curiously, no site has yet been discovered on the eastern slope of the island. It was the same situation in the extreme north, until one of us located *Aepyornis* egg shells, both in the early Quaternary stratum and near the surface in the Diégo–Suarez region (Ambolobozokely and Irodo).

Is there a basis for thinking that the disappearance of these subfossils is connected with human action? Alfred Grandidier was the first to claim the contemporaneity in certain sites of vestiges of subfossils associated with traces of human industry—pottery, notched bones, and pierced aye-aye teeth; he saw there the action "of ancient colonies of an unknown race" (Grandidier and Grandidier, 1908–1928, p. 1, n. 2, and p. 12–13).

Guillaume Grandidier, in a general article on the extinct animals of Madagascar (1905, p. 115–16), studied the most characteristic cases of association in a certain number of sites, including Ampasambazimba and Lamboharano. At Ampasambazimba in Itasy, in a site discovered by Raybaud,

> The bones are immediately under the stratum of humus, generally 30–40 cm, sometimes 1 m deep; they rest on the bank of sedimentary deposits, in which they are sometimes so encrusted that it is impossible to free them. A certain number of skeletal pieces of *Hippopotamus* and of other animals are welded among them by calcareous concretions. The remains are sparse in all directions; it has not been possible, until now, to bring together the elements of a single skeleton, not only the bones belonging to individuals, but even to different types scattered pell mell under the vegetal earth stratum.
>
> M. Raybaud, to whom we owe the discovery of this layer, adds that certain pieces seem to have undergone the effects of fire; shapeless slag and volcanic ash mixed in calcareous formations sometimes imprison a bone, and, on this debris half burnt to a cinder, iridescent metallic traces are perfectly visible.

Seeds and fruit pits in great number have been found, mixed with the bones and with pieces of wood, of which some are admirably well preserved.

Subsequently, the excavations were taken up again by Fontoynont. This author points out a find of great interest:

> In fact, right in the fossil layer, in the middle of fragments of *Aepyornis* and of numerous bones of lemurs, there have been found worked wooden sticks and a bone implement accompanying an absolutely intact earthenware jar, which are undeniable proof of the contemporaneity of man and of the subfossils. These objects were found at a depth of $2\frac{1}{2}$ m, in a soil where they certainly could have been neither willfully nor recently buried. The presence of tangled bones and especially the very context of these objects are themselves sufficient proof.

The attached pictures (Fig. 2) represent the pieces in question, which are evidence of already rather advanced signs of civilization among the first inhabitants of the island, who were contemporary with *Aepyornis*, *Megaladapis*, *Propithecus*, and other now extinct lemurids.

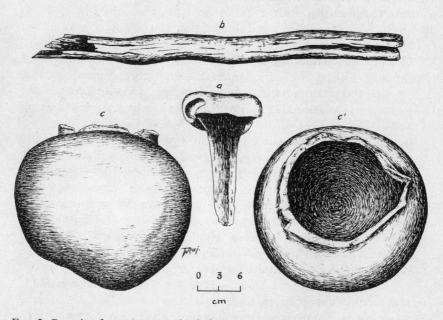

FIG. 2. Remains from Ampasambazimba: a, ax of *Mullerornis* tibio-tarsus; b, its wooden handle; c, earthenware jar, side view; c′, same from above (from Académie Malgache).

The piece of cut wood, whose species it was impossible to determine, is very clearly a tool handle. It is easy to see on one of the extremities the traces produced by a cutting instrument that shaped it slightly into a blunt point. It is 52 cm long and 3 cm in diameter. The other end was broken, probably by accident, for the maker or the owner must have rounded it off. This weapon, which could serve equally as a tool, is shown with an ax cut out of a tibio-tarsus of *Mullerornis*, and it was found right next to the wooden tip described above. This ax, pictured opposite, is very well made. It has a bevelled blade that must have been quite sharp.

The hypothesis that an accidental fracture determined this appearance is inadmissible, for there was a real hollowing out of a special part of the bone. A fracture could have given neither the curvature nor the polish that one can still easily see. The handle and the ax are hinged easily. They seem well made for each other.

(Fontoynont, 1908, p. 6)

Farther on (p. 7), Fontoynont describes the earthenware jar found in the excavation:

The jar is as shown in Plate 1 [Fig. 2 in this paper]. This kind of utensil is still used by the women of the central plateau to fetch water from the fountains and keep it in the house; it is the Madagascan *siny*. It is earthenware and has undergone a firing, even an excellent firing. It presents no varnish, nor any interior or exterior coating. The inner surfaces, thicker, however, than those of jars now in use, are 1 cm thick. It is dished, but crudely made, in the sense that the bottom is not very stable, because it is angular, a fact that makes the whole object a little asymmetrical. The dished part has a diameter of 0.31 m. The interior and the whole lower part are blackened by smoke and by burned contents, indicating that it was an object of everyday use. One notices also on each of the faces the traces of the hand of the modeler, in the form of inequalities. The orifice has a diameter of 0.19 m. This object was neither made on a wheel nor polished.

Some fragments of pottery had already been found in 1904 at the entrance to the basin, in the same strata, but without any clear association.

At Lamboharano, about 40 km south of Morombe on the southwest coast of Madagascar, Guillaume Grandidier found fossil aye-aye teeth "mixed with bones of *Aepyornis*, *Megaladapis*, and *Palaeopropithecus;* they had been cut and pierced, in order to be worn by men contemporary with this extinct fauna, as ornaments" (1928, p. 106–07).

The pierced tooth discovered by Grandidier (Fig. 3) is larger than those of the present *Cheiromys* (*Daubentonia*).

It is essential to note that remains of zebu bones have been found in

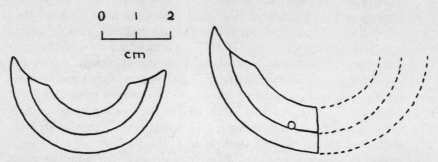

FIG. 3. Pierced lower (left ?) tooth of subfossil *Cheiromys*, on right, compared with lower left incisor of modern *Cheiromys madagascariensis*, on left (from Académie Malgache).

the beds of subfossils. G. Grandidier, who reports this fact (1905, p. 123), states,

> The skeleton of this animal presents, however, a peculiarity that is important to point out—the very clear bifurcation of the dorsal apophysis from the dorsal vertebrae. This characteristic does not occur in as pronounced a manner in any of the domestic or wild cattle (that is to say, cattle that have gone wild) that are still encountered in Madagascar.

Other sites, without possessing any traces of association with human industries, reveal nonetheless an apparent intervention by man. This is the case with beds in which skeletons, instead of being found whole, have fractured and dispersed elements. Thus Jully (1899) notes that at Antsirabe

> The state of preservation differs in certain respects. While in the Vallée des Eaux, heads of hippopotamus and large bones are found whole and in perfect condition, in the Andraingy Valley they are broken and in a state of advanced decomposition that often does not permit them to be removed. On the other hand, while small bones are entirely lacking in the first valley, they are found in abundance in the second and well preserved in the middle of the remains of hippopotamus bones. It is fair to remark that the bones coming from a single individual are always found grouped in a relatively limited space; those of small species, for example, have been found within an area of 1 sq m.

At Itampolo, on the southwestern coast about 200 km south of Tuléar, the bones appear fragmented under a sandy clay layer of about 20 cm. The absolute age of *Hippopotamus lemerlei* bone, according to the radiocarbon method, is 980 ± 200 years (Kigoshi, 1964: Gak-350b).

We know of no account of the excavation on the Taolambiby site, but P. S. Martin, and A. Walker, during a recent visit, found there a pierced piece of tortoise shell (*Geochelone grandidieri*) beneath femur bones associated with potsherds in the upper levels (for similar cases, see Fig. 4).[2] In certain collections at the British Museum, Walker (1967) has noticed other lemur skulls that had been deliberately pierced.

At Amparihingidro, near Majunga, the accumulation of remains of extinct animals seems to have occurred in a natural manner, probably effected by running water; C^{14} dating gives an age of $2,850 \pm 200$ years (Gif-sur-Yvette).

There must exist, then, three types of subfossil beds: natural deposits, such as those of Amparihingidro; accumulations whose origin is exclusively human; and mixed sites, of which only the upper horizons, perhaps, were the result of human action.

These data must be confronted with what is known of the oldest Madagascan past. There is no real prehistory on Madagascar, but it is certain that there existed a widespread and implanted population at the end of the first millennium. The study of the Talaky bed (in the extreme south) has proved that nine centuries ago people of a maritime way of life populated the regions of the mouth of the Manambovo River. The relics of their habitats contain, in addition to shell and iron objects, remains of *Aepyornis* shells; it is not yet known whether the *Aepyornis* were consumed or whether the shells were simply used as containers. One carbon sample taken from a hearth had an age of 840 ± 80 years (Battistini et al., 1963). In the extreme north, the lower stratum of the Irodo archaeological site proved to be slightly older: Antanimena, 980 ± 100 (Kigoshi, 1964, Gak-280b), and Tafiantsirebika, 1090 ± 90 (Gak-692). Hypotheses formulated through linguistic work and study of archaeological beads suggest that research now in progress will probably reveal still older traces of human occupation.

Oral tradition seems to have brought back to us from a less remote period the memory of ancient animals. *Hippopotamus* has left his old name, *lalomena*, to present generations. The eggs of *Aepyornis*, attributed by the inhabitants of the south to the *vazoho*, an enigmatic term, or to the *fanany*, a monstrous snake, are also recognized as coming from the *vorombe* (literally, large bird). In the seventeenth century Flacourt could write (p. 165): "*Vouroun patra* is a large bird that haunts the Ampatres and lays eggs as the ostrich; it is a kind of ostrich; people of the above

---

2. In Figure 4 are reproduced two instances from Ampasambazimba and Beloha Anavoha. The first one seems definitely man-made, but the second could have also been pierced by a Crocodile.

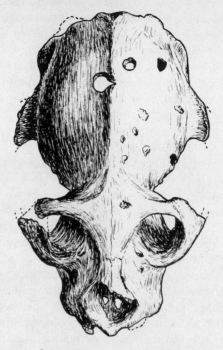

Ampasambazimba, XVI

0   3
|___._._._|
cm

Beloha-Anavoha, VIII
1932

FIG. 4. Skulls of *Archeolemur* showing perforations (collections of Académie Malgache).

places cannot take him; he seeks the most deserted places." This statement suggests that the animals had not yet completely disappeared.

Flacourt describes another animal (p. 154), *Tretretretre* or *Tratratratra*,
probably a large lemur now extinct:

> It is an animal as large as a two-year-old calf that has a round head
> and the face of a man: forefeet like a monkey, and hind feet also;
> he has frizzy hair. He resembles the *Tanacht* described by Ambroise
> Paré. One of them lived near the Lipomani pond, around which is
> his lair. He is a strong, solitary animal; the people of the country are
> afraid of him and flee him as he also does them.

THE INTERDEPENDENCE OF THE CHANGES IN
FAUNA, FLORA, AND CLIMATE

Certain authors have managed to attribute the destruction of the fauna
of subfossils to disasters of volcanic origin; thus, Jully (1899) wrote about
Antsirabe:

> It seems, in a word, that all these creatures were struck at the same
> place where they were found. Volcanic debris, slag, and charred
> pieces of wood, branches and plants recount eloquently this disaster,
> this upheaval, this enormous swelling that suddenly burst in the
> middle of Madagascar. It made Ankaratra a formidable carbuncle,
> where from 70 mouths (the number of extinct craters counted by M.
> Mullens) flowed torrents of lava and mud to the raised bottoms of
> the large lakes, into the grooves cut by the flow of their waters,
> burying the plants and animals under accumulations of liquid soil
> and congealing life into a matrix where corruption itself cannot
> reach it completely.

This opinion is not shared by most other authors. At the Ampasambazimba site it is obvious that the subfossils accumulated in a depression
behind an obstruction created by a flow of lava. Perrier de la Bathie
(1914) is of the same opinion in regard to the Marotampona site, when he
writes, "The Marotampona depression is later than the outflows of the
Antsifotra, one of the most recent volcanoes of the island, and there, as
elsewhere, volcanic phenomena have nothing to do with the disappearance of the subfossil species."

For other authors, the disappearance of the subfossils is linked with
climatic change. Decary, in his monograph on Androy (1930, p. 15–16),
thinks that it is a result of the increasing aridity of the climate, although it
might also have been the result of human action:

> The fauna could not become accustomed to the new conditions of
> life. Lemurids, *Aepyornis*, and *Hippopotamus*, after being concen-
> trated around the spots of water that subsisted longest (Ambovombe,

Antsirasira, Ambovotsimahay, and Ampotaka), perished one after
the other at the very edge of those ponds that should have constituted their last refuge; their bones are found there scattered around
the bottom, or buried in the soil at a depth scarcely more than 1 m.
The giant tortoises also perished. Only the crocodiles subsisted, all
the while being slightly modified .... Thus, a change in climate,
bringing a progressive desiccation, and as a direct consequence, a
modification in flora and fauna, an establishment of the sub-desert
order; such is the process that led the extreme south to the present
stage.

But the fauna of the large subfossils has been found, as we have seen, in
other extremely varied regions of Madagascar—for example, around
Antsirabe and in Itasy.

It seems difficult to explain the very recent disappearance of these
subfossils from the central highlands by a modification of the climate. At
present, 1,448.2 mm of rain per year fall at Antsirabe, and 1,879.9 mm at
Soavinandriana in Itasy. We are therefore forced to admit a more general
cause, even if in the extreme south the drying up of the ponds and the
rivers could also have played a role.

On the other hand, it is apparent that the disappearance of the forest
was a determining factor in the annihilation of certain species, among
them the large lemurs. The disappearance of the flora and the extinction
of the fauna are two phenomena that are probably linked and are at least
parallel, as Perrier de la Bathie had already seen for Marotampona.

Because the subfossil species characteristic of a forest biotope, such as
the large lemurs, have been found in already deforested regions of the
highlands, it is logical to think that these species disappeared with the
forest. The destruction of the tropophile forest, which covered most of
the highlands and the west, occurred at the beginning of the accelerated
erosion of soils, such as the lavaka, over considerable areas. The aquatic
environment in which the dwarf hippopotamus and the fossil crocodile
lived was then suddenly modified by an important increase in turbidity.
The iron oxides in suspension, limonite and hematite, which were probably delivered to the rivers only in a limited quantity when the forest
cover existed, gave the rivers the red color that is visible today when they
are swollen. Apparently, the large aquatic subfossils could not resist this
modification of environment.

In reality, there was, in a variable degree according to species, a combination of direct and indirect destruction, man remaining the essential
agent. For example, one can suppose that in certain regions, because the
fauna flocked together inside the residual boundaries of the vegetation or
was confined to a biotope of a lesser area, it might have been more easily

exterminated by hunters. The *fosa* (*Cryptoprocta ferox*) seems to be becoming extinct before our very eyes. It is becoming more and more rare in the deforested zones where it is easily spotted, but it remains abundant on the edge of the large forest that is not especially its habitat. In the same way a drying climate in the extreme south of the island might have led to a concentration of the fauna around spots of water also making it more vulnerable.

To explain the disappearance of *Aepyornis*, it is not necessary to bring in ecological changes. In the extreme south the bush still covers expanses vast enough to offer a refuge today as it did in the past. It is curious that *Aepyornis*, the remains of whose eggs are relatively rare in the base of the early Aepyornian and more numerous in the upper strata, seems to have multiplied especially during the Karimbolian and the Flandrian, whose dunes are often literally stuffed with broken eggshells, and that just at the time when this large animal was most numerous, which implies optimal living conditions, it disappeared with astonishing rapidity, probably in less than several centuries. Decary (1930, p. 16) doubts that man, with the feeble means of attack at his disposal, could have exterminated these great "birds." But is it not more plausible that he plundered the eggs, a relatively easy feat, for the disposition of the beds suggests that the *Aepyornis* nested in great colonies in the littoral region? We could then understand how the species managed to disappear so quickly and so completely.

The gathering of eggs could have been done to obtain vessels as well as food. Hébert (personal communication) has suggested that the Arabs, coming to barter on the west coast, had managed to stimulate this gathering by trading for the eggs. However, no site on the east coast of Africa contains any vestige of *Aepyornis* eggs.

Whatever reasons be proposed, direct or indirect, certain cases remain hardly understandable. Thus it is difficult to comprehend why the large lemurs, for example, did not endure in certain vast impenetrable zones in the eastern forest. It is equally difficult to conceive that *Hippopotamus* completely disappeared from all the most inaccessible marshlands of Madagascar. One would be tempted to believe that the first inhabitants were extremely active and effective hunters, much more so than present populations. F. Martin (1920, p. 503) indicates that in the seventeenth century, inhabitants of the forest, not far from Alaotra Lake, hunted "monkeys" and used the bow, a practice unknown at the present time.

Among the species of extinct subfossil fauna, the largest, in particular, disappeared, although the ancient Madagascans also attacked smaller species, as the contents of most sites show. In fact, one also finds with them the bones of species generally not extinct. Therefore, one must conclude in favor of a fragility peculiar to the largest species, a conclusion that is explained not by giantism but by a greater vulnerability (longer reproductive cycle and lesser density). Giantism of the species is frequently invoked in paleontology to explain their disappearance. This is not applicable in Madagascar, for it does not explain why so many species would have disappeared at the same time and in such a brief period.

In the process of animal depopulation at the hand of man, cultural factors can intervene. The great Indris (*babakoto*) was perhaps saved by the common Eastern prohibition (*fady*) against killing it. The aye-aye (*Daubentonia*) is tabu today, but it was not always so, as the pierced teeth found at Lamboharano indicate. It must be pointed out that the *fadys* relating to animals are very often local, being applied only by a particular ethnic unit.

Madagascar is not unique with regard to the rapid extinction of a great number of animal species. New Zealand saw the Maoris exterminate the *Dinornis* (*Moa*) in a very short time, between about 900 and 1600 of this era. A complex fauna containing mastodon, mammoth, certain bison, antelope, and others disappeared from North America around the tenth millennium B.C., in a brief period apparently corresponding to the human intrusion into the American continent. The rhythm of destruction is all the more accelerated when the territory is exiguous: thus at Maurice the dodo disappeared very rapidly after the arrival of the Europeans in the seventeenth century.

So it seems that comparative works on the Pleistocene extinction of animals could lead to a quasimathematical law of duration relative to the surface of the sites concerned and to the population density. For Madagascar, investigations are still not advanced enough to furnish such figures, but here and now we can presume that 500 to 1,000 years sufficed to complete the disappearance of most of the large species. The rhythm of regression of the flora is much more irregular because of the play of numerous factors as much human as natural (e.g. karstic zones unsuitable for agriculture or rearing stock, or slopes too steep).

References

Bastian, G., 1964, La forêt d'Ambohitantely, Madagascar: *Rev. Géographie*, no. 5, p. 1–42

Battistini, R., and Vérin, P., 1966, Vohitrandriana, haut-lieu d'une ancienne culture du lac Alaotra: Paris, Ed. Cujas, *Civilisation malgache*, vol. *1964*, p. 53–90

Battistini, R., Vérin, P., and Rason, R., 1963, Le site archéologique de Talaky, cadre géologique et géographique, premiers travaux de fouilles; Notes ethnographiques sur le village actuel proche du site: Paris, Ed. Cujas, Univ. Madagascar, sér. Lettres et Sciences Humaines, *Annales*, no. *1*, p. 111–34

Dandouau, A., and Chapus, G. S., 1952, *Histoire des populations de Madagascar:* Paris, Larose, 317 p.

Decary, R., 1930, *L'Androy*: Paris, Soc. d'editions géographiques, maritimes et coloniales, v. *1*, 224 p.; v. *2*, 268 p.

Flacourt, E. de, 1661, *Histoire de la Grande Isle Madagascar:* Paris, Gervais Clouzier, 471 p.

Fontoynont, M., 1908, Les gisements fossiles d'Ampasambazimba, compte rendu des fouilles effectuées par l'Académie Malgache à Ampazambazimba en 1908: *Académie Malgache, Bull.*, v. 6, p. 3–12

Grandidier, G., 1905, Les animaux disparus de Madagascar, gisements, époques et causes de leur disparition: *Paris, Rev. Madagascar*, p. 112–28

——— 1928, Une variété du Cheiromys madagascariensis actuel et un nouveau Cheiromys fossile: *Académie Malgache, Bull.*, n.s., v. *11*, p. 101–07

Grandidier, A., and Grandidier, G., 1908–1928, *Ethnographie de Madagascar*, en 5 vol.: Paris, Union Coloniale

Griveaud, P., 1962, Rapport sur le gisement de subfossiles d'Amparihingidro: Tananarive-Tsimbazaza, Inst. Recherche Scientifique de Madagascar, 21 p.

Hervieu, J., 1962, Site géologique et pédologique du gisement de subfossiles d'Amparihingidro: Inst. Recherche Scientifique de Madagascar, 21 p.

Humbert, H., 1927, La destruction d'une flore insulaire par le feu: *Académie malgache, Mem.*, no. *5*, 79 p.

Jully, A., 1899, Les tourbières d'Antsirabe et les animaux disparus de Madagascar, dans notes, reconnaissances et explorations: *Tananarive, Imprim. gouvernement*, p. 1175–83

Mahé, J., 1965, Les subfossiles malgaches (collection de l'Académie malgache): *Tananarive, Revue de Madagascar*, n.s., no. *29*, p. 51–58

Martin, F., 1920, Mémoires sur l'Ile de Madagascar (1665–1668), *in* Grandidier, A., et al., *Editors, Collection des ouvrages anciens concernant Madagascar:* Paris, Union coloniale, p. 427–633

Palynology in Africa, 1962, Seventh report covering the years 60/61: Madagascar, Bloemfontein

Perrier de la Bathie, 1914–1921, Au sujet des tourbières de Marotampona: *Académie malgache, ·Bull.*, n.s., v. *1* (1914), p. 137–38; La végétation malgache: *Musée Colonial de Marseille, Annales*, 29ème année, 3ème ser., v. *9* (1921)

Rainihifina, 1958, *Lovan-tsaina*, vol. I. Tantara betsileo: Fianarantsoa, Imprimerie catholique, 211 p.

Riquier, J., 1951, Etude sur les "lavaka": *Recherche Scientifique de Madagascar, Mém.*, sér. D, v. *3*, p. 113–26

Rougerie, M. G., 1965, Les "lavaka" dans l'évolution des versants à Madagascar: *Géog. Franç. Bull.*, p. 15–28

de Waard, H., and Straka, H., 1961, C-14 Datierung zweier Torfproben aus Madagaskar: *Naturwissenschaften*, v. *45*, no. 2, p. 1–2

Walker, Alan, 1967, Patterns of extinction among the subfossil Madagascan lemuroids (this volume)

ALAN WALKER

Department of Anatomy
Makerere University College
Kampala, Uganda

PATTERNS OF EXTINCTION

AMONG THE SUBFOSSIL

MADAGASCAN LEMUROIDS

Abstract

Direct and indirect evidence for an association between man and the subfossil lemuroids of Madagascar is reviewed. A new association from the deposit of Taolambiby is the occurrence of pottery in beds yielding extinct Palaeopropithecus, Archaeolemur, *and* Lemur (Pachylemur). *A skull of* Archaeolemur majori *from the British Museum shows what appears to be a man-made wound. Indirect evidence regarding presumed locomotor habits and activity rhythms indicates the extinct lemuroids were large, slow-moving, ground-living, and diurnal compared with the living species, which are generally smaller, more active, pre-eminently arboreal, and in many cases nocturnal.*

There is no doubt that extinction took place after prehistoric man reached the island. The inferred habits of the extinct species would have made them more easily hunted and killed. The only mechanism for such a selective extinction pattern seems to be the intervention of predatory man.

Only in Madagascar have the primates known as lemurs survived in abundance into recent times. Various authors have concluded that the large extinct lemuroids, indriids, the aye-aye, and typical lemurs of Madagascar disappeared through the direct or indirect intervention of man (Hill, 1953; Mahe, 1965; Paulian, 1961; Romer, 1945). No detailed effort has been made to assess the data for such a conclusion. During a study of the locomotor adaptations of the fossil forms, a certain amount of direct evidence has been gathered for the association between the extinct forms and man, as well as some indirect evidence for the behavior, size, and habitat of the fossil forms.

Only two radiocarbon dates on beds containing extinct lemurs have been published (Mahe, 1965). These are the deposit of Amparihingidro, dated 2,850 ± 200 years and containing bones of *Archaeolemur*, *Megaladapis*, *Lemur* (*Pachylemur*), and *Lemur* (*s.s.*), and that of Itampolo, dated at about 1,000 years (see Battistini and Vérin, this volume), which has yielded *Archaeolemur*. A humerus of *Palaeopropithecus* has been submitted by Dr. K. P. Oakley of the British Museum (Natural History) for a dating of the bone itself, but the result is not yet available.

Dating of the arrival of man on the island is considered in the article in this volume by Battistini and Vérin. A complete checklist of the Madagascan lemurs, recent and subfossil, is given in Table 1. With the

TABLE 1. *Madagascan Lemuroids, Recent and Subfossil, Including Known Skull Lengths*

After Hill (1953)

Living species	Skull length (mm)	Extinct species	Skull length (mm)
Lemuridae		Lemuridae	
Cheirogaleinae		Lemurinae	
*Microcebus murinus**	34.2	*Lemur* (*Pachylemur*) *insignis*	114.0
*M. coquereli**	50.0	*L.* (*Pachylemur*) *jullyi*	126.0
*Cheirogaleus major**	56.3	Indriidae	
*C. medius**	44.8	Indriinae	
*Phaner furcifer**	53.0	*Neopropithecus globiceps*	94.5
Lemurinae		*Mesopropithecus pithecoides*	101.0
*Hapalemur griseus**	58.0	*Archaeoindris fontoynonti*	260.0
H. simus†	81.0	*Palaeopropithecus ingens*	not known
*Lemur catta**	75.0		
*L. variegatus**	110.0	*P. maximus*	186.0
*L. macaco**	101.0	Archaeolemurinae	
*L. fulvus**	90.5	*Archaeolemur edwardsi*	157.0
*L. mongoz**	82.0	*A. majori*	133.0
*L. rubriventer**	85.0	*Hadropithecus stenognathus*	141.0
Lepilemur mustelinus†	59.0	Megaladapidae	
L. ruficaudatus†	56.0	*Megaladapis edwardsi*	310.0
Indriidae		*M. grandidieri*	290.0
Indriinae		*M. madagascariensis*	250.0
Propithecus diadema†	92.0	Daubentoniidae	
P. verreauxi†	86.2	*Daubentonia robusta*	not known
Avahi laniger†	54.5		
Indri indri†	103.3		
Daubentoniidae			
Daubentonia madagascariensis	85.0		

* Active arboreal quadrupedal forms.
† Active arboreal leaping forms.

exception of a few subfossil species of doubtful validity, it follows Hill's classification (1953), including his measurements of skull length.

MAN AND THE EXTINCT FAUNA

During my recent visit to Madagascar with P. S. Martin, preliminary examination of a deposit containing subfossil lemurs yielded evidence regarding the association of man with the extinct forms. The site, Toalambiby, is in the southwest of the island and lies east of the town of Betioky. The fossil deposits are a series of sandy travertines and sands, forming and filling a basin below an intermittent waterfall of one of the tributaries of the River Sakamena. A layer of black humic sand 50 cm thick yielded specimens of the extinct *Palaeopropithecus sp.*, *Archaeolemur majori*, and *Lemur* (*Pachylemur*) *sp.*, and cranial and postcranial material of the extant *Propithecus verreauxi*, *Lepilemur ruficaudatus*, *Lemur sp.*, and *Cheirogaleus sp.*, as well as tenrec, small tortoise, and bird bones. In this same bed were also found four fragments of coarse earthenware pottery 8–15 mm thick. The underlying sands contained bones of *Geochelone grandidieri*, *Hippopotamus lemerlei*, and *Crocodilus nilocticus*. This site has provided stratigraphic evidence of an association in time between man and three species of extinct lemuroids, as well as emphasizing the important fact (already known from other deposits), that representatives of the living species of lemuroids coexisted with the extinct species. This verifies the conclusion of Standing (1908): "One may at any rate from a biological point of view regard all these subfossil Malagasey Lemuroids as the contemporaries of extant species in other parts of the island."

A great number of the first fossil lemuroids to be described came from the caves of Androhomana, near Fort Dauphin in the south of the island. Caves formed in thick calcareous sands overlie a gneissic basement. Collections made by Sikora in the early part of this century from the largest of these caves were bought by the British and Viennese museums. In the collections of the British Museum (Natural History) is an almost complete skull of *Archaeolemur majori* (B.M., M. 7374) that has a depressed fracture of the left frontal bone, which could only have been made by an ax-like instrument with an edge roughly 30 mm long by 5 mm wide (Fig. 1). That the wound was inflicted while the flesh was intact is apparent by the folded and adherent edges of the depression. The zygoma seems to have been broken at a later date. From the same collection is a femur of *Archaeolemur majori* (B.M., M. 7974), which shows traces of burning. Many of the bones exhibit marks of cutting along or across their length. Although it is possible that these were made

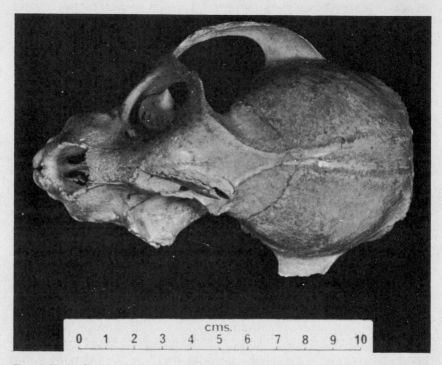

FIG. 1. Skull of *Archaeolemur majori* from Androhomana, showing left frontal fracture apparently made by an ax-like instrument (British Museum, Natural History, M. 7374).

during excavation of the cave, it is less likely than in the case of the diggings in the marsh deposits, where local unskilled help was employed and where excavations were often flooded.

Extinction According to Size and Habits

The size of the extinct large lemuroids has frequently been exaggerated. The largest species, *Megaladapis edwardsi*, has a skull length of about 300 mm and a total head and body length of 1,300 mm. *Archaeoindris* has a skull length of 260 mm; the large species of *Palaeopropithecus* has a skull of over 180 mm. *P. maximus* must have been the size of a small chimpanzee. *Archaeolemur* and *Hadropithecus* were about the same size as females of the living baboons of the genus *Papio*, whereas *Mesopropithecus* and *Neopropithecus* were about the size of modern African monkeys of the genus *Cercopithecus*. Within the living genera *Lemur* and *Daubentonia*, extinct species were greater in size by 15 to 30 per cent. In the absence of complete skeletons, a rough assessment of relative sizes can be based only on skull length, a method that can be misleading because body size and skull length are not necessarily constant among different species. The maximum skull lengths of the living and extinct forms are given in Table 1. It can be seen that all the large forms are now extinct; of the medium-sized animals, only the living indris slightly exceeds the skull length of *Neopropithecus*, the smallest of the extinct species.

The living Madagascan lemuroids can be classified into two basic locomotor types (Petter, 1962). There is an arboreal quadrupedal group and an active leaping group that Petter calls the "specialised group." The latter will shortly be the subject of a more complete account (Napier and Walker, in preparation). At present it is sufficient to say that these animals are almost wholly arboreal leaping forms which prefer vertical supports for resting, takeoff, and landing. On the ground they progress by a series of bipedal hops. This adaptation is in contrast to that of the quadrupedal group which prefer horizontal supports and run on all fours through the trees or on the ground. Surprisingly, the leaping form of locomotion is well adapted for avoiding predators. Petter (1962) states, "Les sauts du Propithèque pourraient peut-être constituer un moyen d'échapper à la poursuite d'un petit carnassier grimpeur, mais il n'existe aucun mammifère dangereux de ce type dans les forêts mal-gaches." Notes of the various species included in the two groups are given in Table 1, based essentially on the observations of Petter.

The locomotor adaptations of extinct forms can be inferred by study of the limb bones, in particular their relative proportions, allied to a comparative study of locomotion among living related forms. In rare cases it has proved necessary to search outside the order for a locomotor equivalent. So far there is no indication of the extinction of an active leaping form. Of the arboreal quadrupeds there are five extinct species as against thirteen living species. The extinct forms are *Daubentonia robusta*, *Lemur* (*Pachylemur*) *jullyi*, *L.* (*Pachylemur*) *insignis*, *Meso-propithecus pithecoides*, and *Neopropithecus globiceps*.

Three species, *Archaeolemur edwardsi*, *A. majori*, and *Hadropithecus stenognathus*, show many features shared by the ground-living quadru-peds among the Old World monkeys. *H. stenognathus* has slightly longer, more gracile limb bones and probably occupied the same sort of locomotor relationship to *Archaeolemur* that the African *Erythrocebus patas* does to the baboons of the genus *Papio*.

The genus *Megaladapis* is comprised of three species that have every indication in the postcranial skeleton of being a modified leaping type.

Many earlier authors were reluctant to consider a large animal (the size of an Alsatian dog) to have been arboreal, but recently Zapfe (1963) inferred that *Megaladapis* led an arboreal way of life, and studies of the huge grasping hands and feet leave no doubt that his diagnosis is correct. The nearest locomotor equivalent is to be found not among the Primates but among the marsupials. Studies of the skeletons and of moving pictures of the koala (*Phascolarctos cinereus*) suggest that *Megaladapis*, like the koala, was probably a slow-moving vertical-clinging form capable of making short leaps between vertical stems. On the ground it may have progressed in a series of short, frog-like hops.

Palaeopropithecus probably had a brachiating (Napier, 1963), or arm-swinging, mode of arboreal progression. Although there is very little postcranial material of *Archaeoindris*, the same may be true for this genus, which is a closely related form. *Palaeopropithecus* has its nearest locomotor equivalent in the modern orang-utan (*Pongo pygmaeus*). The hands of *Palaeopropithecus* were long, strongly curved, hook-like structures that were used, at the end of long slender arms, to support the body weight from above. Progression through the trees would have been leisurely but efficient, whereas ground locomotion, if indeed this animal ever came to the ground, would have been clumsy and very slow.

None of the vertical-clinging group is extinct, but 28 per cent of the arboreal quadrupeds and all the ground quadrupeds, brachiators, and koala-like types have disappeared.

Among the living forms there are eleven diurnal and nine nocturnal species. An attempt has been made to separate these two groups by the relative size of the orbits, on the assumption that the nocturnal forms would have larger eyes. It is possible, within broad limits, to demonstrate the validity of this assumption (see Fig. 2, where orbital diameter is plotted against skull length). The two groups of definite nocturnal and definite diurnal species are separate, whereas the crepuscular forms lie between the two. Of the fossil species, all have relatively smaller orbits, compared with living forms. Thus, except for *Daubentonia robusta*, whose skull is unknown, it appears that all the extinct lemuroids of Madagascar were diurnal.

Information gathered from the dentition is not always reliable, but for the ground-living quadrupeds it suggests the nature of the habitat. Nearly all the adult skulls of *Archaeolemur* and *Hadropithecus* show great wear of the cropping incisors and the molars. Lamberton's complete skull of *H. stenognathus* has the complicated molar cusp pattern, reminiscent of that of *Simopithecus*, worn quite flat. This compares favorably with the condition of modern baboon teeth, especially of those forms that

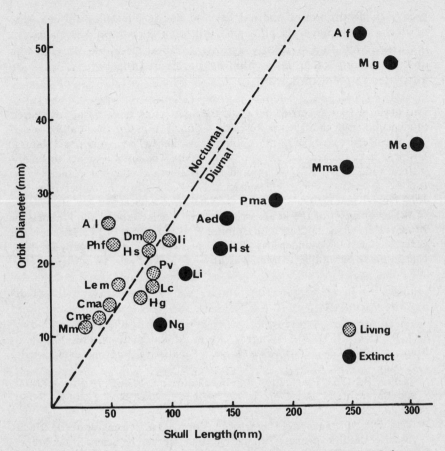

FIG. 2. Division between living nocturnal and diurnal forms among the Madagascan lemuroids. Key: Aed, *Archaeolemur edwardsi;* Af, *Archaeoindris fontoynonti;* Al, *Avahi laniger;* Cma, *Cheirogaleus major;* Cme, *Cheirogaleus medius;* Dm, *Daubentonia madagascariensis;* Hg, *Hapalemur griseus;* Hs, *Hapalemur simus;* Hst, *Hadropithecus stenognathus;* Ii, *Indri indri;* Lc, *Lemur catta;* Li, *Lemur insignis;* Lem, *Lepilemur mustelinus;* Me, *Megaladapis edwardsi;* Mg, *Megaladapis grandidieri;* Mma, *Megaladapis madagascariensis;* Mm, *Microcebus murinus;* Ng, *Neopropithecus globiceps;* Pma, *Palaeopropithecus maximus;* Pv, *Propithecus verreauxi;* Phf, *Phaner furcifer.*

habitually eat grasses (Warwick-James, 1960). In this case the dental evidence supports the ground-living locomotor assignation.

The assembled morphological and inferred behavioral evidence reveals that extinct Madagascan lemuroids were all relatively large, diurnal, and either slow-moving or ground-living. This sort of extinction pattern could have been brought about only by selective and excessive predation. Restriction of the habitat and minor climatological changes, of course, could have played a part, but alone they would not account for a selection that is clearly against a more accessible prey. In the absence of any other large hunting carnivore, the pattern must be attributed to prehistoric man.

Part of this work was carried out while the author was in receipt of a Research Studentship from the Science Research Council, London. The collections in the Sub-Department of Anthropology in the British Museum (Natural History) have formed the basis for this study. The author wishes to record his gratitude to Dr. E. I. White and Dr. Kenneth Oakley for permission to use the collections. He would particularly like to thank his supervisor, Dr. John Napier, and Dr. Oakley for all their help and advice. Moving picture film of the locomotion of the koala was made available through the kindness of Mr. Gerald Durrell and Mr. Christopher Parsons of the B.B.C. Through the generosity of Mr. Alexander Russell the writer was able to accompany Dr. P. S. Martin on an expedition to Madagascar.

References

Hill, W. C. O., 1953, *The Primates*, Vol. 1: Edinburgh Univ. Press, 789 p.

Mahe, J., 1965, *Les subfossiles malgaches:* Tananarive, Imprimerie Nationale, p. 1–11

Napier, J. R., 1963, Brachiation and brachiators, p. 183–95, *in* Napier, J. R., and Barnicott, N. A., *Editors, The primates:* Zool. Soc. London, Symp. no. *10*, 285 p.

Paulian, R., 1961, *Faune de Madagascar*, Vol. 13, La zoogéographie de Madagascar et des îles voisines: Tananarive, Inst. Recherche Scientifique Publ.

Petter, J. J., 1962, Recherche sur l'écologie et l'ethologie des lémuriens malgaches: *Mém. Mus. Hist. nat., Paris*, n.s., v. *27*, p. 1–146

Romer, A. S., 1945, *Vertebrate paleontology:* Univ. Chicago Press, 687 p.

Standing, H. F., 1908, On recently discovered subfossil primates from Madagascar: *Zool. Soc. London, Trans.*, v. *18*, p. 59–162

Warwick-James, W., 1960, *The jaws and teeth of primates:* London, Pitman Medical, 328 p.

Zapfe, H., 1963, Lebensbild von *Megaladapis edwardsi* (Grandidier): Basle, Karger, Folia primat., v. *1*, p. 178–87

NAME INDEX

SUBJECT INDEX

Acacia, 255
Acer, 232
Acheulian, 110–11
Acheulian man. *See* Prehistoric man
Acheulian–Mousterian horizon, 373
Acratocnus, 9, 18, 403
Adams local fauna, Kans., 321, 323, 324–25
Adaptation, 87, 164–65, 355, 357–58
Aden Crater, N.M., 90, 97, 249, 261
Adiantum, 261
Aenocyon. See *Canis*
Aepyornis, 4, 111, 410, 412–15, 417, 419, 421
Afotroa, 411–12
Africa, 109–11, 123, 125, 133–34, 144–45, 149, 156, 210, 263
Afrochoerus, 44
Aftonian, 332
Agate Basin, 280, 283–84
Agave, 261
Age curves, floral, 225–27, 230
Alabama, U.S., 96
Alaska, 88, 217, 218, 227
Alca, 172
Alces, 11, 52, 360, 388, 391
Algae, 207
Allen Site, Nebr., 183
Allerød, 67, 237, 355, 357
Allosorex, 351
Altithermal. *See* Hypsithermal
Altonian, 276
Ambystoma, 2, 324–25, 331
American hemisphere, 142, 267–84
American Indian. *See* Man
Androhomana, Madagascar, 427
Anisus, 326
Anodonta, 326
Anomalopteryx, 4, 102

Antelope, 337–39, 341–42, 346. See also *Myotragus*, *Saiga*
Anticosti Island, 240–41
Antilocapra, 11, 55, 174, 257, 295
Apium, 241
Aplexa, 326
Archaeoindris, 9, 16, 426, 428, 430
Archaeolemur, 9, 15, 413, 425–30
Archaic period, 280
Archidiskodon, 11, 38
Arctic–Alpine flora, 239
Arctodus, 10, 30, 83, 89, 183, 293
Arctotherium, 10, 30, 107
Ardops, 14, 174
Argentina, 125
Aristida, 261
Ariteus, 9, 14, 175
Arizona, U.S., 63, 96–97, 156, 228, 248, 250–52, 262
Armadillo. See *Chlamytheres*, *Dasypus*, *Holmensina*, *Pampatherium*
Armeria, 241
Art: Neolithic, 385–86; Paleolithic, 355, 357, 376, 379, 384
Artemisia, 250, 253, 261, 324
Arvicola, 10, 26, 350
Asia, 349, 365–66, 371–72, 388; central, 387; middle, 369, 389; west, 232–33
Asian nomad. *See* Man
Atriplex, 250, 253, 261
Aurochs, 360–62. See also *Bos*
Australia, 105–07, 134–35
Avahi, 9, 15, 426
Aye-aye, 425. See also *Daubentonia*
Azolla, 214, 231, 233, 351

Baboon. See *Papio*, *Simopithecus*
Badger. See *Meles*, *Taxidea*
Baidzherakh terrain, 393

450